PRACTICE MAKES PERMANENT

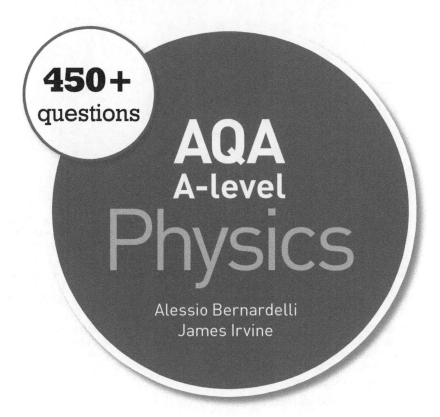

450+ questions

AQA
A-level
Physics

Alessio Bernardelli
James Irvine

HODDER
EDUCATION
AN HACHETTE UK COMPANY

Although every effort has been made to ensure that website addresses are correct at time of going to press, Hodder Education cannot be held responsible for the content of any website mentioned in this book. It is sometimes possible to find a relocated web page by typing in the address of the home page for a website in the URL window of your browser.

Hachette UK's policy is to use papers that are natural, renewable and recyclable products and made from wood grown in well-managed forests and other controlled sources. The logging and manufacturing processes are expected to conform to the environmental regulations of the country of origin.

Orders: please contact Hachette UK Distribution, Hely Hutchinson Centre, Milton Road, Didcot, Oxfordshire, OX11 7HH. Telephone: +44 (0)1235 827827. Email education@hachette.co.uk Lines are open from 9 a.m. to 5 p.m., Monday to Friday.
You can also order through our website: www.hoddereducation.com

ISBN: 978 1 5104 7641 7

© Alessio Bernardelli and James Irvine 2020

First published in 2020 by
Hodder Education,
An Hachette UK Company
Carmelite House
50 Victoria Embankment
London EC4Y 0DZ
www.hoddereducation.co.uk

Impression number   10 9 8 7 6 5 4

Year   2024 2023 2022

Cover photo © pixbox77 – stock.adobe.com

Typeset in India.

Printed and bound by CPI Group (UK) Ltd, Croydon, CR0 4YY

A catalogue record for this title is available from the British Library.

MIX
Paper from
responsible sources
FSC™ C104740

Contents

Acknowledgements

Every effort has been made to trace the copyright holders of material reproduced here. The authors and publishers would like to thank the following for permission to reproduce copyright material:

Page 3 Figure 4 © N. Feather/Science Photo Library; page 118 Figure 4 © Photopictures/shutterstock.com.

References

Page 95, question 13: the stated size of the Very Large Array is drawn from Amazing Space (https://amazing-space.stsci.edu/).

Page 97, question 13: magnitudes for Arcturus and Deneb were obtained from 'The Constellations and their Stars' (C. Dolan, www.astro.wisc.edu/~dolan/constellations/constellations.html)

Page 98, question 16: apparent magnitude and estimated distance for SN 1993J were obtained from *Wikipedia, The Free Encyclopedia*, https://en.wikipedia.org/wiki/List_of_supernovae

Page 110, Table 4: data from the Hubble Space Telescope Key Project Team, as quoted in 'Hubble's Law and the age of the Universe' (ASSIST, 2015, https://assist.asta.edu.au/).

Page 123, question 1: the rate of evaporation for a swimming pool is from The Engineering ToolBox (www.engineeringtoolbox.com).

Introduction

Practice Makes Permanent is a series that advocates the benefits of answering lots and lots of questions. The more you practise, the more likely you are to remember key concepts; practice does make permanent. The aim is to provide you with a strong base of knowledge that you can automatically recall and apply when approaching more difficult ideas and contexts.

This book is designed to be a versatile resource that can be used in class, as homework, or as a revision tool. The questions may be used in assessments, as extra practice, or as part of a SLOP (Shed Loads of Practice) teaching approach.

How to use this book

This book is suitable for the AQA A-level Physics course. It covers all the content that you will be expected to know for the final examination. For Paper B Section 3, the option covered is Astrophysics.

The content is arranged topic-by-topic in the order of the AQA specification, so areas can be practised as needed. Within each topic there are:

- **Quick questions** – short questions designed to introduce the topic
- **Exam-style questions** – questions that replicate the types, wording and structure of real exam questions, but highly targeted to each specification point
- **Topic reviews** – sections of exam-style questions that test content from across the entirety of the topic in a more synoptic way.

These topic questions are tagged with the following:

- Page references to the accompanying Hodder Education student books, for example **SB1 p3** refers you to page 3 of *AQA A-level Physics Student Book 1* (ISBN 978 1471 80773 2) and **SB2 p8** refers to page 8 of *AQA A-level Physics Student Book 2* (ISBN 978 1471 80776 3)
- AQA specification references such as **3.2.1.4**, which can be used if you want to select questions to practise a specific area of the specification
- **MS** indicates questions that test maths skills
- **AT** indicates questions that ask you to use practical knowledge of apparatus and techniques
- **PS** indicates questions that ask you to use knowledge of practical skills
- **RP** indicates questions that test understanding of the required practicals.

At the end of the book there is a full set of practice exam papers. These have been carefully assembled to resemble typical AQA question papers in terms of coverage, marks and skills tested. We have also constructed each one to represent the typical range of demand in the A-level Physics papers as closely as possible.

Full worked answers are included at the end of the book for quick reference, with awarded marks indicated where appropriate.

If you are using a different textbook

Some students will be using a different Hodder Education student book for their course: *AQA A-level Physics (Year 1 and Year 2)*, ISBN 978 1510 46983 9. If you have this book, there is a mapping grid available at https://tinyurl.com/qmbn6u8 to show how each question in *Practice Makes Permanent* links to the content of your student book.

Particles and radiation

Particles

Quick questions

1 Describe what is meant by two isotopes of the same element.

2 Name the **four** fundamental interactions.

3 What happens when a particle and its antiparticle collide?

4 What exchange particles mediate electromagnetic (EM) forces?

5 Name the **two** classes of hadrons and describe their structure.

6 Describe the change of quark character in a β+ decay.

Exam-style questions

7 **Figure 1** represents the decay of particle X into particle Y and two other particles.

The diagram shows the quark structure of particles X and Y.

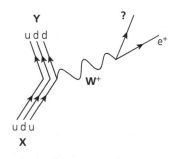

Figure 1

7–1 Deduce the names of particles X and Y. *[2]*

7–2 State what class of particles W+ belongs to and what type of interaction this decay is. *[2]*

7–3 Deduce the name of the particle represented by the question mark (?) in **Figure 1**. *[1]*

7–4 Explain how charge and baryon number are conserved in this interaction. Make reference to all the particles involved. *[2]*

7–5 Explain why the particle represented by the question mark (?) in **Figure 1** is formed in this interaction. *[2]*

7–6 Describe what eventually happens to all unstable baryons. *[1]*

Total: 10

8 In a pair production event, a photon is converted to an electron (e–) and a positron (e+).

Both particles generated have an energy of 3.512 MeV.

8–1 Calculate the wavelength of the photon in this process. *[3]*

8–2 State **two** quantities that are conserved in pair production processes. *[2]*

8–3 Explain why a photon of the same wavelength you calculated in **Question 8–1** could not produce a muon–antimuon pair. [2]

8–4 Calculate the minimum frequency needed for a photon to produce a muon and an antimuon through pair production. [3]

Total: 10

9 Carbon-14 is an unstable isotope of carbon.

SB1 p4 | 3.2.1.1

9–1 State what the term isotope means of carbon. [2]

SB1 p4 | 3.2.1.1

9–2 A carbon atom has 6 electrons orbiting its nucleus. Identify how many nucleons a carbon-14 atom has. [1]

SB1 p4 | 3.2.1.1

9–3 State how many neutrons a nucleus of carbon-14 contains. [1]

SB1 p31 | 3.2.1.4

9–4 Carbon-14 eventually decays into another element through beta decay. Copy and complete the nuclear equation below by filling in the gaps for the proton and nucleon numbers. [3]

$$^{14}_{}C \rightarrow\ ^{}_{}X + ^{}_{-1}e + \bar{v}_e$$

SB1 p31 | 3.2.1.4

9–5 State the name of the particle $\bar{v}_e$ and explain why it is emitted in a beta decay. [3]

Total: 10

SB1 p24–5 | 3.2.1.4

10 **Figure 2** shows a β-decay event.

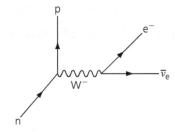

Figure 2

10–1 Add the exchange particle that mediates this interaction on **Figure 2**. [1]

10–2 Name the type of fundamental force that mediates β-decays. [1]

10–3 Name another nuclear decay that is **not** mediated by this type of interaction. [1]

10–4 Give a reason for the charge you chose in the exchange particle you added in **Figure 2**. [1]

10–5 Identify the range of action in metres of the fundamental force mediating the interaction in **Figure 2**. [1]

Total: 5

SB1 p30–2 | 3.2.1.7

11 Look at the following interactions and show whether they are possible or not.

11–1 $p + p \rightarrow p + p + K^- + K^+$ [3]

11–2 $K^- \rightarrow \pi^- + \pi^- + \pi^+$ [3]

11–3 $p \rightarrow n + \beta^- + v_e$ [3]

11–4 $n + e^+ \rightarrow e^- + \Sigma^+ + K^+$ (Σ^+ has quark structure uus) [3]

Total: 12

12   **Figure 3** represents the decay of a Λ⁰ particle into particle X and particle Y.

The diagram shows the quark structure of particles X and Y.

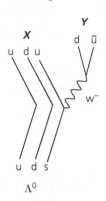

Figure 3

SB1 p31 3.2.1.4

12–1 Deduce the names of particles X and Y.   [2]

SB1 p30–2 3.2.1.7

12–2 By considering the conservation of charge, baryon number and strangeness, explain why this interaction is possible.   [3]

SB1 p22–4 3.2.1.4

12–3 The distribution of charge in the quarks that make up particle X is uneven. Explain how particle X is held together despite its uneven distribution of charge.   [3]

Total: 8

13   **Figure 4** shows the traces left by α-particles in a cloud chamber.

Figure 4

SB1 p8 3.2.1.2 AT i

13–1 Suggest a method that would allow you to accurately measure the range of α-particles in the cloud chamber.

In your method include a description of how you would determine the uncertainty in your results.   [6]

SB1 p6–7 3.2.1.2

13–2 Describe how the traces left by α-particles would change if the cloud chamber were immersed in a perpendicular magnetic field 'entering' the page.   [3]

SB1 p9 3.2.1.1

13–3 Radon is often used as an alpha source for small cloud chambers like the one in **Figure 4**. Copy and complete the nuclear equation below by filling in the gaps.   [2]

$$\frac{}{88}Ra \;\rightarrow\; ^{222}_{\underline{}}Rn + \;^{\underline{}}_{\underline{}}He$$

Total: 11

14 In a PET scanner, the gamma photons used for imaging are produced when a positron meets an electron inside a patient's body.

14–1 Name the process by which these gamma rays are produced. [1]

14–2 Describe what happens to a positron and an electron when they meet inside the patient's body. [2]

14–3 Calculate the frequency and wavelength of the gamma photons produced by the process you described above, when the electron and positron are at rest. [4]

14–4 Explain why PET scans should be avoided unless they are necessary. [2]

Total: 9

15 In a β-decay, a u quark of a neutron changes character to a d quark and this interaction generates a proton, a β-particle and an antineutrino.

15–1 Describe the nature of the β-particle. [2]

15–2 This interaction is mediated by the exchange particle W⁻. Name the type of interaction responsible for this nuclear decay. [1]

15–3 Describe how the W⁻ particle is responsible for the force between the nucleus of the decaying atom, the β-particle and the antineutrino. In your answer consider the properties of all particles involved to explain their behaviour in this interaction. [6]

Total: 9

16 The nuclear interaction below shows the production of a K^0 and Σ^+ (sigma) particle.

$\pi^0 + p \rightarrow K^0 + \Sigma^+$

16–1 The Σ^+ particle is made up of a strange quark and two other quarks. Use the conservation laws to show that this interaction is possible. [3]

16–2 Name the type of interaction in **Question 16–1**. Justify your answer. [2]

16–3 Deduce the quark structure of Σ^+ and show that its relative charge is +1. [3]

16–4 Explain why the interaction below would not be possible. [2]

$\pi^0 + p \rightarrow \pi^0 + \Sigma^+$

16–5 Σ^+ can decay into a proton and a pion in the interaction shown below.

$\Sigma^+ \rightarrow \pi^0 + p$

Explain why this can only be a weak interaction. [2]

Total: 12

Electromagnetic radiation and quantum phenomena

Quick questions

SB1 p61 | 3.2.2.1

SB1 p43 | 3.2.2.1

SB1 p45 | 3.2.2.2

SB1 p47 | 3.2.2.2

SB1 p49 | 3.2.2.3

SB1 p51 | 3.2.2.2

1   Define the work function of a metal.

2   What is a photon?

3   Give the definition of electron volt (eV).

4   What is the ionisation energy of an atom?

5   Describe how the emission spectrum for an element is formed.

6   Describe how thermionic emission of electrons occurs.

Exam-style questions

SB1 p60–1 | 3.2.2.1 | MS 1.1, 2.3

7   A TV manufacturer is testing a range of metals to use in their TV phototransistor as a receiver for the remote control infrared signals. **Figure 5** shows a graph of the maximum energy $E_{k(max)}$ of the photoelectrons emitted by the phototransistor for different frequencies of incident photons.

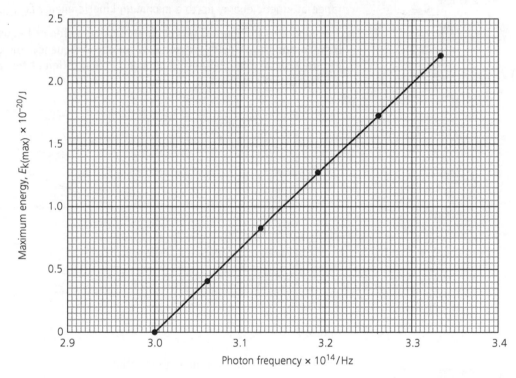

Figure 5

7–1   The infrared LED in the remote control emits photons with wavelength of 940 nm. Show whether the metal from the graph in **Figure 5** is suitable to use with this wavelength. [3]

7–2   Calculate the Planck constant using the graph. You must show your calculations. [3]

7–3   State the threshold frequency for this metal. [1]

7–4   Calculate the work function φ for this metal. [2]

7–5   Copy the graph in **Figure 5** and draw a line on it to show a metal with a greater work function. [2]

Total: 11

8 **Figure 6** shows a photoelectric cell connected to a variable power source and a high sensitivity ammeter. A voltmeter is connected in parallel to the photocell to measure the p.d. across it.

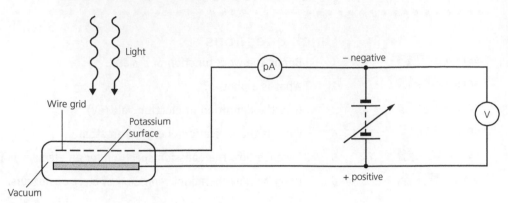

Figure 6

SB1 p57–61   3.2.2.1   PS 3.2
MS 2.3

8–1 When white light is shone on the potassium surface, a small current is registered by the picoammeter. Explain how this process happens.   [3]

SB1 p57–61   3.2.2.1   PS 3.2
MS 2.3

8–2 Explain why the photoelectrons emitted by the potassium plate can have a range of kinetic energy up to a maximum kinetic energy $E_{k(max)}$.   [3]

SB1 p57–61   3.2.2.1   PS 3.2
MS 2.3

8–3 The apparatus in **Figure 6** was calibrated using different frequencies of light to find the stopping potential V_s for each frequency. The graph in **Figure 7** shows the relationship between V_s and different frequencies of electromagnetic radiation.

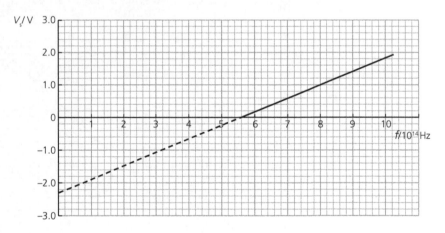

Figure 7

State what is meant by stopping potential.   [1]

SB1 p57–61   3.2.2.1   PS 3.2
MS 2.3

8–4 Suggest a method to use the apparatus in **Figure 6** and the data in **Figure 7** to measure the frequencies of visible light emitted by a gas lamp.   [6]

SB1 p49   3.2.2.3   MS 0.1, 0.2

8–5 **Figure 8** shows the emission spectrum of hydrogen.

Wavelength, λ/nm

Figure 8

Identify the wavelength(s) of the hydrogen spectrum that could be detected with the apparatus in **Figure 6**. Use calculations to support your answer.   [4]

SB1 p49   3.2.2.3   MS 0.1, 0.2

8–6 Suggest how the apparatus in **Figure 6** could be adapted to be able to detect the wavelengths emitted by hydrogen that could not be detected before.   [2]

Total: 19

SB1 p61 3.2.2.1 PS 3.2
           MS 2.3

9 A calcium surface is hit by light emitted by a violet LED of wavelength 410 nm. The threshold frequency of calcium is 6.5×10^{14} Hz.

9–1 Calculate the maximum kinetic energy of the photoelectrons emitted by the calcium plate in eV. *[6]*

9–2 Calculate the maximum velocity of the photoelectrons emitted by the calcium plate. *[3]*

Total: 9

10 **Figure 9** represents the kinetic energy levels in a hydrogen atom.

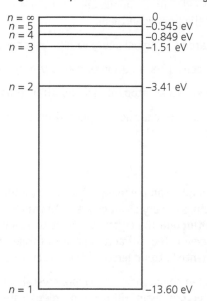

Figure 9

SB1 p48–50 3.2.2.3 MS 0.1, 0.2

10–1 Calculate the wavelength of the photon the atom would need to absorb to excite an electron from the ground state to the 3rd energy level. *[4]*

SB1 p48–50 3.2.2.3 MS 0.1, 0.2

10–2 The electron that was excited to the 3rd energy level eventually drops to the 2nd energy level and then back to the 1st energy level. Calculate the frequencies of the two photons emitted in this process and state in what region of the electromagnetic spectrum they can be found. *[5]*

SB1 p47 3.2.2.2

10–3 State the definition of ionisation energy. *[2]*

SB1 p47 3.2.2.2

10–4 Give the ionisation energy of the hydrogen atom in joules. *[2]*

Total: 13

11 **Figure 10** shows the emission spectrum of a sodium lamp.

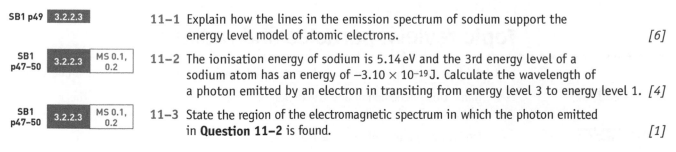

Figure 10

SB1 p49 3.2.2.3

11–1 Explain how the lines in the emission spectrum of sodium support the energy level model of atomic electrons. *[6]*

SB1 p47–50 3.2.2.3 MS 0.1, 0.2

11–2 The ionisation energy of sodium is 5.14 eV and the 3rd energy level of a sodium atom has an energy of -3.10×10^{-19} J. Calculate the wavelength of a photon emitted by an electron in transiting from energy level 3 to energy level 1. *[4]*

SB1 p47–50 3.2.2.3 MS 0.1, 0.2

11–3 State the region of the electromagnetic spectrum in which the photon emitted in **Question 11–2** is found. *[1]*

SB1 p47–50   3.2.2.3   MS 0.1, 0.2

11–4 The two emission wavelengths shown in **Figure 10** are so close to each other that they need to be represented as a single line in an energy levels diagram.

Draw a diagram of energy levels 1, 2 and 3 for an atom of sodium. Use the information in **Figure 10** and in **Question 11–2**.

Indicate the energy of each level in eV on your diagram. [5]

Total: 16

SB1 p51   3.2.2.2

12 A fluorescent tube is a glass tube coated with phosphor and filled with mercury gas. A cathode and an anode are at either end of the tube and a p.d. of 500 V is applied between them.

12–1 Describe how the atoms in the mercury vapour become excited. [3]

12–2 Describe what happens after an atom of mercury becomes excited. [2]

12–3 Explain what happens to an atom of mercury after ionisation. [3]

12–4 Explain why the phosphor coating is essential for the fluorescent tube to emit visible light. [3]

Total: 11

SB1 p64–5   3.2.2.4   MS 1.1, 2.3

13 Electron crystallography can complement X-ray crystallography when studying the structure of very small crystals. In this method, beams of electrons are shot through the crystal we want to study and electrons will be diffracted on screen. The diffraction pattern formed by the scattered electrons helps scientists to understand the structure of the crystal.

13–1 The distance between the atoms of a sample crystal is 0.025 µm. Calculate the velocity the electron beam will need in order to study this sample crystal. [3]

13–2 Explain why the velocity you calculated is a good choice for achieving electron diffraction through the sample crystal. [2]

13–3 Calculate the kinetic energy of the electrons shot at the sample crystal in eV. [2]

13–4 Explain how the wavelength of the electrons would change, if their kinetic energy is halved. Ignore relativistic effects on the mass of the electron. [3]

Total: 10

SB1 p64–5   3.2.2.4   MS 1.1, 2.3

14 A radioactive source emits α-particles with energy of about 5 MeV. In Rutherford's experiment, α-particles were directed at a foil of gold. When the α-particles hit the gold foil, some were scattered, but did not undergo diffraction.

14–1 Calculate the velocity of an α-particle emitted by this radioactive source. [4]

14–2 Calculate the de Broglie wavelength of one of these α-particles. [2]

14–3 The distance between nuclei in the gold foil is roughly 2×10^{-10} m. Explain why the α-particles from this radioactive source do not undergo diffraction. [2]

14–4 Show that the units of the de Broglie wavelength are metres. Use the de Broglie equation. [2]

Total: 10

Topic review: particles and radiation

1 A 'slow' positron and a 'slow' electron collide with each other, generating a photon pair.

SB1 p12   3.2.1.3

1–1 Name this phenomenon **and** the region of the electromagnetic spectrum the photon pair belongs to. [2]

SB1 p12   3.2.1.3

1–2 Explain why a pair of photons is produced. [1]

SB1 p47 3.2.2.2

1–3 One of the photons produced is absorbed by an electron in the ground state of a helium atom. The ionisation energy of helium is 24.6 eV. The total energy of the electron–positron pair before the collision is approximately 1.6×10^{-13} J. Show whether the helium atom will be ionised by absorbing the photon. [3]

SB1 p49–50 3.2.2.3

1–4 The same helium atom absorbs a photon that causes an electron in its ground state to transition to the 3rd energy level. **Figure 11** shows a diagram of this event.

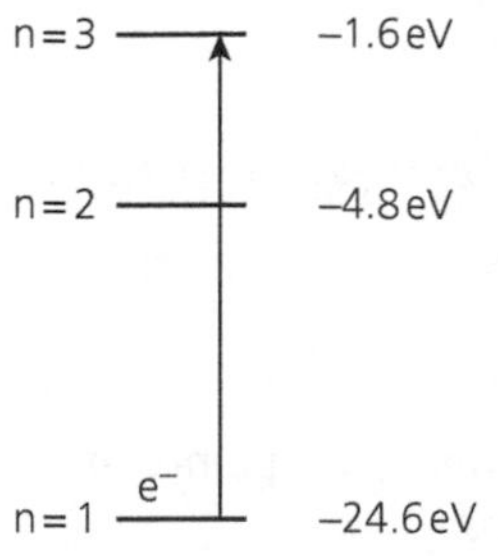

```
n = 3 ———————   −1.6 eV
          ▲
          |
          |
n = 2 ——————   −4.8 eV
          |
          |
          |
          |
          |
n = 1 —— e⁻ ——   −24.6 eV
```

Figure 11

Calculate the wavelength of the photon absorbed by the electron. [4]

SB1 p49–50 3.2.2.3

1–5 After a short time, the electron in **Figure 11** 'drops' from the 3rd to the 2nd energy level. Calculate the wavelength of the photon emitted during this transition and use **Table 1** to suggest which colour it is. [4]

Colour	Wavelength interval
red	~ 625–740 nm
orange	~ 590–625 nm
yellow	~ 565–590 nm
green	~ 500–565 nm
cyan	~ 485–500 nm
blue	~ 440–485 nm
violet	~ 380–440 nm

Table 1

SB1 p30–1 3.2.1.2

1–6 The nucleus of a helium atom is composed of two protons and two neutrons. Describe the sub-atomic composition of a neutron and a proton. [4]

SB1 p30–1 3.2.1.2

1–7 Explain how the helium nucleus is held together despite the electrostatic repulsion between the protons. [3]

Total: 21

2 An electric current passes through a mercury gas to excite electrons in mercury atoms to their fifth energy state.

SB1 p49 3.2.2.3

2–1 When an electron in the fifth energy state de-excites to the ground state, it emits a photon of wavelength 184.9 nm. Calculate the energy of the emitted photon. [1]

SB1 p43 3.2.1.3

2–2 State in what region of the electromagnetic spectrum the photons emitted by the mercury gas are. [1]

SB1 p47 3.2.2.2

2–3 The photons emitted by the mercury gas hit a zinc plate, which emits photoelectrons as a result. Describe the difference between photoelectrons emitted from the surface of the zinc plate and photoelectrons emitted from deeper in the zinc plate. [3]

2–4 The work function of zinc is 4.33 eV. Calculate the maximum energy of the electrons emitted by the zinc plate in joules. [3]

2–5 Explain what would happen if green light was incident to the zinc plate, instead of the light emitted by the mercury gas. [2]

Total: 10

Waves

Progressive and stationary waves

Quick questions

SB1 p73–4 | 3.3.1.1 | MS 0.1, 4.7

1 What is the phase difference between two points on a wave of $\lambda = 34\,cm$ separated by a distance of 85 mm?

SB1 p80 | 3.3.1.2

2 Draw a suitable diagram to explain the difference between an unpolarised and a polarised wave.

SB1 p101–2 | 3.3.1.3

3 Describe how standing waves are formed on a string fixed at its two ends.

SB1 p103 | 3.3.1.3 | MS 4.7

4 What is the wavelength of the first standing wave that can be produced on a string fixed at both ends with length 260 mm?

SB1 p80 | 3.3.1.2

5 Explain why a longitudinal wave cannot be polarised.

SB1 p72 | 3.3.1.1

6 Calculate the frequency of a sound wave in air travelling at $340\,m\,s^{-1}$ and with a wavelength of half a metre.

Exam-style questions

SB1 p101–4 | 3.3.1.3 | MS 4.7 | PS 1.2, 2.1 | AT i

7 A student sets up the apparatus in **Figure 1** to investigate the stationary waves generated along the thin cord when she changes the frequency of the signal generator.

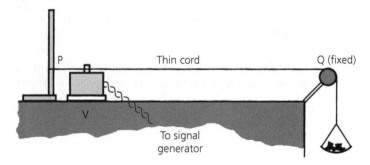

P   Thin cord   Q (fixed)

V

To signal generator

Figure 1

7–1 Describe what stationary waves are. [3]

7–2 In one of her measurements, the student sets the frequency to 50 Hz and calculates a velocity for the standing waves on the string of $17.5\,m\,s^{-1}$. What is the distance between the nodes on the string? [3]

7–3 When the student suspends 500 g masses over the pulley at point Q, she generates the first harmonic stationary wave on the thin cord at a frequency of 61.43 Hz. The length of the thin cord is 1.2 m. Calculate the mass of the cord. [5]

7–4 Calculate the percentage and absolute uncertainty for the mass of 1 m of thin cord. Assume that the absolute uncertainty on the tension is ±0.1 N, the uncertainty on the length is ±0.001 m and on the frequency ±0.01 s^{-1}. [3]

7–5 Suggest a more accurate way to find the mass of 1 m of thin cord and the uncertainty. [3]

Total: 17

8   **Figure 2** shows a microwave transmitter (T), a metal sheet and a microwave sensor (P). The distance between the opened end of T and the metal sheet is 55 cm.

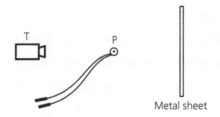

Figure 2

SB1 p101–4 | 3.3.1.1

8–1   The closest distance sensor P can be placed from the metal sheet to detect a maximum value is 5.5 cm. Explain how to determine the wavelength of microwaves between T and the metal sheet.   *[3]*

SB1 p101–4 | 3.3.1.1

8–2   Give the number of wavelengths of microwaves between T and the metal sheet.   *[1]*

SB1 p72–4 | 3.3.1.1 | MS 0.1, 4.7

8–3   Calculate the frequency of the microwaves generated by transmitter T.   *[3]*

SB1 p79–82 | 3.3.1.2 | PS 2.2, 2.4 | MS 1.2, 3.2, 3.4, 3.5 | AT i

8–4   The frequency of the microwaves transmitted by T is changed to 1.2×10^{10} Hz and the metal screen is replaced with an oven grid with metal bars separated by 2.5 cm gaps. Explain what happens to the microwaves that go through the oven grid.   *[2]*

8–5   A second identical oven grid is placed in front of the one in **Question 8–4** but with the metal bars perpendicular to each other. Explain what the sensor P would detect beyond the two grids.   *[2]*

Total: 11

SB1 p71–7, 101–2 | 3.3.1.1, 3.3.1.3 | MS 0.1, 4.7 | AT a, b

9   **Figure 3** shows the diagram of a Kundt's tube where the vibrating disc generates a train of sound waves inside the tube. The sound waves are reflected at the stationary disc and stationary waves can be set up at certain frequencies of the vibrating disc. The copper disc can also be moved inside the tube to change the distance between discs. Assume the speed of sound in air to be 330 m s⁻¹.

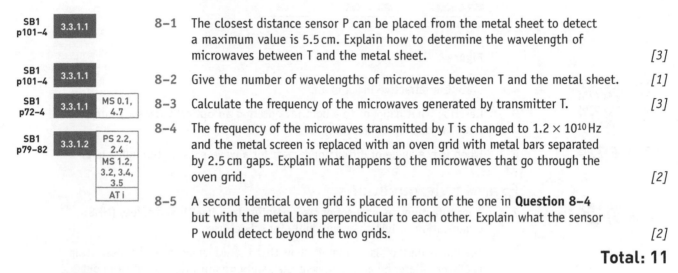

Figure 3

9–1   When small polystyrene balls are placed inside the tube and a stationary sound wave is set up in the tube, the balls are forced in the pattern shown in **Figure 3** where N is a node and A an antinode. Suggest a reason why the polystyrene balls arrange themselves in this way.   *[4]*

9–2   When the copper disc is vibrating with a frequency of 119 Hz, the polystyrene balls collect in the centre of the tube in a single peak and move away from both discs. Deduce the distance between the discs using appropriate calculations.   *[3]*

9–3   The copper disc is now set to vibrate at a different frequency. When the distance between discs is 1.2 m, four nodes can be seen inside the tube. Calculate the frequency of vibration of the copper disc.   *[3]*

Total: 10

Refraction, diffraction and interference

Quick questions

SB1 p96 3.3.2.1

1   What is resultant displacement of two waves meeting at the same point?

SB1 p97 3.3.2.1

2   Describe what is meant when two waves are coherent.

SB1 p101–2 3.3.2.2

3   Which diagram in **Figure 4** shows the situation where the greatest diffraction will occur?

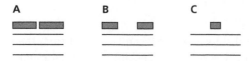

Figure 4

SB1 p83–5 3.3.2.3

4   State the refractive index in air.

SB1 p83–5 3.3.2.3

5   Describe what happens to a light ray entering an optically denser medium than the one it was travelling in.

SB1 p95 3.3.2.2 AT i

6   When is the angle of diffraction largest for a wave passing through a gap?

Exam-style questions

SB1 p99–100 3.3.2.1 MS 4.7 AT i

7   A manufacturer sends a set of laser pointers to a health and safety (H&S) organisation to test their suitability for school use.

7–1 The labels on the laser pointers show that the wavelength of the laser beam is 650 nm. Describe an experiment the safety organisation could carry out to measure the wavelength of the laser pointers. You should use a diagram to support your answer.   [6]

7–2 Through their measurements, the H&S organisation determined a mean wavelength of 630 nm. In their calculations they used the relationship $\lambda = \frac{sw}{D}$. Their percentage uncertainty on s was 2.5%, on w it was 1.2% and on D it was 0.1%. Explain why the manufacturer can still claim their laser pointers' wavelength is 650 nm. Use a calculation in your answer.   [3]

Total: 9

SB1 p99–100 3.3.2.1 MS 4.7 AT i

8   **Figure 5** shows a Young's double slit experiment using a laser beam shining through a double slit. An interference pattern with bright and dark fringes appears on the screen.

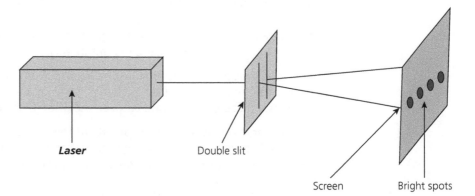

Laser   Double slit

Screen   Bright spots

Figure 5

8–1 Explain why a laser beam is suitable for this experiment.   [2]

8–2 Describe what happens to the fringes when the distance between the screen and the double slit is increased.   [2]

8–3 When a laser of wavelength 635 nm is used and the distance between screen and double slits is set to 1.54 m, the fringe spacing is 2.4 mm. Calculate the spacing between the slits. [3]

8–4 Describe **two** safety precautions that should be in place when using a laser beam in the science lab. [2]

8–5 The laser beam is replaced with a coherent source of white light. Describe the interference pattern that can now be observed on the screen. [2]

Total: 11

9 **Figure 6** shows two loudspeakers, L_1 and L_2, (not to scale) used on a concert stage. A group of sound engineers is testing the sound along a parallel line to L_1 and L_2. Both speakers are connected to the same signal generator playing a sound frequency of 265 Hz. Consider the speed of sound in air to be 340 m s^{-1}.

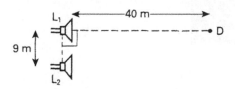

Figure 6

9–1 Calculate the path difference between the sound waves generated by L_1 and L_2 at point D. [3]

9–2 Calculate at what distance from the central maximum the first minimum will occur. You might find it useful to draw another diagram to help you. [4]

9–3 The sound engineers decide to put another speaker at point D. Explain why they will need to delay the sound played by the speaker at point D. [3]

9–4 Calculate the delay needed for the speaker at point D from the moment speaker L_1 plays a sound. [2]

Total: 12

10 **Figure 7** shows a beam of coherent microwave radiation passing through the two slits in the metal sheet. A microwave detector is moved parallel to the metal sheet at 30 cm from the slits.

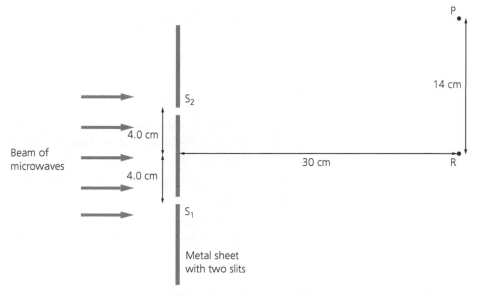

Figure 7

10–1 The second maximum is detected at point P. Calculate the frequency of the microwaves using the information from **Figure 7**. [3]

10–2 The microwave detector is now moved parallel to the metal sheet, but at a distance of 50 cm. Calculate the fringe spacing in the interference pattern that is detected. [2]

10–3 Describe how the fringe spacing changes when the distance between slits is reduced to 4 cm. [2]

Total: 7

SB1
p108–9   3.3.2.2

11   **Figure 8** shows a diagram of the diffraction pattern of monochromatic light going through a single pin hole in a sheet of aluminium foil.

Figure 8

11–1 Describe how the central maximum in **Figure 8** changes when the wavelength of light decreases. [1]

11–2 Describe how the central maximum in **Figure 8** changes when the slit width decreases. [1]

11–3 The monochromatic light is replaced with a source of white light. Describe the appearance of the central maximum and fringes when white light is used. [2]

11–4 A patient is called by a doctor for his surgery appointment. The patient is in the waiting room and the doctor is in her room with the door open. Explain why the patient can hear the doctor but cannot see her. [3]

Total: 7

SB1
p109–10   3.3.2.2   MS 4.7
                     AT i

12   A student shines a red laser beam of wavelength 650 nm through a diffraction grating of unknown spacing between slits. **Table 1** shows the measurements the student recorded.

Order of maxima	Angle, ϑ/rad	Slit spacing/m
1	0.216	
2	0.480	
3	0.752	
4	1.076	

Table 1

12–1 Estimate the number of slits per millimetre for the diffraction grating used by the student. Use the information in **Table 1** and a suitable method for your calculations. [4]

12–2 Estimate the absolute uncertainty on the value of slit spacing you used in **Question 12–1**. [2]

12–3 Describe how the angles for the orders of maxima would change, when a green laser beam is used instead of a red laser. Explain your answer. [2]

Total: 8

SB1
p109–11   3.3.2.2   MS 4.7
                     AT i

13   **Figure 9** shows the diffraction pattern of a green laser beam of wavelength 540 nm passing through the barbs of a feather. The diagram is drawn in its true dimensions.

Figure 9

13–1 The diffraction pattern in **Figure 9** was formed on a screen about 3.5 m away from the feather. Estimate the distance between the barbs in the feather. Use the diagram in **Figure 9** and a suitable method in your calculations. [4]

13–2 Derive the number of barbs per millimetre in this feather. [2]

13–3 A student suggests moving the screen further away from the feather to obtain 'better' measurements. Suggest **one** advantage and **one** disadvantage of such a change. Give a reason for each point you make. [4]

Total: 10

14 The light ray in **Figure 10** is parallel to the base of the prism AB. The refractive index of the prism is 1.5.

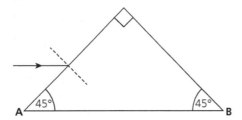

Figure 10

SB1 p86   3.3.2.3   MS 0.6, 4.1

14–1 Calculate the critical angle for the glass of the prism. [2]

SB1 p83–5   3.3.2.3   MS 0.6, 4.1

14–2 Calculate the angle of refraction for the light ray. [2]

SB1 p83–5   3.3.2.3   MS 0.6, 4.1

14–3 Copy **Figure 10** and draw the path of the refracted ray inside the prism. You only need to show the path of the light ray inside the prism in your answer. You will need a protractor to complete the path of the light ray. [2]

Total: 6

SB1 p88–9   3.3.2.3

15 Optical fibres have a core made of materials of high refractive index. The core is usually clad with a different material of low refractive index.

15–1 Explain why the cladding must have a lower refractive index than the core. [2]

15–2 Optical fibres would work even without cladding. Explain **two** advantages of using cladding in optical fibres. [4]

15–3 Explain why pulse broadening can affect the function of optical fibres. [2]

15–4 Describe the processes that could cause this effect. [2]

Total: 10

SB1 p83–5   3.3.2.3   MS 0.6, 4.1

16 The refractive index of a material is a concept used to compare the speed of waves in different materials.

16–1 Prove that the speed of light in air is (virtually) the same as the speed of light in a vacuum, using a suitable equation. [3]

16–2 Explain why the refractive index of a material cannot be smaller than 1. [2]

16–3 The refractive index of water is 1.33. Calculate the speed of light in water. [2]

16-4 The *relative refractive index* between two media can be defined as $_1n_2 = \dfrac{c_1}{c_2}$.

The *relative refractive index* between medium 1 and medium 2 is $_1n_2 = 0.4$ and the speed of light in medium 1 is $\dfrac{1}{3}$ of the speed of light in air. Calculate the speed of light in medium 2.

[3]

Total: 10

Topic review: waves

1 **Figure 11** shows a guitar string clamped to a table and held in tension by a set of masses. A movable bridge can be used to change the length of the guitar string that can be plucked to produce a sound.

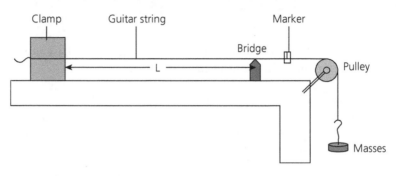

Figure 11

SB1 p75–7 3.3.1.2

1–1 State what type of waves sound waves are. [1]

SB1 p75–7 3.3.1.2

1–2 State what type of waves are set along the guitar string when it is plucked. [2]

SB1 p101–3 3.3.1.3

1–3 Sketch the third harmonic of the guitar string when it is being plucked. [2]

SB1 p101–3 3.3.1.3

1–4 The mass per unit length of the guitar string is $2.3 \times 10^{-3}\,\text{kg m}^{-1}$. A mass of 320 g is attached to the string. Calculate the distance from the clamp the bridge should be placed to play the first harmonic frequency of 147 Hz. [2]

SB1 p97–9 3.3.2.1

1–5 The same frequency of the first harmonic in **Question 1–4** is generated by a signal generator and played by two speakers placed next to each other at a distance of 1.2 m apart. Students are asked to walk along a parallel line to the two speakers at a distance of 8 m from the speakers. Explain what the students will notice. Your answer should include a calculation. You can use $330\,\text{m s}^{-1}$ for the speed of sound in air. [6]

Total: 13

SB1 p83–6 3.3.2.3

2 A laser beam enters the water in the fish tank in **Figure 12** at an angle of 70° from the normal. The speed of the laser beam in the water is approximately $2.3 \times 10^8\,\text{m s}^{-1}$.

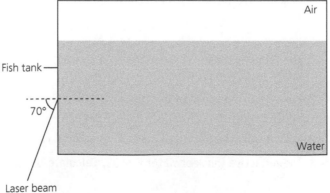

Figure 12

2–1 Calculate the angle of refraction of the light beam entering the fish tank. [2]

2–2 Copy and complete the diagram by showing the beam inside the tank. [1]

2–3 Describe what will happen to the laser beam as it hits the boundary between the water and the air above the tank. You should justify your answer with a calculation. [3]

2–4 A bag of sugar is poured into the bottom of the tank and left overnight to dissolve in the water. When the laser beam is shone inside the tank, the light beam follows the path shown in **Figure 13**. Explain why this happens. [3]

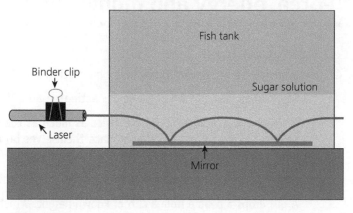

Figure 13

Total: 9

3 A student carries out an investigation to find the wavelength of a laser beam. She shines her laser pointer through a double slit of separation s = 0.4 mm and measures the mean fringe spacing w = 3.6 mm on a screen 2.3 m from the double slit.

SB1 p99 3.3.2.1

3–1 Calculate the wavelength of the laser beam. [2]

SB1 p108–11 3.3.2.2

3–2 The student uses the same laser pointer in a diffraction investigation using a diffraction grating with 330 slits per mm. Calculate the spacing between each slit. [1]

SB1 p108–11 3.3.2.2

3–3 In her diffraction investigation, the student calculates a wavelength of 635 nm. Give a reason to explain why her two results are different. [2]

SB1 p108–11 3.3.2.2

3–4 Calculate the uncertainty on the value of the wavelength that could be deduced from these two investigations. [1]

Total: 6

Mechanics and materials

Force, energy and momentum

Quick questions

SB1 p117 3.4.1.1

1   Name **three** vector and **three** scalar quantities.

SB1 p121 3.4.1.1

2   State what is meant when the forces acting on an object are in equilibrium.

SB1 p123 3.4.1.2

3   A child and an adult who weighs twice as much as the child sit on a seesaw. Suggest how they should place themselves on the seesaw so that it is in equilibrium.

SB1 p123 3.4.1.2

4   State the principle of moments.

SB1 p137 3.4.1.3 | MS 20.5, 2.2, 2.3, 2.4

5   A girl drops a pebble from a cliff of known height. Which equation should she use to calculate how long it will take the pebble to fall to the bottom of the cliff?

A: $v = u + at$

B: $s = \left(\dfrac{u + v}{2}\right)t$

C: $s = ut + \dfrac{a}{2}t^2$

D: $v^2 = u^2 + 2as$

SB1 p141 3.4.1.4

6   What are the factors that affect the air resistance on an object?

Exam-style questions

SB1 p117–20 3.4.1.1 | MS 0.6, 4.2, 4.4, 4.5 | PS 1.1

7   In **Figure 1** a newton meter is pulled vertically downward. The newton meter is hooked to a string at point C and the string is pinned on a wall at points A and B.

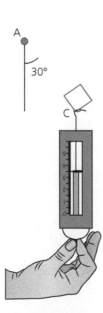

Figure 1

On a copy of **Figure 1**, draw and label the force of tension on AC and BC using the information in it. You will need to calculate each force from the information provided. [2]

Total: 2

8   During a science show, a presenter places a few pellets of dry ice (solid CO_2) inside a deflated balloon and then seals the balloon. The balloon is placed on top of a digital mass scale and the measurement of the mass of the balloon decreases on the scale, as the dry ice sublimates into gaseous CO_2 inflating the balloon.

8–1   Explain why the mass of the balloon appears to decrease on the digital scale using your understanding of forces. [3]

8–2   When all the dry ice has sublimated into CO_2 gas, the presenter fills another balloon with the same volume of air (mixture of 80% N_2 and 20% O_2 gas). He then drops both balloons side by side. Explain which balloon will fall to the floor first. [2]

Total: 5

9   **Figure 2** shows a box at rest on a ramp. The forces on the box are in equilibrium and the mass of the box is 5 kg.

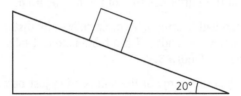

Figure 2

9–1   Copy **Figure 2** and draw all the forces applied on the box. Make sure the forces show clearly that the box is in equilibrium. [1]

9–2   Calculate the value of all the forces acting on the box. [3]

9–3   A lubricant is now placed on the ramp and the box starts to move along the ramp with an acceleration of 0.52 m s^{-2}. Calculate the forces acting on the box in this situation. [2]

9–4   The box continues to slide to the bottom of the ramp for a total distance of 3.4 m. When the box reaches the bottom of the ramp, it decelerates along the horizontal floor until it stops after 1.2 s. Calculate the force needed to stop the box. [2]

9–5   State what assumption you had to make about the stopping force in **Question 9–4**. [1]

Total: 9

10 A ball bearing is moving at constant velocity on a table, as shown in **Figure 3**. At point P a constant force $F = 0.13\,\text{N}$ is applied to the ball bearing.

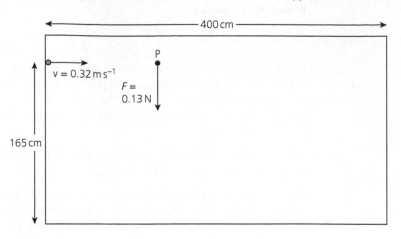

Figure 3

SB1 p117–18 | 3.4.1.1 | MS 3.5, 3.6

10–1 On a copy of **Figure 3**, sketch the trajectory of the ball bearing. [2]

SB1 p142–3 | 3.4.1.3, 3.4.1.4 | MS 0.5, 2.2, 2.3, 2.4

10–2 The ball bearing takes 2.25 s to reach point P. Calculate the distance between P and the initial position of the ball bearing. [1]

SB1 p142–3 | 3.4.1.3, 3.4.1.4 | MS 0.5, 2.2, 2.3, 2.4

10–3 The mass of the ball bearing is 23 g. Calculate the distance travelled by the ball bearing along the length of the table before it falls off the table, using the information in **Figure 3**. [4]

SB1 p172 | 3.4.1.3 | MS 0.3 / PS 3.3, 4.1

10–4 Calculate the kinetic energy of the ball bearing just before it falls off the table. [3]

Total: 10

11 **Figure 4** shows a tennis ball and a ping pong ball joined by an elastic band. The two balls are pulled in opposite directions and held 1.84 m above the floor, so that the elastic band attached to each ball is in tension. F_t is the force of the elastic band pulling the tennis ball and F_p is the force of the elastic band pulling the ping pong ball.

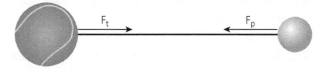

Figure 4

SB1 p153–7 | 3.4.1.5 | PS 4.1 / MS 0.5, 3.2 / AT a, b, d

11–1 The mass of the ping pong ball is 2.7 g and the mass of the tennis ball is 57.7 g. The two balls are dropped simultaneously and at the instant they are dropped, the acceleration on the tennis ball is 7.28 m s^{-2}. Deduce the acceleration of the ping pong ball at that instant. [2]

SB1 p134–7 | 3.4.1.3 | AT d

11–2 Describe the motion of the ping pong ball and tennis ball when they are released and allowed to drop. [3]

SB1 p171–2 | 3.4.1.8 | MS 0.4, 2.2

11–3 Calculate the velocity of the tennis ball as it hits the floor. [2]

SB1 p173–4 | 3.4.1.7 | MS 0.3 / PS 3.3, 4.1 / AT a, b, f

11–4 A group of students determines that the efficiency of the tennis ball bounce is 43%. Calculate the height of bounce of the tennis ball. [1]

Total: 8

SB1
p134–5,
171–4

3.4.1.3,
3.4.1.7,
3.4.1.8

MS 0.3,
0.5, 2.2,
2.3, 2.4

12 **Figure 5** shows the velocity–time graph of a basketball dropped from a certain height.

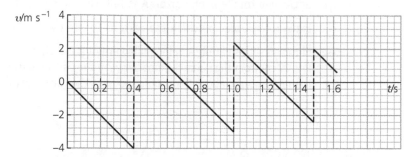

Figure 5

12–1 Calculate the efficiency of the ball on its first bounce. *[3]*

12–2 The ball weighs 5.17 N. Calculate the kinetic energy of the ball at the start of its second bounce. *[2]*

12–3 Sketch the displacement–time graph and acceleration–time graph for the bouncing basketball. *[2]*

Total: 7

SB1
p109–11

3.4.1.3,
3.4.1.4

MS 4.7

AT i

13 A cricket ball is batted to a velocity of 92 km h⁻¹ and at an angle of 45°. The ball is hit by the bat when it has bounced to a height of 1.30 m.

13–1 Calculate the maximum height the ball reaches. *[3]*

13–2 Calculate the horizontal velocity of the ball. *[1]*

13–3 Calculate the time taken by the ball to reach its highest point after being hit. *[1]*

13–4 Calculate how long it takes for the ball to reach the ground after reaching its highest point. *[3]*

13–5 Calculate how far the ball travels from the bat before it hits the ground. *[2]*

13–6 Suggest how the distance travelled by the ball would change, if the angle at which it was hit changed. *[1]*

Total: 11

SB1
p184–7

3.4.1.6

MS 2.2,
2.3

14 Cars have safety features to reduce the risk of fatal injury to the driver and passengers.

14–1 Explain how the safety features of a car reduce the risk of fatal injury to the driver and passengers. *[3]*

14–2 The mass of an average human head is 4.5 kg and it is estimated that the stopping force on it during a crash should not exceed 130 N, if a person is to survive the crash. Estimate the minimum time for the airbag to decelerate the driver's head safely when the car is travelling at a speed of 83 km h⁻¹ when it crashes. *[2]*

14–3 State what assumption you had to make to calculate the time in **Question 14–2**. *[1]*

Total: 6

15 Three identical juice cartons are pushed together from the top, so they tilt until they topple over. Carton A is full, carton B is half-full and carton C is empty.

15–1 Explain the order in which the cartons topple over. [3]

15–2 **Figure 6** shows carton A being pulled at point P with a constant force $F = 0.83\,N$. P is 27 cm from the contact point between the carton and the table. Calculate the moment of F about the contact point of the carton with the table. [1]

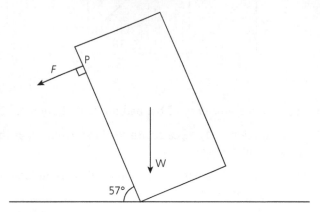

Figure 6

15–3 The total height of carton A is 0.30 m and the total anticlockwise moment about the contact point of the carton with the table is 1.2 N m. Calculate the mass of the carton. [3]

15–4 Force F is suddenly removed when carton A is tilted as shown in **Figure 6**. Explain whether the carton will topple over or return to its upright position. [1]

Total: 8

16 A boy and a girl are pushing a playground roundabout with a force of 56 N, but in opposite directions and at opposite ends of the roundabout's diameter.

16–1 State the name of the two forces considered together. [1]

16–2 The radius of the roundabout is 1.30 m. Calculate the moment of the two forces. [1]

16–3 The boy trips and falls, so the girl is now the only one pushing one end of the roundabout with half her original force. Calculate the moment of the forces now. [1]

16–4 For the roundabout to keep spinning with only the girl pushing it, another force must be applied. State where this second force is applied, its magnitude and its direction. [2]

Total: 5

Materials

Quick questions

SB1 p204 3.4.2.1
1   Describe what the elastic limit of a material is.

SB1 p201–2 3.4.2.1
2   Explain why a solid sphere of steel sinks in water, but an oil tanker can float.

SB1 p208–9 3.4.2.1
3   How can the elastic strain energy of a material that does not obey Hooke's law be calculated?

SB1 p208 3.4.2.1
4   Show how the energy stored by a spring can be expressed in terms of the spring constant. Start from the equation *energy stored* $= \frac{1}{2}F\Delta L$.

SB1 p205 3.4.2.1
5   The same mass is hooked to different spring-systems, each using identical springs. Choose the system that will show the lowest extension.

A: two springs in series

B: two springs in parallel attached to a single spring

C: two springs in parallel

D: a single spring

SB1 p211–12 3.4.2.2
6   Show that the units of the Young modulus of a material are $N\,m^{-2}$.

Exam-style questions

7   **Figure 7** shows the apparatus used by a school to investigate the elastic properties of a copper wire. The pulley is slightly above the height at which the copper wire is clamped on the opposite side of the desk. This results in the wire being stretched at an angle of 2° compared to the plane of the desk.

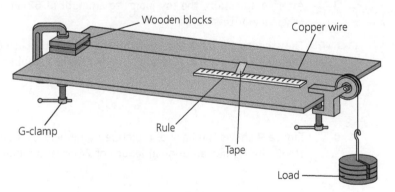

Figure 7

SB1 p117–20 3.4.2.1 | MS 0.6, 4.2, 4.4, 4.5 | PS 1.1

7–1   Estimate the actual extension when an extension of 2.3 mm is recorded by the students. *[2]*

SB1 p206–7 3.4.2.1 | MS 0.2, 4.2, 4.3 | PS 3.3, 4.1 | AT e

7–2   Calculate the percentage uncertainty in the extension measured introduced by this slight tilt in the wire. *[2]*

SB1 p206–7 3.4.2.1 | MS 0.2, 4.2, 4.3 | PS 3.3, 4.1 | AT e

7–3   Explain whether the school should ignore the effects of the tilt in their results, or not. *[1]*

SB1 p206–7 3.4.2.1 | MS 0.2, 4.2, 4.3 | PS 3.3, 4.1 | AT e

7–4   The spring constant of the copper wire is 6125 N m⁻¹. Calculate the extension the wire should experience for a load of 2.68 kg. *[2]*

Total: 7

8   A student measures the mass of the spring toy in **Figure 8** on a digital scale and then presses the toy down until the sucker is in contact with the base of the toy. This causes a compression of the spring of 3.2 cm. The mass of the toy was 14.8 g and the reading on the scale when the toy was pushed down was 985.5 g.

Figure 8

SB1 p207–8	3.4.2.1	MS 0.4, 4.3

8–1   Calculate the spring constant of the spring toy.   *[2]*

SB1 p207–8	3.4.2.1	MS 0.4, 4.3

8–2   Calculate the energy stored by the spring toy when it is compressed and about to jump.   *[1]*

SB1 p173–4	3.4.1.7	MS 0.3
		PS 3.3, 4.1

8–3   After compression, the toy jumps to a height of 62 cm. Calculate the efficiency of this toy.   *[2]*

SB1 p207–8	3.4.2.1	MS 0.4, 4.3

8–4   The student stretches the spring of the toy, so that it now can be compressed by 4.0 cm. Explain how this change would affect the height of jump of the toy.   *[2]*

Total: 7

9   **Figure 9** shows the back of a picture frame held by a nail at point P. The steel wire AB has an original length of 24 cm and a diameter of 1 mm.

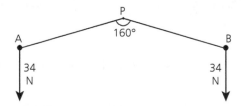

Figure 9

SB1 p117–20	3.4.1.1	MS 0.6, 4.2, 4.4, 4.5
		PS 1.1

9–1   Calculate the tension on the steel wire.   *[1]*

SB1 p211–12	3.4.2.2	MS 1.1

9–2   Calculate the tensile stress in the steel wire.   *[2]*

SB1 p211–12	3.4.2.2	MS 1.1

9–3   The Young modulus of the steel wire is 210 GPa. Estimate the extension of the steel wire when the picture frame is held by the nail at point P.   *[2]*

Total: 5

Topic review: mechanics and materials

1 A sphere of mass 320 g is suspended by two strings of negligible mass at point C, as shown in **Figure 10**. The distance between A and B is 52 cm.

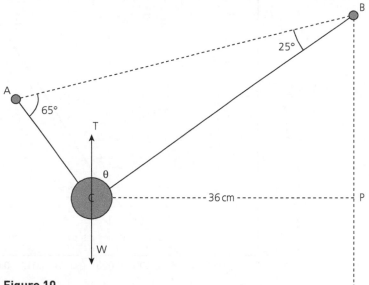

Figure 10

SB1 p118–21	3.4.1.1	MS 0.6, 4.2, 4.4, 4.5

1–1 Calculate angle θ.   [2]

SB1 p118–21	3.4.1.1	MS 0.6, 4.2, 4.4, 4.5

1–2 Calculate the tension on string AC and string BC.   [3]

SB1 p118–21	3.4.1.1	MS 0.6, 4.2, 4.4, 4.5

1–3 String AC is cut. Calculate the initial acceleration of the ball.   [1]

SB1 p118–21	3.4.1.1	MS 0.6, 4.2, 4.4, 4.5

1–4 After string AC is cut, the ball will swing towards the vertical line PB. A sharp razor blade is placed at point P, so that string BC is cut as the ball reaches the vertical line. Calculate the velocity of the ball when string BC is cut.   [2]

SB1 p133–9	3.4.1.3	MS 0.5, 2.2, 2.3, 2.4

1–5 Describe the motion of the ball after point P.   [1]

SB1 p133–9	3.4.1.3	MS 0.5, 2.2, 2.3, 2.4

1–6 After string BC is cut, the ball falls from a height of 85 cm. Calculate the horizontal displacement of the ball from the moment string BC is cut.   [2]

Total: 11

2 **Figure 11** shows an arrow loaded on a bow at the instant when the string has just been released. The mass of the arrow is 0.060 kg.

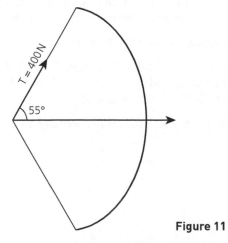

Figure 11

The tension of one of the strings is shown in **Figure 11**.

SB1 p118–21	3.4.1.1	MS 0.6, 4.2, 4.4, 4.5

2–1 Calculate the resultant horizontal force on the arrow.   [1]

SB1 p135 | 3.4.1.3 | MS 3.6, 3.7

2–2 The graph of velocity against time from the moment the arrow is released is shown in **Figure 12**.

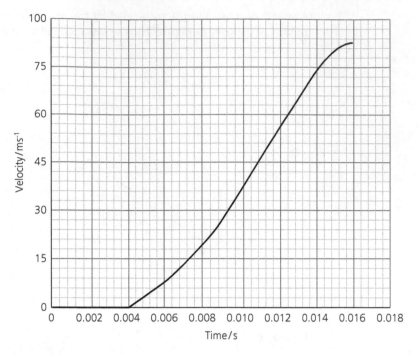

Figure 12

Explain why the curve in **Figure 12** plateaus at 0.016 s. [2]

SB1 p135 | 3.4.1.3 | MS 3.6, 3.7

2–3 Use a linear section of the curve in **Figure 12** to calculate the acceleration of the arrow. [2]

SB1 p83–6 | 3.4.1.6 | MS 2.2, 2.3

2–4 Calculate the resultant force on the arrow from the linear section of the curve you used in **Question 2–3**. [1]

SB1 p83–6 | 3.4.1.6 | MS 2.2, 2.3

2–5 Explain why the resultant force is different from the force you calculated in **Question 2–1**. [1]

SB1 p83–6 | 3.4.1.6 | MS 2.2, 2.3

2–6 Deduce the impulse of the arrow. Use both values of forces on the arrow. [2]

SB1 p83–6 | 3.4.1.6 | MS 2.2, 2.3

2–7 Calculate the percentage discrepancy between the impulse you calculated in **Question 2–6** and the impulse that can be deduced from the graph in **Figure 12**. [2]

Total: 11

SB1 p118–21 | 3.4.1.1 | MS 0.6, 4.2, 4.4, 4.5

3 A firework of mass 250 g is shot at an angle of 20° from the vertical line. The magnitude of the thrust pushing the firework up is 54.0 N.

3–1 Calculate the resultant force on the firework. [4]

3–2 The firework was faulty and it didn't explode. It reached a maximum height of 45 m with a horizontal velocity of 8 ms⁻¹. Calculate how long it will take for the firework to fall back to the ground. [2]

3–3 Calculate how far the firework travels horizontally after reaching the highest point. [1]

Total: 7

4 Electricity

Current electricity

Quick questions

SB1 p221 3.5.1.1

1 Explain why an ammeter needs to have a very low resistance.

SB1 p227 3.5.1.6

2 Describe the difference between potential difference (p.d.) and electromotive force (e.m.f.).

SB1 p229 3.5.1.3

3 State the properties of superconductors with respect to current flow, resistance, temperature and fields.

SB1 p227–8 3.5.1.1 AT b, f

4 Choose the correct ratio of p.d. shared across the points AB, BC and CD in **Figure 1**.

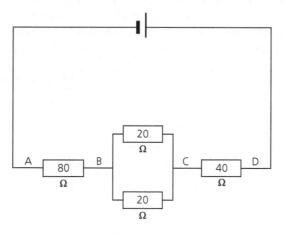

Figure 1

A: $80:40:40$

B: $2:1:1$

C: $8:1:4$

D: $4:2:1$

SB1 p231 3.5.1.2

5 Sketch the graph of a filament lamp with a higher resistance than the one shown in **Figure 2**.

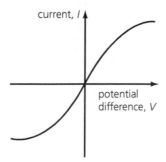

Figure 2

SB1 p246–7 3.5.1.4

6 Express electrical power $P = IV$ in terms of I and R, and in terms of V and R.

Exam-style questions

7   A multiple extension lead with 6 sockets has a built-in trip switch that breaks the circuit if the current flowing through the main cable becomes too large.

7–1   Three students are discussing how the extension lead works. Each statement contains not only valid points about the extension lead but also incorrect points.

Which part of each statement is incorrect?   [3]

Statement 1: Each socket is connected in series with the cable, so that the current from the cable can be shared by each device plugged in.

Statement 2: All sockets are connected in parallel, so the total resistance of the block of sockets increases, causing the wires to become very hot.

Statement 3: The potential difference is shared across each appliance, so as more appliances are connected there will be a build-up of current in the cable.

7–2   Explain why a safety mechanism that opens the circuit of a multiple socket lead when too much current builds at the cable might be needed. Include electrical power in your answer.   [4]

7–3   A builder uses a four-socket extension lead that can operate with a maximum current of 27.5 A before the safety mechanism breaks the circuit. The extension lead is connected to a mains potential difference (p.d.) of 230 V. The builder has plugged in an electric heater already. Use **Table 1** to choose **three** other devices she can plug in the extension lead without activating the safety mechanism. Show your calculations.   [4]

Device	Resistance / Ω
Electric drill	115
Electric kettle	36
Electric heater	15
Grinder	44
Radio	50
Plaster mixing tool	38

Table 1

7–4   Calculate the power drawn by the extension lead when the appliances you selected and the electric heater are plugged in.   [1]

7–5   The builder suggests that the electric drill, the grinder and the radio could be connected in series with the mains potential difference (p.d.). Calculate the p.d. across each appliance in this scenario.   [4]

7–6   The appliances might not operate correctly if they were wired as suggested by the builder in **Question 7–5**. Explain why. Refer to the current in your answer.   [1]

Total: 17

8   The relationship of the resistance of a light-dependent resistor (LDR) as a function of light intensity is shown in **Figure 3**.

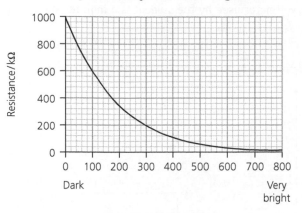

Figure 3

8–1   A manufacturer of light-sensitive switches wants to use this LDR to turn a set of garden lights on when the outdoor light intensity drops below 200 lux. Draw a potential divider that would allow the circuit in the garden lights to turn the lights on when the output potential difference reaches 2.5 V or below. [4]

8–2   The manufacturer wants to use the same LDR in a light probe for schools. Explain what resistance the fixed resistor in the potential divider circuit for this probe should have to collect accurate measurements of light intensity in the range of 300 to 500 lux. [4]

Total: 8

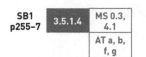

9   **Figure 4** shows a block of resistors connected to a 9 V cell.

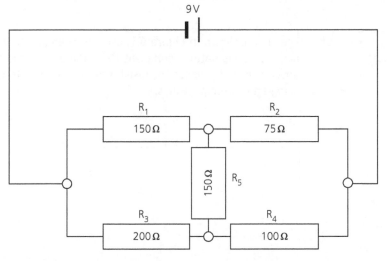

Figure 4

9–1   Explain what current flows through resistor R_5. [4]

9–2   Calculate the resistance of the block of resistors in **Figure 4**. [3]

9–3   Calculate the potential difference across resistor R_1. [2]

Total: 9

10　A student connects a variable resistor as shown in **Figure 5** to make a potential divider. The student's circuit is incorrectly connected. The voltmeter used by the student is a high resistance digital voltmeter.

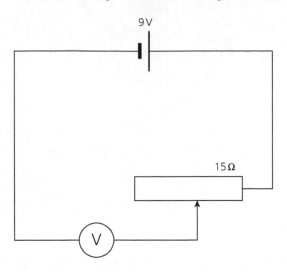

Figure 5

10–1 Explain what values of potential difference the student will read as the sliding contact on the variable resistor is moved. [4]

10–2 Draw a correct potential divider circuit using the same components from **Figure 5**. [1]

10–3 Explain for what value of resistance in the variable resistor to the left of the sliding contact the potential difference measured by the voltmeter is 3 V (in the correct potential divider circuit you drew). Use calculations in your answer. [3]

10–4 Calculate the resistance of the variable resistor to the right of the sliding contact when the reading on the voltmeter is 7.3 V. [3]

Total: 11

11　The circuit shown in **Figure 6** is used as a temperature probe. T_1 is a negative temperature coefficient (NTC) thermistor and the potential difference (p.d.) across the resistor R_2 can be used to calibrate the probe using known temperatures.

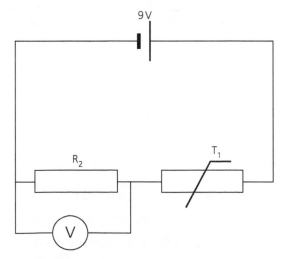

Figure 6

11–1 Suggest an experiment a group of students could carry out to calibrate this circuit to work as a temperature probe. List the equipment needed, describe the layout of the experiment and the method to be used. [6]

11–2 When T_1 is placed inside a cup of hot coffee, the resistance R_2 and the resistance of T_1 are equal. Explain how the p.d. across R_2 will change when T_1 is removed from the cup and is left to cool down. [3]

11–3 At $-12°$ the resistance of T_1 becomes $150\,k\Omega$. Calculate the p.d. across R_2 if its resistance equals $220\,k\Omega$. [3]

11–4 The voltmeter across R_2 is replaced with an analogue voltmeter which has a resistance of $200\,k\Omega$. Calculate the p.d. recorded by the analogue voltmeter. [3]

11–5 Compare the p.d. across R_2 from your answer to **Questions 11–3 and 11–4**. Explain why it is better to use an analogue or a high resistance digital voltmeter in this circuit. [3]

Total: 18

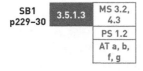

12   The circuit in **Figure 7** shows a superconductor S and two resistors R_1 and R_2 connected to a 6 V cell at room temperature. The critical temperature of the superconducting material is 90 K.

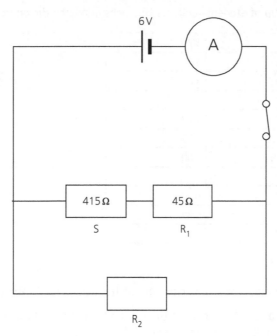

Figure 7

12–1 Describe what happens in a superconductor below its critical temperature. [2]

12–2 State **two** applications of superconductors. [2]

12–3 At room temperature, the current read by the ammeter shown in **Figure 7** is 63 mA. Calculate the resistance of R_2 at room temperature. [4]

12–4 The superconductor S is immersed in liquid nitrogen. The boiling point of liquid nitrogen is $-195.8\,°C$. Calculate the current read by the ammeter in **Figure 7** in this situation. [4]

Total: 12

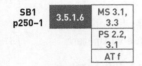

13 A 55 Ω resistor is connected to two dry cells in series. The e.m.f. of each cell is 1.5 V and their individual internal resistance is 1.2 Ω.

13–1 Calculate the current through the circuit. [2]

13–2 Calculate the p.d. across the resistor. [2]

13–3 Calculate the p.d. across **one** of the two cells in series. [2]

Total: 6

14 The circuit in **Figure 8** is used to derive the internal resistance of the cell.

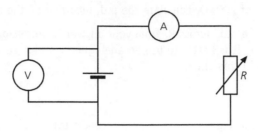

Figure 8

A group of students collects the measurements shown in **Table 2**.

I / A	V / V
0.089	1.50
0.210	1.36
0.310	1.29
0.400	1.23
0.500	1.15
0.615	1.00
0.700	0.99
0.790	0.90
0.910	0.78
1.000	0.73

Table 2

14–1 Plot a graph of p.d. versus current from the data in the table of results. [5]

14–2 Write a suitable equation to represent the line of best fit from the graph you plotted. Use the symbols V, I, ε and r. [1]

14–3 Use your graph to derive the internal resistance of the cell in **Figure 8**. [3]

14–4 Estimate the e.m.f. of the cell from your graph. [2]

Total: 11

SB1
p236–7

3.5.1.3

MS 3.2,
4.3
PS 1.2
AT a, b,
f, g

15 A group of students wants to estimate the resistivity of constantan. The students measure the diameter of a constantan wire at six different places along the wire. The length of the wire is 50.0 cm.

15–1 Five of the six measurements of diameter recorded by the students are shown in **Table 3**.

Diameter / mm	0.27	0.30		0.30	0.32	0.26

Table 3

The missing value from **Table 3** is shown in **Figure 9**, which represents the micrometer used to collect this measurement.

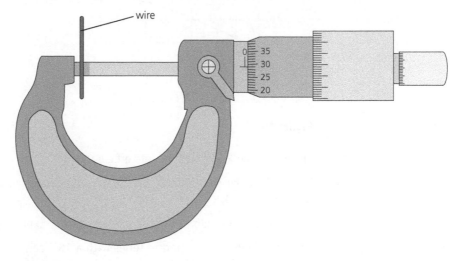

Figure 9

Read the diameter shown in **Figure 9**. [1]

15–2 Estimate the cross-sectional area of the constantan wire used by the students. [3]

15–3 The teacher looks at the micrometer used by the students and tells them that it was not calibrated correctly. The students find a zero error of $+2 \times 10^{-5}$ m. Correct your estimate of the cross-sectional area to account for this systematic error. [2]

15–4 The resistance across the wire measured by the students is 4.0 Ω. Calculate the resistivity of constantan using the students' results. [3]

15–5 The uncertainty on the length of the wire was ±0.1 cm and the uncertainty on the resistance was ±0.1 Ω. Estimate the absolute uncertainty in the value of the resistivity of constantan. [4]

15–6 The accepted value of resistivity of constantan is 4.9×10^{-7} Ω m. Explain whether the value derived from the students' data is acceptable or not. [2]

Total: 15

Topic review: electricity

SB1
p230–3

3.5.1.2

MS 0.3
PS 4.1

1 **Figure 10** shows the I–V characteristics of a 2.5 W light bulb and a 1.0 W light bulb.

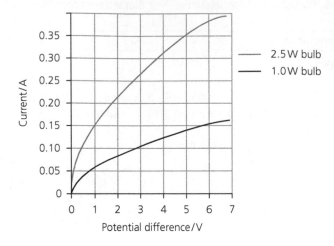

Figure 10

1–1 Show that the 2.5 W bulb is more efficient. [3]

1–2 Explain how the resistance of the 2.5 W bulb changes with the potential difference (p.d.). [2]

1–3 State which light bulb has the higher resistance. Explain your answer. [2]

1–4 The 2.5 W bulb is connected in series with a 16 V cell and an LDR as shown in **Figure 11**. **Figure 12** shows how the resistance of the LDR changes with the light intensity.

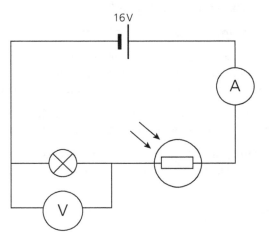

Figure 11

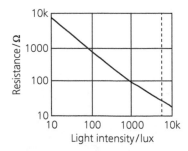

Figure 12

Deduce the current in the circuit when the light intensity shining on the LDR is 1000 lux. Use the information in **Figures 10, 11 and 12**. [5]

Total: 12

2   A group of students moulds a piece of play-dough in the shape of a conical frustum as in **Figure 13**.

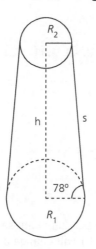

Figure 13

Side s = 25 cm and radii R_1 and R_2 are 0.8 cm and 0.6 cm, respectively.

SB1 p236–7	3.5.1.3	MS 3.2, 4.3
		AT a, b, f, g

2–1   Describe a suitable method that could be used to investigate the resistivity of the play-dough shape. Include a circuit diagram for your suggested investigation. *[6]*

SB1 p236–7	3.5.1.3	MS 3.2, 4.3
		AT a, b, f, g

2–2   The resistivity of play-dough is approximately 0.24 Ω m. Calculate the resistance of the block of play-dough in **Figure 13**.   *[5]*

SB1 p250–1	3.5.1.6	MS 3.1, 3.3

2–3   When the block of play-dough is connected to a cell of e.m.f. 4.5 V, a current of 8.6×10^{-3} A flows through the play-dough. Calculate the internal resistance of the cell.   *[2]*

Total: 13

SB1 p230–3	3.5.1.2	MS 0.3
		PS 4.1

3   **Figure 14** shows a circuit diagram. The current measured by the ammeter is 220mA.

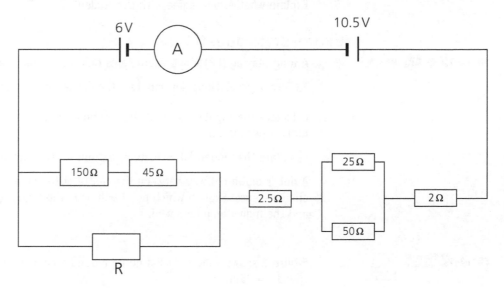

Figure 14

3–1   Calculate the value of the resistance of R.   *[5]*

3–2   Describe how the total current in the circuit would change if resistor R were disconnected from the circuit. Explain why the current would change in the way you suggested.   *[3]*

Total: 8

35

5 Further mechanics and thermal physics

Periodic motion

Quick questions

SB2 p2 · 3.6.1.1

1 Express 2 radians in degrees.

SB2 p18–19 · 3.6.1.3

2 A mass-spring oscillator and a simple pendulum are set to oscillate on the Moon. State whether their time period T will change or not and explain why.

SB2 p229 · 3.6.1.3

3 Describe how the x–t graph for a damped mass-spring system changes in terms of its frequency and amplitude.

SB1 p21–3 · 3.6.1.3 · AT b, f

4 Choose the correct statement about a pendulum reaching the centre of oscillation.

A: The pendulum will have maximum E_k and zero ΔE_p.

B: The pendulum will have maximum E_k and maximum ΔE_p.

C: The pendulum will have minimum E_k and maximum ΔE_p.

D: The pendulum will have zero E_k and maximum ΔE_p.

SB2 p26–7 · 3.6.1.4

5 State the condition needed for resonance to occur in an oscillating system.

SB2 p14–15 · 3.6.1.2

6 A student says:

'The acceleration of a pendulum is maximum in the centre of oscillation, because that's where it is fastest. But the acceleration at maximum displacements is zero, because the pendulum has stopped there.'

Explain whether you agree with the student.

Exam-style questions

SB2 p2–3 · 3.6.1.1 · MS 0.4

7 A long-playing (LP) vinyl record spins on a turntable at an angular speed of $33\frac{1}{3}$ rpm (revolutions per minute). The diameter of an ordinary LP is 30 cm.

7–1 Calculate the angular speed of the turntable in degrees per second and in radians per second. *[1]*

7–2 Calculate the tangential velocity of a point on the circumference of the LP. *[2]*

7–3 A dot is drawn halfway between the centre of the LP and a point on the LP's circumference. Explain which point will have the higher tangential velocity and the higher angular speed. *[2]*

Total: 5

SB2 p3–5 · 3.6.1.1 · MS 3.2 · PS 4.1

8 **Figure 1** shows a motorcyclist going around a circular bend at constant speed $v = 25\,\mathrm{m\,s^{-1}}$.

Figure 1

8–1 Copy and complete the free body diagram by drawing the contact force F, its vertical component N and its horizontal component F_c. [2]

8–2 If the bend were a full circle, it would take the motorcyclist 3.5 s to go around the full circle. Calculate the radius of the bend. [2]

8–3 The combined weight of the motorcycle and rider $W = 2651$ N. Calculate the centripetal force on the motorcycle and rider. [2]

8–4 State what provides the centripetal force for the motorbike and rider in this situation. [1]

8–5 Calculate the angle between the motorcycle's vertical axis and the ground during the bend. [2]

Total: 9

9 **Figure 2** shows a student swinging a beaker with some water placed on a tray suspended by four strings in a uniform circular motion. The angle between radius AF and radius BF is 30°. The mass of the beaker and water is 430 g.

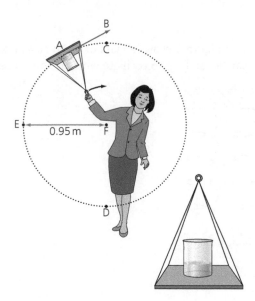

Figure 2

9–1 At point A the contact force N on the beaker from the tray is 23.9 N. Calculate the centripetal acceleration on the beaker–water system. [3]

9–2 Calculate the tangential velocity of the beaker–water system. [2]

9–3 Calculate the period of **one** cycle. [2]

9–4 On a copy of **Figure 2**, sketch and label all the forces on the beaker–water system at points C, D and E. [3]

Total: 10

SB2 p18–19   3.6.1.3   MS 4.6   AT b, c

10 A baby and his sister are being swung on two swings of identical length in the park. Both swings started from the same displacement with a small angle from the rest position. The mass of the sister is three times the mass of her baby brother.

10–1 Explain why the sister will remain in phase with her brother's swing. [2]

10–2 Describe what would happen if one of the swings were swung at large angles from the rest position. [1]

10–3 The length of the chain holding the swings is 2.4 m. Calculate the frequency of the oscillations of the baby's swing. [2]

10–4 Suggest what changes the sister should make to swing with a time period $T = 2.5$ s. Show your calculations. [2]

Total: 7

11 A bicycle engineer wants to test the front fork suspension for a new bike. She compresses the suspension spring with a compression $x = 2.5$ cm and records the time of 5 oscillations. Her results are shown in **Table 1**.

T_1/s	T_2/s	T_3/s	T_4/s	T_5/s
0.32	0.29	0.31	0.32	0.33

Table 1

SB2 p14–15 | 3.6.1.2 | MS 3.6, 3.8, 3.9, 3.12 | AT i, k

11–1 Calculate the maximum acceleration of the spring system. [2]

SB2 p24–5 | 3.6.1.4 | AT g, i, k

11–2 The suspension of the front fork is damped. Explain what effect this damping has on the oscillations of the spring and why the maximum acceleration you calculated can only apply on the initial compression. [4]

SB2 p14–15 | 3.6.1.2 | MS 3.6, 3.8, 3.9, 3.12 | AT i, k

11–3 The engineer measures a damping in amplitude of approximately 8.5% per oscillation. Estimate the maximum velocity of the fork after three oscillations. [2]

SB2 p24–5 | 3.6.1.4 | AT g, i, k

11–4 Explain why it is important to damp the oscillations of bicycle suspensions. [3]

Total: 11

SB2 p13–15 | 3.6.1.2 | MS 3.6, 3.8, 3.9, 3.12 | AT i, k

12 The diagram in **Figure 3** shows a mass of 850 g suspended by two identical springs in parallel and set to oscillate with an amplitude of 10 cm.

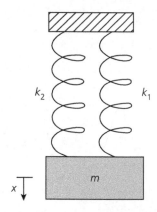

Figure 3

12–1 **Figure 4** represents the displacement–time graph for this mass. Draw the velocity–time and acceleration–time graphs using the information in **Figure 4**. [4]

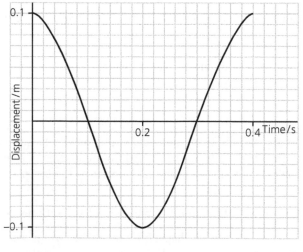

Figure 4

12–2 Calculate the spring constant for each spring. [3]

Total: 7

SB2
p15–23 3.6.1.2, 3.6.1.3

MS 3.6, 3.8, 3.9, 3.12, 4.6
AT b, c, i, k

13 The total energy and the changes in potential (E_p) energy and kinetic (E_k) energy over time of a system oscillating in simple harmonic motion is represented by **Figure 5**.

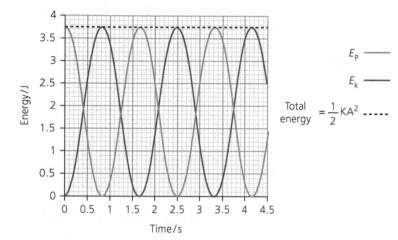

Figure 5

The potential energy can be expressed as $E_p = \frac{1}{2}kx^2$.

13–1 Explain, from the graph, where the starting position of the system is. *[1]*

13–2 Calculate the frequency of the oscillations. *[1]*

13–3 The system was set to oscillate with an amplitude of 4.8 cm. Calculate the constant k. *[2]*

13–4 Calculate the mass of the oscillating object. *[2]*

13–5 Calculate the values of x at which the graph of E_p intersects the graph of E_k. *[4]*

Total: 10

Thermal physics

Quick questions

SB2 p57 3.6.2.1 **1** What is meant by heat?

SB2 p57 3.6.2.1 **2** What is meant by internal energy?

SB2 p68 3.6.2.2 **3** Give the value in Celsius of absolute zero (0 K).

SB2 p74 3.6.2.2 **4** What is meant by an ideal gas?

SB2 p74 3.6.2.3 **5** Brownian motion can be observed using a smoke cell under the microscope. Describe what you would see. Explain your answer in terms of the movement of molecules.

SB2 p71 3.6.2.3 **6** What is the value of Avogadro's constant?

SB2 p58 3.6.2.1   MS 2.1, 2.2 **7** The formula that links thermal energy used and the specific heat capacity is:

$$Q = mc\Delta\theta$$

Copy and complete **Table 2** to show what the symbols mean for this formula and what the units are for each term.

Symbol	Meaning	Units
Q		
m		
c		
$\Delta\theta$		

Table 2

SB2 p71 3.6.2.2 MS 2.2

8   The formula that links pressure and temperature is:

$$pV = nRT$$

Copy and complete **Table 3** to show what the symbols mean for this formula and what the units are for each term.

Symbol	Meaning	Units
p		
V		
n		
R		
T		

Table 3

SB2 p64 3.6.2.1 MS 0.1, 0.2

9   Calculate how much energy is needed to boil 0.55 kg of water to steam.

The specific latent heat of vaporisation of water is $2.26 \times 10^6 \, J \, kg^{-1}$.

SB2 p58 3.6.2.1 MS 0.1, 0.2, 2.1

10   Calculate the energy flow when 0.25 kg of olive oil is heated from 20 °C to 45 °C.

The specific heat capacity of olive oil is $1790 \, J \, kg^{-1} \, K^{-1}$.

SB2 p60 3.6.2.1 MS 0.1, 0.5, 2.3

11   A china cup of mass 0.225 kg is filled with 0.25 kg water which has a temperature of 15 °C. It is placed in a microwave oven, which transfers energy to the cup at a rate of 800 W. Calculate how long it takes the water to heat up until it just starts to boil.

The specific heat capacity of water is $4200 \, J \, kg^{-1} \, K^{-1}$.

The specific heat of china is $1000 \, J \, kg^{-1} \, K^{-1}$.

SB2 p60 3.6.2.1 MS 0.1, 0.5, 2.3

12   A lump of iron of mass 2.5 kg has a temperature of 600 K. It is dropped into 15 kg water which has a temperature of 300 K. Neglecting the heat capacity of the container, and assuming that the mass of water lost to steam is negligible, what is the final temperature of the iron and water?

c for iron $= 438 \, J \, kg^{-1} \, K^{-1}$. c for water $= 4200 \, J \, kg^{-1} \, K^{-1}$.

SB2 p77 3.6.2.3 MS 0.5

13   The speed (in $m \, s^{-1}$) of seven particles is measured as follows:

3.2; 4.7; 3.4; 5.8; 7.2; 9.3; 2.6

a)   Calculate the mean speed, $\bar{c}$.

b)   Calculate the mean speed squared, $\bar{c}^2$.

c)   Square the speeds and calculate the mean square speed, $\overline{c^2}$.

d)   Now calculate the root mean square speed, $\sqrt{\left(\overline{c^2}\right)}$.

e)   What is the difference between the root mean square speed and the mean speed?

SB2 p77 3.6.2.3 MS 2.2, 2.3

14   The density of air at 15 °C and $1.013 \times 10^5 \, Pa$ is $1.23 \, kg \, m^{-3}$.

Show that the root mean square speed of the molecules is about $500 \, m \, s^{-1}$.

SB2 p70 3.6.2.2 MS 2.2

15   High flying aeroplanes fly at a height of about 10 000 m. The air pressure at this height is $2.64 \times 10^4 \, Pa$. The cabin pressure is $1.00 \times 10^5 \, Pa$. Calculate the force acting on a cabin door, which has an area of $1.2 \, m^2$.

SB2 p77 3.6.2.3 MS 2.2, 2.3

16   An air molecule has an average mass of $4.8 \times 10^{-26} \, kg$.

What is its RMS speed at a temperature of 500 K?

Exam-style questions

17 Look at the graphs in **Figure 6**:

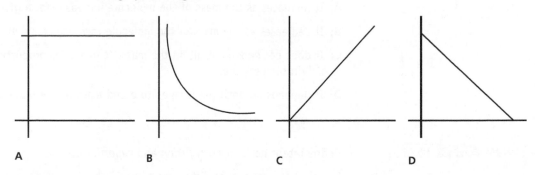

Figure 6

SB2 p66 | 3.6.2.2 | MS 3.1
17–1 Which one of these graphs shows Boyle's law? *[1]*

SB2 p67 | 3.6.2.2 | MS 3.1
17–2 Which one of these graphs shows Amonton's law (pressure–temperature law)? *[1]*

Total: 2

SB2 p74 | 3.6.2.2
18 Which one of these gases is an ideal gas? *[1]*

A: Nitrogen

B: Xenon

C: Propane

D: Hydrogen chloride

Total: 1

SB2 p78 | 3.6.2.3
19 Which one of the graphs in **Figure 7** correctly shows the relationship between the temperature and the average kinetic energy of a molecule of an ideal gas? *[1]*

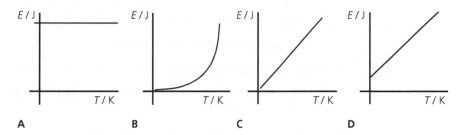

Figure 7

Total: 1

41

SB2 p74 | 3.6.2.2
20 Which one of the graphs in **Figure 8** correctly shows the relationship between the temperature and the RMS speed of an ideal gas? *[1]*

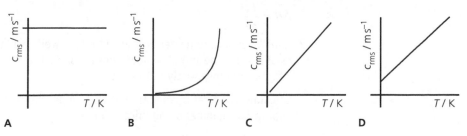

Figure 8

Total: 1

SB2 p78 3.6.2.3

21 Which one of these statements is correct about the translational kinetic energy of a molecule of an ideal gas? [1]

   A: It increases as the mass of the molecule increases for a given temperature.

   B: It decreases as the mass of the molecule increases for a given temperature.

   C: It does not depend at all on the mass. It increases proportionally with Kelvin temperature.

   D: It depends on both the temperature and mass of the molecule.

Total: 1

SB2 p64 3.6.2.1 MS 2.1

22–1 Define latent heat for both *fusion* and *vaporisation*. [2]

22–2 Explain how *latent heat* differs from *specific heat*. [2]

22–3 A 2.7 kW kettle contains 0.67 l of water. It is accidentally left open, so that the kettle does not switch off when the water boils. The kettle boils dry.

   1. Show that the energy used, from when the water just starts to boil, is about 1.5×10^6 J. The specific latent heat of vaporisation of water is 2.26×10^6 J kg^{-1}. [1]

   2. Calculate the time needed for the kettle to boil dry. [2]

   3. Once the kettle has boiled dry, explain what would happen to the kettle. A calculation is not expected. [1]

Total: 8

SB2 p64 3.6.2.1

23 In a certain industrial process, vaporised silicon is condensed onto components. In order to do this, silicon is melted, then the liquid silicon is heated until it boils and evaporates. 0.50 kg silicon is needed for each batch.

   • Specific latent heat of fusion of silicon = 1790 kJ kg^{-1}

   • Specific heat capacity of liquid silicon = 712 J kg^{-1} K^{-1}

   • Specific latent heat of evaporation of silicon = 12 800 kJ kg^{-1}

   • Melting point of silicon = 1414 °C

   • Boiling point of silicon = 3265 °C

23–1 Calculate the energy needed to melt the 0.50 kg of silicon. [2]

23–2 Calculate the energy to raise the temperature of 0.50 kg silicon from 1414 °C to 3265 °C. [2]

23–3 Calculate the energy required to boil the 0.50 kg of silicon. [2]

23–4 Determine the total energy required for each batch. [2]

Total: 8

SB2 p63 3.6.2.1 MS 2.3

24–1 A material is being heated at a constant rate. Show that the temperature change in a material is linked to the specific heat capacity by the expression:

$$\Delta\theta = \frac{Pt}{mc}$$ [5]

24–2 A 48 W laboratory heater is inserted into a well-insulated block of copper, mass 1.0 kg. The initial temperature is 21 °C. The heater is turned on and left running for exactly 5.0 minutes.

Calculate the final temperature of the copper block. Give your answer to an appropriate number of significant figures. [3]

Specific heat capacity of copper = 385 J kg^{-1} K^{-1}.

Total: 8

SB2 p66 `3.6.2.2` `AT a`
`PS 2.1,`
`2.2, 2.3`

25 A simple method of measuring how volume varies with pressure (Boyle's law) is shown in **Figure 9**.

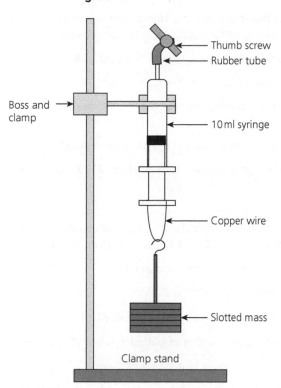

Figure 9

The syringe is full of air.

25–1 Describe how you would carry out the experiment to investigate how the volume of the air varies with pressure. [5]

25–2 State how you would process the data, in order to obtain values for pressure and volume. [5]

25–3 Sketch a graph to show what results you might expect from this experiment. [2]

25–4 Evaluate the effectiveness of investigating Boyle's law in this way. [3]

Total: 15

SB2 p77 `3.6.2.3` `MS 2.1,`
`2.2`

26–1 In the model of molecular kinetic theory, this equation is derived:

$$pV = \frac{1}{3}Nm\left(c_{rms}\right)^2$$

Explain how the term $\frac{1}{3}$ is reached in this equation. [3]

26–2 Pressure, p, and density, ρ, are related by:

$$p = \frac{1}{3}\rho\left(c_{rms}\right)^2$$

Stating the quantities involved, show how this equation is derived from the equation given in **Question 26–1**. [4]

26–3 At normal temperature and pressure (293 K and 1.013×10^5 Pa), helium has a density of 0.166 kg m⁻³. Calculate the root mean square speed of helium atoms at normal temperature and pressure. [4]

Total: 11

SB2 p78 3.6.2.3

27 The density of air at a temperature of 20 °C and pressure of 1.013×10^5 Pa is 1.2 kg m^{-3}. The volume of 1 mole of gas is 0.0244 m^3 at this pressure.

27–1 Show that the mass of a single molecule of air is about 5×10^{-26} kg. Assume all air molecules have the same mass. [3]

27–2 Calculate the average energy of an air molecule.

Boltzmann's constant $k = 1.38 \times 10^{-23}$ J K^{-1}. [1]

27–3 Using the density, show that the root mean square speed of an air molecule is about 500 m s^{-1}. [3]

27–4 Calculate the kinetic energy of an air molecule. [2]

27–5 Compare your answers to **Questions 27–2 and 27–4**. Explain your answer. [2]

Total: 11

SB2 p77 3.6.2.3

28 Molecules of a particular gas are confined to a box which is a cube. The length of each side of the box is 0.15 m. The density of this gas is 0.937 kg m^{-3}.

The pressure inside the container is 1.013×10^5 Pa.

The temperature is 273 K.

The volume of 1 mole of gas is 0.0224 m^3.

28–1 Show that the number of particles is about 9×10^{22}. [3]

28–2 Calculate the mass of a single particle. [2]

28–3 Calculate the root mean square speed of the molecules of this gas. [3]

28–4 Calculate the change of momentum of each molecule of gas as it hits a side. [3]

Total: 11

SB2 p77 3.6.2.3

29 Nitrogen (N_2) is a diatomic gas. Each atom has a relative atomic mass of 14.0.

29–1 For nitrogen:

1. Give the mass of one mole. [1]

2. The volume of 1 mole of the gas is 0.0224 m^3. Show that the density of the gas is 1.15 kg m^{-3}. [2]

29–2 The nitrogen gas is stored at a temperature of 20 °C and a pressure of 1.013×10^5 Pa, making the volume of the gas 0.0244 m^3 mol^{-1}.

1. Calculate the root mean square speed of the molecules. [3]

2. Determine the temperature at which the molecules travel twice as fast. [4]

29–3 As a car goes along a road, its tyres get warm, with the temperature in the tyre getting warmer by about 10 K.

1. Describe **one** way in which the molecular motion of the air in the tyre when warm is **similar** to the molecular motion in the tyre when it is cold. [1]

2. Describe **one** way in which the molecular motion of the air in the tyre when warm is **different** to the molecular motion in the tyre when it is cold. [1]

SB2
p76–7 | 3.6.2.3 | MS 3.1

30 Argon is an inert gas that forms about 1% of the Earth's atmosphere.

30–1 1. Sketch **two** graphs on the same axes that show the distribution of the speeds of molecules within an ideal gas at 100 K and 1000 K. [3]

2. State what these graphs are called and explain what they show. [5]

30–2 Argon has a density of 1.66 kg m^{-3} at a temperature of 20 °C at a pressure of 1.013×10^5 Pa.

1. Calculate the root mean square speed of the argon atoms. [3]

2. Determine the mass of an argon atom. (1 mole occupies 0.0244 m^3 at this temperature and pressure.) [4]

3. The root mean square speed doubles. Calculate the new temperature. [3]

Total: 18

Topic review: further mechanics and thermal physics

● ●

SB2 p67 | 3.6.2.2 | MS 3.1, 3.2, 3.3, 3.4
PS 3.1, 3.2

1 In an experiment to investigate the Pressure Law, the pressure of a fixed volume of gas is 101 kPa at 0 °C. The apparatus is cooled to −20 °C, before being heated steadily to a temperature of 100 °C. The pressure and temperature are recorded to two significant figures.

Table 4 gives the data recorded.

Temp / °C	Pressure / kPa
−20	94
−10	97
0	100
10	110
20	110
30	110
40	120
50	120
60	120
70	130
80	130
90	130
100	140

Table 4

1–1 Plot this data as a graph of pressure against temperature. [5]

1–2 Determine the gradient of your graph. Give the units. [4]

1–3 Write an equation that describes the graph, using the term θ for the Celsius temperature. [5]

1–4 Use the equation to determine the *x*-axis intercept. Give the correct unit. [3]

Total: 17

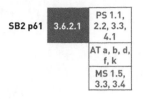

2–1 Define specific heat capacity. [1]

2–2 Describe a simple experiment using an electric heater that would allow you to determine the specific heat capacity of a block of aluminium. Include the basic measurements you need to make. [6]

2–3 The graph of such an experiment is shown in **Figure 10**. A 48 W heater was used.

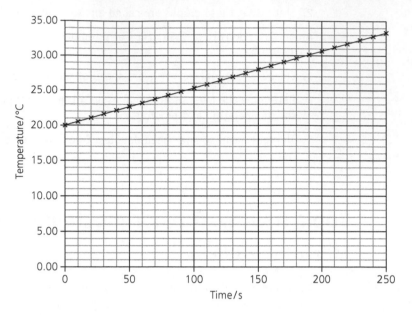

Figure 10

A student says that the graph shows *direct proportionality*. Comment on the student's answer. [2]

2–4 Use the graph to determine the specific heat capacity of aluminium. [4]

2–5 State and explain how uncertainty can be minimised for this experiment. [3]

Total: 16

3–1 Define the specific latent heat of vaporisation. [1]

Figure 11 shows a very simple refrigerator made with a cylinder of porous material. The cylinder is soaked before 1.5 kg of extra water is placed in the cylinder. After that, food or drink that needs to be kept cool is placed in a container that stands in the water in the cylinder.

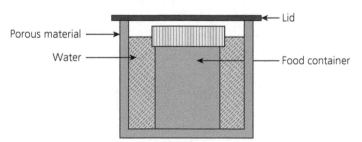

Figure 11

- The rate of evaporation of the water is found to be $1.5 \times 10^{-4}\,\text{kg s}^{-1}$.
- The specific latent heat of evaporation of water = $2.264 \times 10^{6}\,\text{J kg}^{-1}$.
- The specific heat capacity of water = $4200\,\text{J kg}^{-1}\,\text{K}^{-1}$.

3–2 Explain how water evaporating on a surface keeps the water inside cool. [2]

3–3 Show that the rate of energy transfer due to the evaporation of the water is about 340 W. [2]

3–4 State what provides the energy source for the evaporation. [1]

3–5 Calculate the initial rate of change of temperature, assuming that the rate of evaporation remains constant. Give your answer to an appropriate number of significant figures. [3]

3–6 Discuss whether the rate of change of temperature within the water will remain constant. [3]

Total: 12

4 In an investigation into the suspension of a car, the car can be modelled as a block mounted on four identical springs, as shown in **Figure 12**.

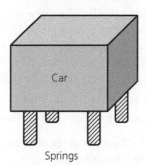

Car

Springs

Figure 12

The block has a mass of 1500 kg. Each spring is 0.30 m long when unloaded.

When the block is lowered on to the four springs, the springs are compressed by 2.0×10^{-2} m.

The block is then pushed down so that the length of the springs is now 0.27 m. It is then released. The block oscillates with simple harmonic motion.

4–1 Explain what is meant by simple harmonic motion. [1]

4–2 Show that the spring constant for each spring is about 1.8×10^5 N m^{-1}. [1]

4–3 Show that the frequency of the oscillations is about 3.5 Hz. [2]

4–4 A real car has worn dampers, whose damping effect is very much reduced. It drives along a road where the bumps occur at a frequency that is about the same as the answer to **Question 4–3**. Explain what will happen to the car. [3]

4–5 Calculate the maximum value of the kinetic energy. [1]

4–6 In the model, a single damper is added to the system. It consists of a piston in a cylinder of length 0.25 m, which is filled with nitrogen gas. It has an area of 1.52×10^{-3} m^2. The displacement of the damper is 0.010 m. The damper removes 10% of the energy of the first half-oscillation.

The work done on a gas (J) = pressure (Pa) × change in volume (m^3).

Show that the pressure in the nitrogen gas is about 2.4×10^5 Pa. [1]

4–7 The block that represents the car moves to its rest position. Assume that the temperature stays constant.

Calculate the pressure in the damper when it is fully extended. [2]

4–8 The temperature in the room where the experiment was done was 293 K.

Nitrogen molecules have a molar mass of 0.028 kg mol^{-1}.

1. Calculate the number of moles of nitrogen molecules. [1]

2. Determine the density of the nitrogen. [1]

3. Calculate the root mean square speed of the nitrogen molecules. [1]

Total: 14

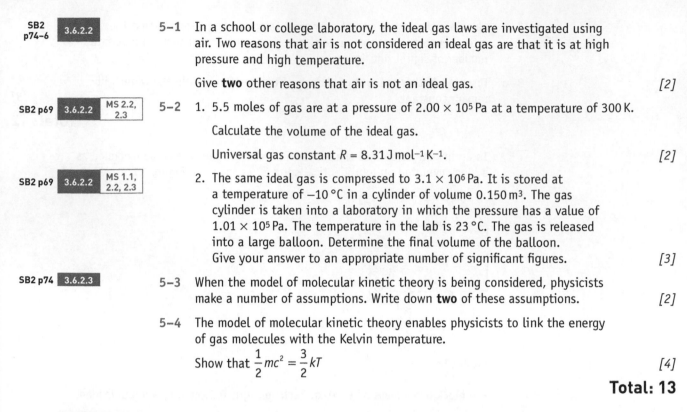

SB2
p74–6   3.6.2.2

5–1   In a school or college laboratory, the ideal gas laws are investigated using air. Two reasons that air is not considered an ideal gas are that it is at high pressure and high temperature.

Give **two** other reasons that air is not an ideal gas.   [2]

SB2 p69   3.6.2.2   MS 2.2, 2.3

5–2   1. 5.5 moles of gas are at a pressure of 2.00×10^5 Pa at a temperature of 300 K.

Calculate the volume of the ideal gas.

Universal gas constant $R = 8.31$ J mol^{-1} K^{-1}.   [2]

SB2 p69   3.6.2.2   MS 1.1, 2.2, 2.3

2. The same ideal gas is compressed to 3.1×10^6 Pa. It is stored at a temperature of $-10\,°C$ in a cylinder of volume 0.150 m³. The gas cylinder is taken into a laboratory in which the pressure has a value of 1.01×10^5 Pa. The temperature in the lab is 23 °C. The gas is released into a large balloon. Determine the final volume of the balloon. Give your answer to an appropriate number of significant figures.   [3]

SB2 p74   3.6.2.3

5–3   When the model of molecular kinetic theory is being considered, physicists make a number of assumptions. Write down **two** of these assumptions.   [2]

5–4   The model of molecular kinetic theory enables physicists to link the energy of gas molecules with the Kelvin temperature.

Show that $\frac{1}{2}mc^2 = \frac{3}{2}kT$ [4]

Total: 13

SB2
p14–15   3.6.1.2   MS 3.6, 3.8, 3.9, 3.12

6   A cyclist is riding a bicycle of mass 820 g. When she hits a pothole in the road, she lifts her front wheel and rides the bike on the back wheel. The back suspension begins to oscillate with frequency 3.2 s^{-1}, maximum amplitude 2.8×10^{-2} m and maximum kinetic energy of 12 J.

6–1   Deduce the spring constant k of the back suspension and the mass of the cyclist.   [5]

6–2   What assumptions did you make about the mass of the bike?   [1]

6 Fields and their consequences

Gravitational fields

Quick questions

SB2 p38 3.7.1.1

1 What is meant by a field in physics?

SB2 p36 3.7.2.1

2 State Newton's law of gravitation.

SB2 p36 3.7.2.2

3 The equation for Newton's law of gravitation is $F = G\dfrac{m_1 m_2}{r^2}$.

Copy and complete **Table 1** with the meaning and units of each term.

Term	Meaning	SI unit
F		
G		
m_1		
m_2		
r		

Table 1

SB2 p36 3.7.2.1

4 Some textbooks show that the gravitational force is negative. What does this mean?

SB2 p36 3.7.2.1 MS 0.1, 0.2, 2.3

5 Two objects each have a mass of 75 kg and are 1.25 m apart. Calculate the gravitational force between the two objects.

SB2 p36 3.7.1.1

6 Write down two properties of gravitational forces.

SB2 p36 3.7.2.1 MS 0.1, 0.2, 2.2, 2.3

7 Two large rocks floating in space have a mass of 1.5×10^7 kg and 2.5×10^7 kg. Calculate the distance apart they are when the force between them is 2.7 N.

SB2 p39 3.7.2.2 MS 2.3

8 The mass of the Sun is 1.99×10^{30} kg and its average radius is 6.96×10^8 m. Calculate the gravitational field strength at the Sun's average radius.

SB2 p38 3.7.2.2

9 Sketch the field between the Earth and the Moon, showing the neutral point where the overall gravitational field strength is zero.

SB2 p40 3.7.2.3 MS 2.1

10 An object of mass 5.0 kg is in a gravity field close to a planet which has gravitational field strength 4.0 N kg⁻¹. Calculate the potential energy every 20 m from the surface to a height of 100 m.

SB2 p42 3.7.1.3 MS 3.8

11 Sketch a graph that shows the gravitational field strength against the distance from the centre of the planet. On the graph, show the potential difference, V_g between points r_1 and r_2.

Exam-style questions

SB2 p38 3.7.2.2

12 Which one of these correctly describes gravitational field strength? *[1]*

A: It is defined as mass per unit force.

B: It is defined as force per unit mass.

C: It is defined as the energy per unit mass.

D: It is defined as the mass per unit energy.

Total: 1

SB2 p45 | 3.7.2.4

13 A satellite is orbiting a planet at a constant height. This means that:

A: the required centripetal force is provided by the gravity of the planet

B: the gravitational force and the centripetal force cancel each other out

C: there are no forces at this height

D: a continuous thrust is needed from the satellite's motors to keep the satellite in orbit. *[1]*

Total: 1

SB2 p44 | 3.7.2.4

14 Which one of these is NOT true about the escape speed, v, of a rocket escaping from a planet? *[1]*

A: $v^r = \dfrac{2GM}{r}$

B: The mass of the planet does not need to be known.

C: The speed depends on the potential difference between the surface and infinity.

D: The mass of the rocket does not need to be known.

Total: 1

SB2 p43 | 3.7.2.3

15 An object moves along the same gravitational equipotential line in a radial field. Which one of these statements is true about the work done, W? *[1]*

A: $W = 2\pi m v^2$

B: $W = m v^2$

C: $W = 2\pi G M m$

D: $W = 0$

Total: 1

SB2 p38 | 3.7.2.2

16–1 Define gravitational field strength. *[1]*

SB2 p38 | 3.7.2.2

16–2 Write an equation to calculate gravitational field strength. *[1]*

SB2 p38 | 3.7.2.2

16–3 A certain planet has a gravitational field strength of $18\,\text{N}\,\text{kg}^{-1}$.

  1. Calculate the force due to gravity acting on an object of mass $250\,\text{kg}$. *[1]*

  2. What is this force called? *[1]*

SB2 p39 | 3.7.2.2

16–4 Draw a diagram to show the idea of a radial field of a planet. Mark on your diagram where the field strength has the highest value and the lowest value. Show the direction of the field. *[3]*

Total: 7

SB2 p39 | 3.7.2.2

17–1 Sketch a graph that shows how the gravitational field strength varies with the distance from the centre of a planet. Show where the surface is on your graph. *[3]*

SB2 p39 | 3.7.2.2 | MS 2.3

17–2 The radius of the Earth is $6.37 \times 10^6\,\text{m}$. Its mass is $5.97 \times 10^{24}\,\text{kg}$. Show that the gravitational field strength at the surface of the Earth is $9.8\,\text{N}\,\text{kg}^{-1}$. *[2]*

SB2 p38 | 3.7.2.2

17–3 Using the definition of gravitational field strength, show that gravitational field strength is the same as acceleration. *[2]*

Total: 7

SB2 p40–1	3.7.2.3	MS 2.1	**18–1** Define gravitational potential. [1]
SB2 p40–1	3.7.2.3	MS 2.1	**18–2** Give the unit of gravitational potential. [1]
SB2 p40–1	3.7.2.3	MS 2.1	**18–3** Write down an equation for potential difference in a uniform gravity field. [1]

SB2 p41 | 3.7.2.3 | MS 2.1, 2.2

18–4 A 5.0 kg mass is at a height of 120 m above the surface of a planet of which the gravitational field strength is 4.0 N kg^{-1}. Calculate the gravitational potential of the mass. [2]

SB2 p42–3 | 3.7.2.3

18–5 Explain why the use of the equation for potential energy ($E_p = mg\Delta h$) in a gravity field is not appropriate for calculating potential energy over long distances, for example, between the Earth and the Moon. [2]

Total: 7

SB2 p42 | 3.7.2.3

19–1 Write down an equation that gives the gravitational potential at a point r that is a long way from the centre of a planet of mass m_1. [1]

SB2 p41 | 3.7.2.3 | MS 2.1, 2.3

19–2 In a uniform gravity field of 8.5 N kg^{-1}, an object moves from 250 m above the surface to 110 m above the surface.

    1. Calculate the potential difference. [3]

    2. Explain whether the work done results in an increase or decrease in gravitational potential as the object is moved. [1]

Total: 5

20 A rocket of mass 4500 kg is orbiting the Earth (mass 5.97 × 10^{24} kg) at a height of 450 km. It fires its engines to rise to an orbit of 550 km.

SB2 p42 | 3.7.2.3 | MS 0.1, 0.2, 2.1, 2.3

20–1 Calculate the change in potential. Include the correct unit. [4]

20–2 Which data item do you not need for this calculation? [1]

SB2 p42 | 3.7.2.3 | MS 0.1, 0.2, 2.1

20–3 Define gravitational potential energy at a point. [1]

20–4 Calculate the energy needed to raise the rocket from low orbit to high orbit. [1]

SB2 p42 | 3.7.2.4

20–5 Explain what is meant by the escape velocity of a rocket. [1]

SB2 p44 | 3.7.2.4 | MS 2.1

20–6 By considering the equation for potential: $V = -\dfrac{GM}{r}$

show that the escape speed of a rocket is: $v = \sqrt{\left(\dfrac{2GM}{r}\right)}$

where M is the mass of a planet. [3]

SB2 p44 | 3.7.2.4 | MS 0.1, 0.2, 2.3

20–7 Calculate the escape velocity of a rocket from Jupiter. Give your answer to an appropriate number of significant figures.

Mass of Jupiter = 1.90 × 10^{27} kg; radius of Jupiter = 6.99 × 10^7 m. [2]

Total: 13

SB2 p45 | 3.7.2.4 | MS 2.1

21–1 By considering the centripetal force exerted on a satellite of mass m by a planet of mass M, show that the orbital speed is given by:

$$v = \sqrt{\left(\dfrac{GM}{r}\right)}$$

[2]

SB2 p45 | 3.7.2.4 | MS 2.1

21–2 A space probe is exploring Jupiter's atmosphere. It is 22 500 km above the surface of Jupiter. (For the purposes of this question, assume that the surface is solid.)

Calculate the speed of the probe to ensure that it stays in orbit. [3]

Mass of Jupiter = 1.90 × 10^{27} kg; radius of Jupiter = 6.99 × 10^7 m

Total: 5

SB2 p46–7 | 3.7.2.4 | MS 2.2

22–1 State the equation that links the linear speed, v, of an object moving in a circular orbit around a central point, the angular velocity, ω, and the radius, r. [1]

SB2 p46–7 | 3.7.2.4 | MS 2.2

22–2 Modify this equation to determine the time period, T, for **one** orbit. [1]

SB2 p46–7   3.7.2.4   MS 2.2

22–3 Hence show that: $T^2 = \left(\dfrac{4\pi^2}{GM}\right)r^3$ [2]

SB2 p48   3.7.2.4   MS 0.2, 2.2

22–4 A satellite is required to have an orbital period of 30 hours exactly.

Calculate the height at which it orbits around the Earth.

Mass of the Earth = 5.97×10^{24} kg; radius of the Earth is 6.37×10^6 m.   [3]

Total: 7

23   A planet has a radius of 3.5×10^6 m. It has an average density of 5500 kg m^{-3}.

SB2 p38   3.7.2.2   MS 0.2, 2.2

23–1 Show that the mass of the planet is about 9.9×10^{23} kg.   [2]

SB2 p38   3.7.2.2   MS 0.2, 2.2

23–2 1. State the definition of gravitational field strength.   [1]

     2. Calculate the gravitational field strength of this planet. Give your answer to an appropriate number of significant figures.   [2]

23–3 A space probe has a mass of 250 kg.

SB2 p42   3.7.2.3

     1. Calculate the gravitational potential at the surface of the planet. Give the correct unit.   [2]

SB2 p44   3.7.2.4   MS 0.2, 2.2

     2. Calculate the escape velocity of the probe.   [2]

Total: 9

24   **Figure 1** shows the gravitational potential against the radius from the centre of a planet.

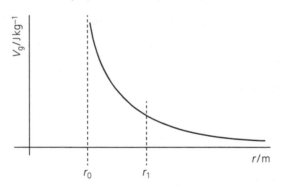

Figure 1

SB2 p42   3.7.2.3   MS 3.6

24–1 State what is shown by the distance r_0.   [1]

SB2 p42   3.7.2.3   MS 3.6

24–2 Explain how you would determine the gravitational field strength at point r_1 from the graph.   [2]

SB2 p39   3.7.2.2   MS 2.3

24–3 The International Space Station orbits 408 km above the surface of the Earth. Show that the gravitational field strength at this height is about 8.7 N kg^{-1}.   [2]

Total: 5

25   At a certain point between the Earth and Moon, there is a neutral point where the gravitational field strength is zero. This is shown in **Figure 2**.

Distance between the Earth and the Moon = 384 000 km between centres

Mass = 5.97×10^{24} kg   Mass = 7.35×10^{22} kg
Radius = 6.37×10^6 m

Figure 2

SB2 p39 3.7.2.2

25–1 1. On a copy of the diagram, draw the field lines of the gravitational fields of both the Earth and the Moon, especially around the neutral point.   [2]

     2. Explain why the gravitational field is zero at this point.   [2]

SB2 p39 3.7.2.2 MS 0.3, 2.2, 2.3

25–2 The radii of the Earth and the Moon are r_E and r_M, respectively.

    The masses of the Earth and the Moon are M_E and M_M, respectively.

    1. By considering the gravitational field strengths of the Earth and Moon, show that: [3]

$$\frac{M_E}{M_M} = \frac{r_E^2}{r_M^2}$$

    2. Show that the neutral point is about 9 times further from the Earth than it is from the Moon.   [2]

SB2 p39 3.7.2.2 MS 0.3, 2.2, 2.3

25–3 1. Calculate the distance from the centre of the Moon to the neutral point.   [2]

    2. Calculate the distance from the centre of the Earth to the neutral point.   [1]

    3. Calculate the distance, d, above the Earth's surface to the neutral point.   [1]

SB2 p39 3.7.2.2

25–4 Explain what, if any, would be the effect on the force at the neutral point of including the mass, m_S, of a space probe.   [2]

Total: 15

Electric fields

Quick questions

SB2 p86 3.7.3.1

1   Explain in terms of charges how the following arise:

a)   a positively charged body

b)   a negatively charged body

c)   a neutral body.

SB2 p86 3.7.1

2   State the main difference between a gravity field and an electric field.

SB2 p86 3.7.3.1

3   State Coulomb's law.

SB2 p86 3.7.3.1

4   The equation for Coulomb's law of electrostatic force is:

$$F = \frac{1}{4\pi\varepsilon_0}\frac{Q_1 Q_2}{r^2}$$

Copy and complete **Table 2** to show the meanings of the terms with their units.

Term	Meaning	SI unit
F		
ε_0		
Q_1		
Q_2		
r		

Table 2

SB2 p87 3.7.3.1

5   Some textbooks show the constant in Coulomb's law as k. So the equation is:

$$F = k\frac{Q_1 Q_2}{r^2}$$

Determine the value of k and give its units.

SB2 p86-7 3.7.3.1 MS 0.1, 0.2, 2.3

6   Two point charges of +75 nC are 1.25 m apart. Calculate the electrostatic force between the two objects, and state whether they attract or repel.

SB2 p88 3.7.1.1

7   Write down the properties of electrostatic forces.

SB2 p88 | 3.7.3.1 | MS 0.1, 0.2, 2.2, 2.3

8 Two spherical charges floating in space have a charge of 1.5×10^{-7} C and -2.5×10^{-6} C. Calculate the distance apart they are when the attractive force between them is 0.27 N.

SB2 p36, p101 | 3.7.1

9 Electric fields and gravitational fields have many features in common, but there are many differences.

Copy and complete **Table 3** to give a summary of the similarities and differences. One line has been completed as an example.

Property	Gravitational field	Electric field
Acts on	All objects with mass	Charged objects only
Exchange particle		
Direct contact?		No
Range	Infinite	
Attractive/repulsive	Attractive only	
Can be shielded?		Yes

Table 3

Exam-style questions

SB2 p94 | 3.7.3.2

10 Which one of these correctly describes the electrical field between two far distant charged objects? [1]

A: The field is radial, and can be attractive or repulsive.

B: The field is uniform, and is attractive only.

C: The field is radial, and is attractive only.

D: There is no electrical field between objects that far apart.

Total: 1

SB2 p90 | 3.7.3.2

11 Which one of these correctly describes electric field strength? [1]

A: It is defined as charge per unit force.

B: It is defined as force per unit charge.

C: It is defined as the energy per unit charge.

D: It is defined as the charge per unit energy.

Total: 1

SB2 p95 | 3.7.3.2

12 A positive charge is placed in a hollow metal sphere. Which one of the diagrams in **Figure 3** shows correctly the electric field outside the sphere? [1]

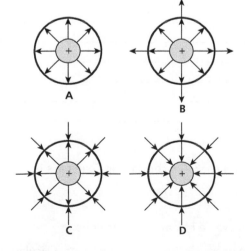

Figure 3

Total: 1

SB2 p94 3.7.3.2

13 **Figure 4** shows a uniform electric field of field strength E set up between two parallel plates.

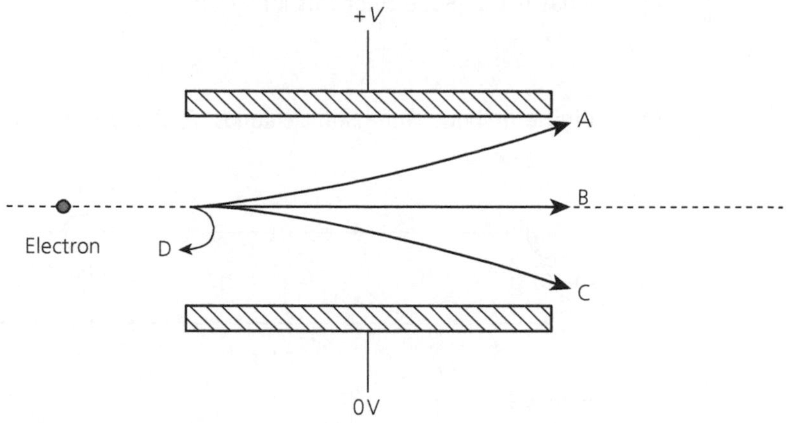

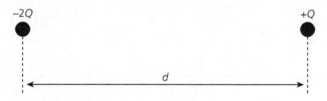

Figure 4

An electron passes between the plates. Which of the paths **A–D** does the electron follow? [1]

Total: 1

SB2 p87 3.7.3.2

14 **Figure 5** shows two charged particles at a distance d apart.

$-2Q$ $+Q$

d

Figure 5

The two charged particles exert an attractive force of F.

The charge of each particle is changed by $+Q$. The distance between the two charges is increased to $2d$. What is the force between the two charges now? [1]

A: attractive force of $\dfrac{F}{4}$

B: repulsive force of $\dfrac{F}{4}$

C: attractive force of $\dfrac{F}{2}$

D: repulsive force of $\dfrac{F}{2}$

Total: 1

SB2 p89 3.7.3.2

15–1 Define electric field strength. [1]

SB2 p89 3.7.3.2

15–2 Write an equation to calculate electric field strength. [1]

SB2 p94–5 3.7.3.2

15–3 Draw a diagram to show the idea of a radial electric field of a positive point charge. Mark on your diagram where the field strength has the highest value and the lowest value. Show the direction of the field. [2]

15–4 Draw the electric field between two parallel plates. The bottom plate is at 0 V. The top plate is at +V. [2]

15–5 What is the property of this kind of field? [1]

Total: 7

16 **Figure 6** shows three point charges, *P*, *Q* and *R*, in a line.

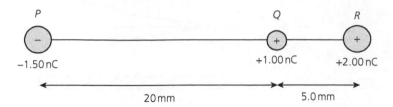

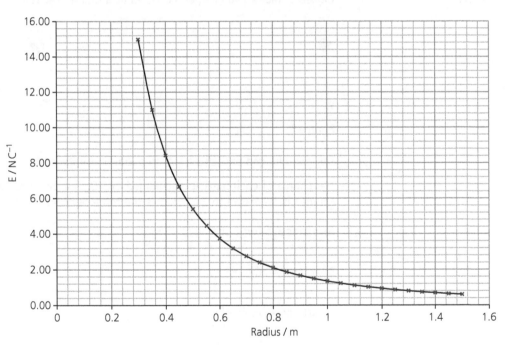

Figure 6

The charges and distances are shown on the diagram.

16–1 Calculate the force acting on charge *Q*. State its direction. [3]

16–2 Calculate the electric field strength at *Q*. [1]

16–3 A sphere of radius 0.30 m carries a charge of 1.50×10^{-10} C.

Figure 7 shows how the electric field strength changes with the radius.

Figure 7

Calculate the change in potential from a radius of 0.40 m to 0.80 m. Give your answer to an appropriate number of significant figures. [3]

Total: 7

17 A sphere of radius 0.30 m carries a charge of 6.50×10^{-10} C.

Figure 8 shows how the potential changes with radius.

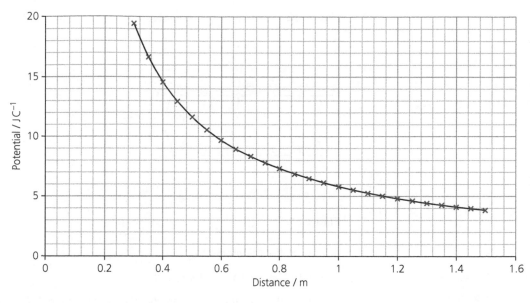

Figure 8

SB2 p94–5 3.7.3.2 MS 3.7

17–1 Use the graph to calculate the electric field strength at 0.60 m. *[2]*

SB2 p94–5 3.7.3.2 MS 3.7

17–2 Explain what the gradient of the graph shows in terms of the potential and the distance. *[1]*

SB2 p99 3.7.3.3 MS 3.1

17–3 Use the graph to sketch a diagram showing the equipotential lines at 15, 10 and 5 J C⁻¹. Show the values and the distances. Indicate the field lines and the equipotentials. *[3]*

Total: 6

SB2 p99 3.7.3.3 MS 2.1

18 **Figure 9** shows equipotentials in a uniform electric field of field strength +1500 V m⁻¹.

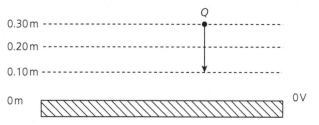

Figure 9

18–1 Give the potential of each equipotential line. *[2]*

18–2 The charge Q is moved from 0.30 m to 0.10 m.

1. Calculate the change in potential. *[1]*

2. Determine the work done if the charge of Q is +1.5 nC. *[1]*

Total: 4

SB2 p97 3.7.3.2 MS 2.1

19 A hydrogen atom consists of a proton and an electron at an average separation of 5.0×10^{-11} m.

19–1 Sketch the electric field pattern between the proton and the electron. *[2]*

19–2 Calculate:

1. the electric potential of the electron at this distance *[1]*

2. the potential energy of the electron *[1]*

3. the field strength. *[1]*

Total: 5

20 **Figure 10** shows two parallel plates with one charged to 2000 V, and one at 0 V.

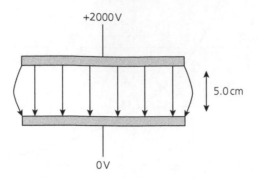

Figure 10

A uniform field is produced.

SB2 p90 3.7.3.2   20–1 Calculate the electric field strength. [1]

SB2 p90 3.7.3.2   20–2 A small particle of dust between the plates has a mass 3.0×10^{-6} kg and carries a charge of +1.5 nC.

Calculate the magnitude of:

1. the force on the particle [1]

2. the acceleration of the particle. State the direction of the acceleration. [2]

SB2 p91 3.7.3.2   20–3 Copy the diagram and draw the equipotential lines every 500 V. [2]

Total: 6

21 Four charges, A, B, C and D, are arranged in a square as shown in **Figure 11**.

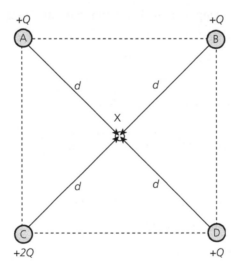

Figure 11

SB2 p100 3.7.3.3 MS 2.1   21–1 Determine the potential at X. [1]

SB2 p100 3.7.3.2 MS 2.1   21–2 Write an expression for the electric field strength at point X. [2]

SB2 p100 3.7.3.2 MS 2.1   21–3 Give the direction of the force due to the resultant field. [1]

Total: 4

SB2 p98 | 3.7.3.3 | MS 2.1, 2.3

22 A positive test charge of +1.25 nC is placed at R, 100 mm from a second charge of +6.3 nC, which is at P. The arrangement is shown in **Figure 12**. The charge is moved from R to S. The distance PS is 115 mm.

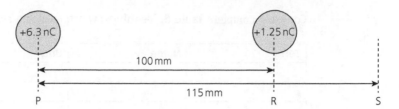

Figure 12

22–1 Calculate the change in potential when the test charge is moved from R to S. [3]

22–2 Calculate the work done on the test charge to move it from R to S. [1]

Total: 4

Capacitance

Quick questions

SB2 p109 | 3.7.4.3

1 Both capacitors and batteries store energy. How are the two different?

SB2 p109 | 3.7.4.2

2 Name the two kinds of capacitor shown in **Figure 13**.

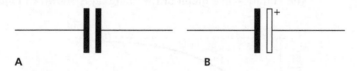

Figure 13

SB2 p109 | 3.7.4.2

3 Explain the advantage and disadvantages of an electrolytic capacitor over a non-electrolytic capacitor.

SB2 p109 | 3.7.4.2

4 Draw a simple diagram to show the construction of a capacitor.

SB2 p109 | 3.7.4.1 | MS 0.1, 0.2

5 A 1 farad capacitor is a large capacity capacitor. Copy and complete **Table 4** to show the common multipliers used in most capacitor calculations.

Prefix	Capacitance/F
μF	
nF	
pF	

Table 4

SB2 p109 | 3.7.4.1

6 The equation for capacitance is: $Q = CV$

Copy and complete **Table 5**.

Term	Meaning	SI unit
Q		
C		
V		

Table 5

7   The value of a capacitor can be worked out using the relationship:

$$C = \frac{\varepsilon_0 \varepsilon_r A}{d}$$

Copy and complete **Table 6**, identifying which term is a constant and which is a variable.

Term	Name	Unit	Constant/variable
C			
ε_0			
ε_r			
A			
d			

Table 6

8   A capacitor of capacitance C is charged to a potential difference of V.
Derive an expression for the energy held by the capacitor.

9   A 2200 µF capacitor holds 3.0 J of energy. Calculate the potential difference
across its plates. Give your answer to an appropriate number of significant figures.

10 A capacitor has a capacitance of 680 µF and is discharged through a resistor of
5.6 kΩ. The capacitor is charged up to 12.0 V. Calculate the time for the voltage to fall to 4.5 V.

Exam-style questions

11 A capacitor is being charged up from a source of potential difference V_0.
The voltage–time graph of the charging is shown in **Figure 14**.

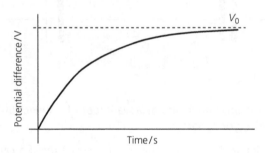

Figure 14

Which one of the graphs in **Figure 15** shows the variation of the current with the time?   *[1]*

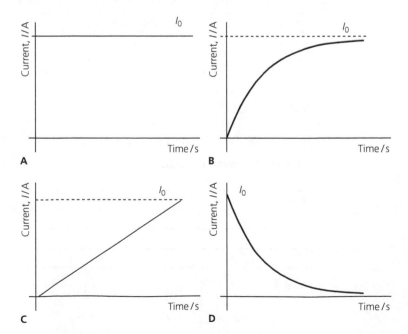

Figure 15

Total: 1

SB2 p115 `3.7.4.3`

12 A source of potential difference, V, provides charge Q to a capacitor of capacitance C. The source provides energy E. Which one of these is the maximum energy that is held by the capacitor? *[1]*

A: $\dfrac{E}{4}$ **B:** $\dfrac{E}{2}$ **C:** E **D:** $2E$

Total: 1

SB2 p119 `3.7.4.3` `MS 3.9`

13 At position 1 of the switch S in **Figure 16**, a capacitor of capacitance C is charged up by a battery of potential difference V.

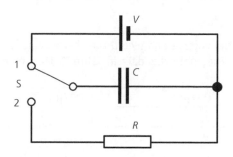

Figure 16

The switch is moved very quickly to position 2 so that the capacitor discharges through the resistor, which has a resistance R. It takes the potential difference 36 ms to fall to $\dfrac{V}{2}$. Once the switch has been changed to position 2, what is the time taken for the energy to fall to $\dfrac{E}{16}$? *[1]*

A: 51 ms   **B:** 72 ms   **C:** 144 ms   **D:** 576 ms

Total: 1

SB2 p115 `3.7.4.1` `MS 2.3`

14 A capacitor of capacitance 2500 µF is charged up by a constant current of 200 µA. What is the potential difference across the plates of the capacitor after 25 s? *[1]*

A: 0.50 V   **B:** 1.0 V   **C:** 2.0 V   **D:** 4.0 V

Total: 1

SB2 p118 `3.7.4.4`

15 Which one of these equations for the discharge of a capacitor is NOT correct? *[1]*

A: $Q = Q_0 e^{-\frac{t}{RC}}$ **B:** $V = V_0 e^{-\frac{t}{RC}}$ **C:** $I = I_0 e^{-\frac{t}{RC}}$ **D:** $C = C_0 e^{-\frac{t}{RC}}$

Total: 1

SB2 p109 `3.7.4.1`

16–1 The unit for capacitance is farad. Give the definition for the farad. *[1]*

SB2 p109 `3.7.4.4`

16–2 Explain how a capacitor charges up when connected to a cell. *[3]*

SB2 p109 `3.7.4.4`

16–3 Explain how a capacitor discharges when connected to a resistor. *[3]*

SB2 p109 `3.7.4.2` `MS 2.2`

16–4 Nylon will no longer act as an insulator if the electric field strength exceeds 12.0×10^6 V m^{-1}. Calculate the thickness of the dielectric if the working voltage of a capacitor is to be 200 V. *[1]*

Total: 8

SB2 p109 `3.7.4.1` `MS 1.1`

17–1 A 4.7 µF capacitor has a potential difference of 2.5 V across its terminals. Calculate the charge held on its plates. Give your answer to an appropriate number of significant figures. *[1]*

SB2 p109 `3.7.4.1` `MS 2.2`

17–2 When a capacitor is charged up to 17.6 V, the charge that has flowed onto the plates is measured as 135 nC. Calculate the capacitance of the capacitor. *[2]*

SB2 p109 `3.7.4.1` `MS 0.1, 0.2, 2.2`

17–3 A 2.2 µF capacitor is connected to a coulombmeter, which reads a charge of 3500 nC. Calculate the potential difference across the terminals of the capacitor. *[2]*

Total: 5

61

18   Some students set up the circuit in **Figure 17** to measure the charge held by a capacitor *C*.

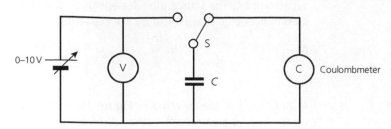

Figure 17

They measure the voltage and record the charge that is shown on the coulombmeter. They note this data in **Table 7**. They repeat the readings.

Voltage/V	Q1/μC	Q2/μC	Average Q/μC
0.0	0	0	
1.0	38.7	39.2	
2.0	77.0	78.2	
3.0	116.3	118.4	
4.0	157.6	153.2	
5.0	194.0	191.9	
6.0	234.8	230.7	
7.0	272.6	339.5	
8.0	313.6	318.3	
9.0	353.0	360.5	
10	393.8	395.1	

Table 7

SB2 p110 3.7.4.1 18–1 Which **one** data item is anomalous? [1]

SB2 p110 3.7.4.1 18–2 What should be done with it? [1]

SB2 p110 3.7.4.1 MS 1.2 18–3 Draw a table of voltage and average charge. Calculate the average charge. [2]

SB2 p110 3.7.4.1 MS 3.1, 3.2 18–4 Plot the data as a graph, charge against voltage. [3]

SB2 p110 3.7.4.1 MS 3.4 18–5 Use the graph to calculate the value of the capacitor. [2]

Total: 9

SB2 p114 3.7.4.2 19–1 Explain how the dielectric of a capacitor works. You may use a diagram to help you with your answer. [4]

SB2 p114 3.7.4.2 19–2 A non-electrolytic capacitor has a mica dielectric that has a dielectric constant of 7.0. It is placed across the terminals of a high-frequency AC supply. Explain why the component starts to get warm. [4]

SB2 p111 3.7.4.2 MS 0.1, 2.3 19–3 A simple capacitor is made of two plates of aluminium foil separated by a layer of polythene. Each plate is 30 cm × 30 cm. The thickness of the polythene is 40 μm. The dielectric constant of polythene is 2.25. Calculate the capacitance of the capacitor. [2]

Total: 10

SB2 p115 3.7.4.3 20–1 Explain how a graph of charge against voltage can be used to derive the equation for energy stored on a capacitor when the voltage is raised from 0 to *V*. [2]

SB2 p115 3.7.4.3 20–2 Show that for a capacitor of capacitance *C* charged with charge *Q*, the energy stored is:

$$E = \frac{1}{2}\frac{Q^2}{C}$$

[3]

SB2 p115 | 3.7.4.3 | MS 0.1 **20–3** A capacitor has a charge of 500 μC with a potential difference of 25 V across its plates. Calculate the energy held by the capacitor. *[1]*

Total: 6

SB2 p115 | 3.7.4.3 | MS 0.1 **21–1** A capacitor of capacitance 45 nF is charged to a potential difference of 4000 V. Calculate the energy held by the capacitor. *[1]*

SB2 p115 | 3.7.4.3 | MS 0.1, 2.2 **21–2** A capacitor holds 3.0 J of energy when the charge on its plates is 5000 μC.

1. Calculate the capacitance of the capacitor. *[2]*

2. Show that the potential difference across the capacitor is about 1200 V. *[1]*

SB2 p117 | 3.7.4.4 | MS 3.6, 3.7 **21–3** A capacitor of capacitance 220 μF is discharged through a resistor of 10 kΩ.

Figure 18 shows the discharge between 0 s and 10.0 s.

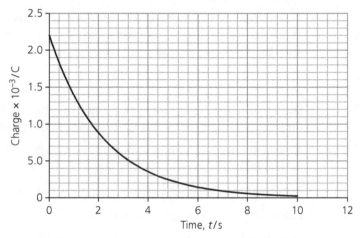

Figure 18

1. Calculate the current that is flowing at 0.40 s. *[3]*

2. Explain how the current changes when $t = 1.6$ s. *[2]*

SB2 p117 | 3.7.4.4 | MS 3.12 **21–4** Sketch a graph to show how the voltage of a capacitor changes with time as it discharges through a fixed value resistor. The initial voltage is V_0. *[3]*

SB2 p117 | 3.7.4.4 | MS 3.12 **21–5** State what this kind of graph is called. *[1]*

Total: 13

SB2 p118 | 3.7.4.4 | MS 3.9 **22** The circuit in **Figure 19** can be used to investigate the current and voltage of a capacitor as it discharges.

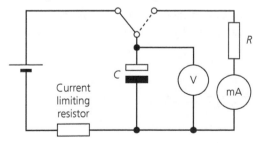

Figure 19

22–1 Write an expression for instantaneous current in terms of charge flow as the capacitor discharges. *[1]*

22–2 Combine this equation with Ohm's law. *[1]*

22–3 State the equation that links charge with capacitance and voltage. *[1]*

22–4 Show that $\dfrac{dQ}{dt} = -\dfrac{Q}{CR}$ [1]

22–5 The equation in **Question 22–4** is a differential equation of the form: $\dfrac{dx}{dt} = -kx$

State what the constant term k corresponds to in the equation in **Question 22–4**. [1]

22–6 The general solution to a differential equation is: $x = x_0 e^{-kt}$

Give the solution to the differential equation: $\dfrac{dQ}{dt} = -\dfrac{Q}{CR}$ [1]

Total: 6

SB2 p119 3.7.4.4

23–1 Show that the units ohms × farad = second. [2]

SB2 p119 3.7.4.4 MS 0.5

23–2 Calculate the time constant, τ, for a 470 µF capacitor discharging through a 1.5 kΩ resistor. [1]

SB2 p119 3.7.4.4 MS 0.5

23–3 A capacitor has a capacitance of 680 µF and is discharged through a resistor of 5.6 kΩ. The capacitor is charged up to 12.0 V. Calculate the voltage 10 s after the discharge begins. [2]

Total: 5

SB2 p119 3.7.4.4 MS 3.9

24–1 Show that when a capacitor is discharged to a time corresponding to the time constant, the charge left is about 37% of the initial charge. [2]

SB2 p121 3.7.4.4 MS 0.5

24–2 A capacitor has a capacitance of 820 µF and is charged through a resistor of 3.3 kΩ. It is connected to a 12.0 V supply.

Calculate the voltage 2.0 s after the supply is connected. [2]

SB2 p119 3.7.4.4 MS 3.9

24–3 Define the half-life of capacitor discharge. [1]

SB2 p119 3.7.4.4 MS 3.9

24–4 Show that the half-life of a capacitor is about 69% of the time constant. [2]

Total: 7

SB2 p120 3.7.4.4 MS 3.9, 3.10

25 A capacitor discharge can be modelled as a straight-line graph rather than an exponential curve.

25–1 Sketch the graph for voltage, labelling the axes. [2]

25–2 State what quantity can be worked out from the gradient. [1]

25–3 State what quantity is given by the y-axis intercept. [1]

25–4 State what quantity can be worked out from the x-axis intercept. [1]

Total: 5

SB2 p112 3.7.4.2

26 **Figure 20** shows a capacitor consisting of two parallel plates, each of area A. They are separated by a distance, d. The plates are separated by dry air.

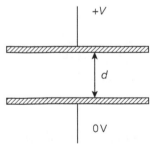

Figure 20

The plates are charged to a potential difference, V. A uniform electric field E is set up.

The plates are disconnected so that the potential difference V is maintained. Then the separation of the plates is increased to $2d$.

Determine an equation for the electric field strength, with respect to charge, of the uniform field at this new separation. [3]

Total: 3

27 A capacitor has plates of area $0.15\,m^2$. These are separated by a layer of dielectric material that is $1.2 \times 10^{-4}\,m$ thick. The relative permittivity of the material is 32.

SB2 p112 3.7.4.2

27–1 1. Explain how a dielectric material increases the capacitance. [3]

2. Many other materials are insulating. Why are they not used as dielectrics? [2]

SB2 p112 3.7.4.2   MS 2.3
SB2 p109 3.7.3.1   MS 2.3

27–2 1. Show that the capacitor has a capacitance of about $350\,nF$. [2]

2. The capacitor is connected to a source of potential difference $6.0\,V$. Calculate the charge on the plates after some time. Give your answer to an appropriate number of significant figures. [3]

27–3 The capacitor has several different voltages applied across its plates, and the charge stored is recorded.

SB2 p115 3.7.4.2   MS 3.1

1. Sketch a graph to show how the charge varies with voltage. [2]

SB2 p115 3.7.4.3   MS 3.1

2. Calculate the energy stored on the plates when the potential difference is $8.0\,V$. [2]

SB2 p115 3.7.4.2   MS 3.4

3. Explain how you would find the capacitance from the graph you sketched in **Question 27–3–1**. [1]

Total: 15

28 A DC power supply of $15\,V$ is connected to a capacitor of capacitance $680\,\mu F$. The capacitor is in series with a resistor of resistance $4700\,\Omega$.

SB2 p122 3.7.4.4   MS 2.3

28–1 Calculate the current immediately after the capacitor is connected to the power supply. [1]

SB2 p119 3.7.4.4   MS 2.3

28–2 Show that the time constant of the circuit is about $3\,s$. [1]

SB2 p119 3.7.4.4   MS 3.1

28–3 **Figure 21** shows how the voltage varies with time as the capacitor charges.

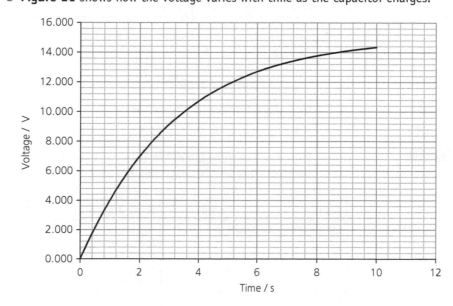

Figure 21

Determine the percentage of the final voltage that is reached when $t = RC$. [1]

SB2 p119 3.7.4.4   MS 3.12

28–4 Sketch the graph of the current when measured against the time. [1]

SB2 p119 3.7.4.4   MS 3.12

28–5 Explain the shape of your graph. [2]

Total: 6

Magnetic fields

Quick questions

SB2 p131 3.7.5.1
1a) Draw a bar magnet and the field lines around it.

b) Show where the field is weak and strong.

SB2 p131 3.7.5.1
2 Show the magnetic field between:

a) a north pole and a south pole

b) two north poles. Show the neutral point and explain what it is.

SB2 p132 3.7.5.1
3 Draw the magnetic flux pattern for a loop of current-carrying wire. Use the convention in **Figure 22** in your diagram.

Current going | Current
into the page | comimg out of
 | the page

Figure 22

SB2 p133 3.7.5.1
4 Draw a field diagram to show how the field around current-carrying wire interacts with a uniform magnetic field between a north pole and a south pole.

SB2 p133 3.7.5.1
5 Explain how the direction of the force acting on a current-carrying wire in a magnetic field is determined by Fleming's Left Hand Rule.

6 The formula for the force on a current-carrying wire is:

$F = BIl$

Symbol	Quantity	Unit	Vector or scalar
F			
B			
I			
l			

Table 8

SB2 p134 3.7.5.1
a) Copy and complete **Table 8**.

SB2 p134 3.7.5.1
b) Give the definition for the unit tesla.

SB2 p134 3.7.5.1
c) If the wire is at an angle θ to the magnetic field, how is the equation above changed?

SB2 p148 3.7.5.3
7 Copy and complete **Table 9** to help you to understand magnetic flux density, magnetic flux and magnetic flux linkage. One line has been done as an example.

Term	Symbol	What it means	Units	Equation
Magnetic flux density	B	The amount of magnetic flux through a unit area taken perpendicular to the direction of the magnetic flux.	Tesla (T) or weber per square metre (Wb m^{-2})	$B = \dfrac{\Phi}{A}$
Magnetic flux				
Magnetic flux linkage				

Table 9

SB2 p148 | 3.7.5.3 | MS 3.12

8   A coil of *N* turns and area *A* is fixed to a shaft that enters the page perpendicularly as shown in **Figure 23**.

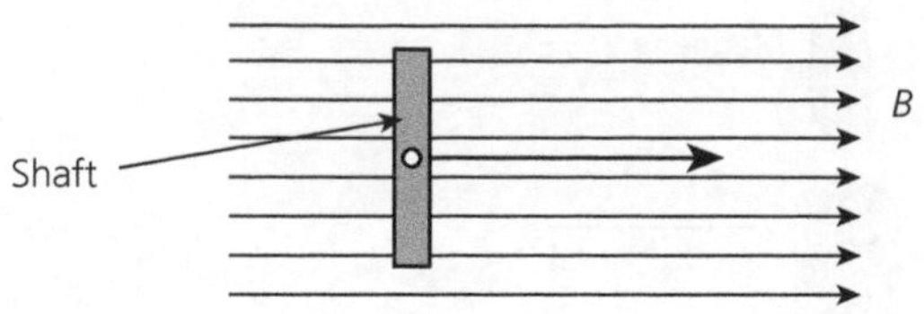

Figure 23

The magnetic field has flux density *B*.

Sketch a graph that shows the flux linkage for 1 complete revolution from 0 rad to 2π rad.

SB2 p155 | 3.7.5.4 | MS 0.6, 4.4, 4.5

9   A coil of 150 turns is rotating at 1200 r.p.m. in a magnetic field of flux density 0.050 T. The area of the magnetic field is 0.15 m^2.

a) What is the maximum e.m.f. E_0 that is generated?

b) What is the e.m.f. generated 0.065 s after the coil leaves the vertical position (when the e.m.f. = 0)?

10 Copy and complete **Table 10** about the controls on an oscilloscope. One has been done as an example.

Control	What it does
On/off switch	Turns the instrument on
Brightness	
Focus	
Time base	
y-gain or voltage gain	

Table 10

Exam-style questions

SB2 p149 | 3.7.5.3

11 Which one of the following is **not** a unit of magnetic flux?   *[1]*

   A: N m A^{-1}

   B: Wb

   C: T m^2

   D: V s^{-1}

Total: 1

SB2 p139 | 3.7.5.2

12 A particle of mass *m* and charge *q* is in a magnetic field of flux density *B*. It is travelling at a speed of *v* in a circular path of radius *r*.

A second particle has a mass 2*m*. It passes into a magnetic field of flux density 2*B* at a speed of 2*v*. What is the radius now?   *[1]*

   A: $\dfrac{r}{2}$ **B:** *r*   **C:** 2*r*   **D:** 4*r*

Total: 1

SB2 p154 3.7.5.4

13 A magnet is dropped through a vertical solenoid as shown in **Figure 24**.

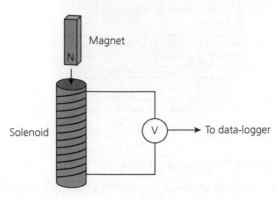

Figure 24

Which graph in **Figure 25** is the graph that the data-logger will plot? *[1]*

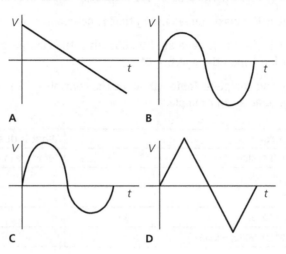

Figure 25

Total: 1

SB2 p169 3.7.5.6

14 In a simple transformer, the primary and secondary coils are linked by a laminated soft iron core. Which one of these statements is correct about the soft iron core? *[1]*

A: The core eliminates energy losses from eddy currents.

B: The core reduces eddy currents but does not eliminate them entirely.

C: The core reduces energy loss by being partly magnetised when the primary voltage is 0.

D: The core is soft and malleable, so that it can be moulded easily into the optimum shape to avoid energy loss.

Total: 1

SB2 p169 3.7.5.6

15 **Figure 26** shows three identical strong cylindrical magnets that are simultaneously dropped down three tubes that are made from different materials, but identical in size.

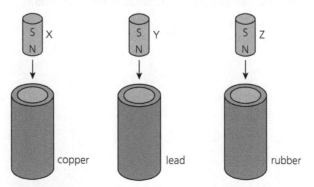

Figure 26

The materials and their resistivities are:

Copper, $1.7 \times 10^{-8}\,\Omega\,\text{m}$; lead, $22 \times 10^{-8}\,\Omega\,\text{m}$; rubber, $50 \times 10^{13}\,\Omega\,\text{m}$

Which row in **Table 11** gives the correct order in which the magnets emerge from the tubes? [1]

	First	Second	Third
A	X	Y	Z
B	Z	X	Y
C	X	Z	Y
D	Z	Y	X

Table 11

Total: 1

16 The apparatus in **Figure 27** can be used to investigate how the force acting on a current-carrying wire varies with a current.

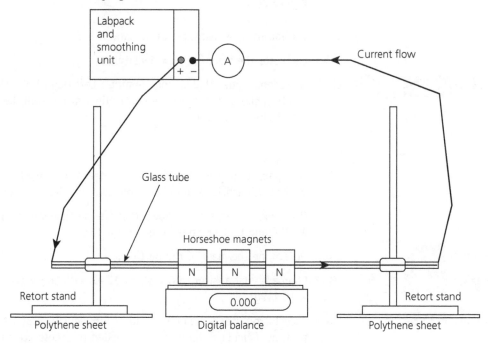

Figure 27

SB2 p136 | 3.7.5.1 | MS 3.4, 3.11 | RP 10

16–1 Describe how you would carry out this experiment to establish the relationship between the force and the current. Include what precautions you need to take to ensure that the experiment is carried out safely. [5]

SB2 p136 | 3.7.5.1 | MS 3.4, 3.11 | RP 10

16–2 Explain how you would collect and process the data to ensure that uncertainties are minimised. [4]

SB2 p136 | 3.7.5.1 | MS 3.4, 3.11 | RP 10

16–3 Sketch a graph to show your expected results. [2]

16–4 Explain how you would calculate the flux density, B. [1]

SB2 p134 | 3.7.5.1 | MS 2.2, 2.3

16–5 A wire is placed between the poles of a magnet that gives a uniform magnetic field that is 5.0 cm long. A current of 4.5 A flows through the wire, giving a force of 0.063 N. Calculate the magnetic flux density acting on the wire. [1]

Total: 13

SB2 p137 | 3.7.5.2

17–1 Show that the relationship between the force acting on a stream of charges of charge q passing through a uniform magnetic field B is given by:

$$F = Bqv$$

[2]

SB2 p138 | 3.7.5.2 | MS 2.2

17–2 Explain why the path of a charged particle in a magnetic field is circular. [1]

SB2 p138 | 3.7.5.2 | MS 2.2

17–3 A charged particle of charge q and mass m enters a uniform magnetic field of flux density B at a speed of v. Derive an expression that can be used to calculate the radius r. [2]

SB2 p138 | 3.7.5.2 | MS 2.2

17–4 Explain what difference, if any, the sign of the charge will make to the path the particle takes. [1]

SB2 p139 | 3.7.5.2 | MS 2.2, 2.3

17–5 An electron is accelerated to travel at $1.6 \times 10^6\,\mathrm{m\,s^{-1}}$ in the apparatus shown in **Figure 28**.

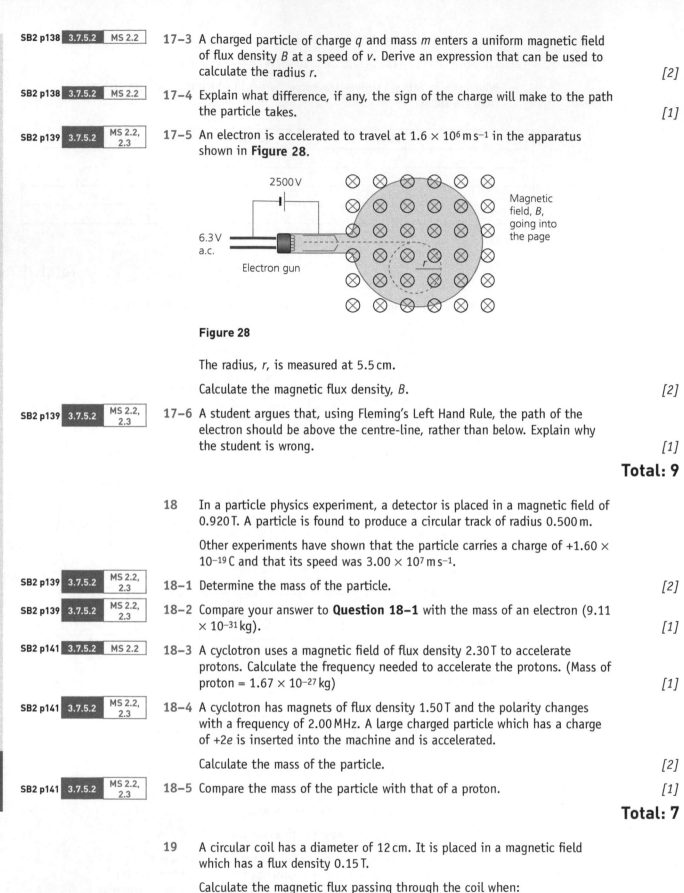

Figure 28

The radius, r, is measured at 5.5 cm.

Calculate the magnetic flux density, B. [2]

SB2 p139 | 3.7.5.2 | MS 2.2, 2.3

17–6 A student argues that, using Fleming's Left Hand Rule, the path of the electron should be above the centre-line, rather than below. Explain why the student is wrong. [1]

Total: 9

18 In a particle physics experiment, a detector is placed in a magnetic field of 0.920 T. A particle is found to produce a circular track of radius 0.500 m.

Other experiments have shown that the particle carries a charge of $+1.60 \times 10^{-19}$ C and that its speed was $3.00 \times 10^7\,\mathrm{m\,s^{-1}}$.

SB2 p139 | 3.7.5.2 | MS 2.2, 2.3

18–1 Determine the mass of the particle. [2]

SB2 p139 | 3.7.5.2 | MS 2.2, 2.3

18–2 Compare your answer to **Question 18–1** with the mass of an electron (9.11×10^{-31} kg). [1]

SB2 p141 | 3.7.5.2 | MS 2.2

18–3 A cyclotron uses a magnetic field of flux density 2.30 T to accelerate protons. Calculate the frequency needed to accelerate the protons. (Mass of proton = 1.67×10^{-27} kg) [1]

SB2 p141 | 3.7.5.2 | MS 2.2, 2.3

18–4 A cyclotron has magnets of flux density 1.50 T and the polarity changes with a frequency of 2.00 MHz. A large charged particle which has a charge of $+2e$ is inserted into the machine and is accelerated.

Calculate the mass of the particle. [2]

SB2 p141 | 3.7.5.2 | MS 2.2, 2.3

18–5 Compare the mass of the particle with that of a proton. [1]

Total: 7

19 A circular coil has a diameter of 12 cm. It is placed in a magnetic field which has a flux density 0.15 T.

Calculate the magnetic flux passing through the coil when:

SB2 p148 | 3.7.5.3 | MS 2.2

19–1 The normal line to the coil is parallel to the field [2]

SB2 p148 | 3.7.5.3 | MS 2.2

19–2 The normal line to the coil is at 50° to the field. [1]

SB2 p148 **3.7.5.3** **MS 2.2** 19–3 A rectangular coil that has dimensions 4.0 cm × 5.0 cm has 120 turns. It is placed in a magnetic field of flux density 0.125 T as shown in **Figure 29**.

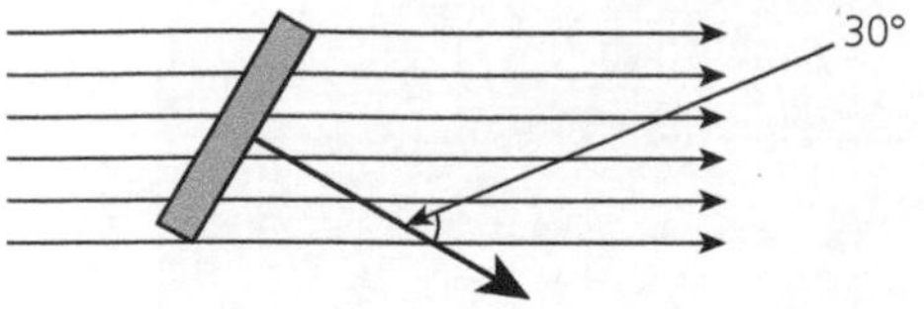

Figure 29

Calculate the flux linkage. Give the correct unit. Give your answer to an appropriate number of significant figures. [2]

Total: 5

SB2 p150 **3.7.5.4** 20–1 State Faraday's law. [1]

SB2 p150 **3.7.5.4** 20–2 State Lenz's law. [1]

SB2 p150 **3.7.5.4** 20–3 Write an expression that summarises both laws. [1]

SB2 p151 **3.7.5.4** **MS 2.1** 20–4 A coil of 500 turns has a diameter of 15 cm. It is perpendicular to a magnetic field of magnetic flux density 0.15 T. The magnetic flux density is reduced from 0.15 T to 0 in a time of 10 s.

Calculate the e.m.f. induced. [4]

SB2 p151 **3.7.5.4** **MS 2.1, 2.2, 2.3** 20–5 A search coil has 2500 turns and an area of 1.5×10^{-4} m². It is placed between the poles of a large horseshoe magnet. It is rapidly pulled out of the field in a time of 0.30 s. A data-logger records an average value for the e.m.f. of 0.75 V. Calculate the flux density between the poles of the magnet. [2]

SB2 p152 **3.7.5.4** 20–6 A small cylindrical magnet is dropped into a vertical PVC tube 12 mm in diameter and 30 cm in length. The magnet drops out of the bottom almost immediately. It is then dropped into a copper tube of exactly the same dimensions. It takes two seconds before it drops out of the bottom.

Explain these observations. [2]

Total: 11

SB2 p155 **3.7.5.4** **MS 0.1** 21 An aeroplane is flying at 150 m s⁻¹ through an area where the average magnitude of the Earth's magnetic field is 6.5×10^{-5} T (65 µT). The angle of the magnetic field is 35° to the vertical.

The wingspan of the aeroplane is 35 m.

21–1 Show that the vertical component of the magnetic field is about 53 µT. [1]

21–2 Calculate the e.m.f. over the end of the wings. [1]

Total: 2

SB2 p163 **3.7.5.5** 22–1 Compare alternating current (AC) to direct current (DC). [2]

SB2 p164 **3.7.5.5** **MS 2.3** 22–2 An AC supply with a peak current of 5.0 A passes through a 10 Ω resistor. Calculate:

1. the RMS current [1]

2. the RMS p.d. across the resistor [1]

3. the peak power [1]

4. the mean power supplied. [1]

SB2 p164 **3.7.5.5** **MS 2.3** 22–3 The RMS value of an alternating voltage is 12 V. Calculate the peak voltage. [1]

Total: 7

SB2 p164 | 3.7.5.5 | MS 2.3

23 The CRO screen in **Figure 30** shows a sinusoidal waveform.

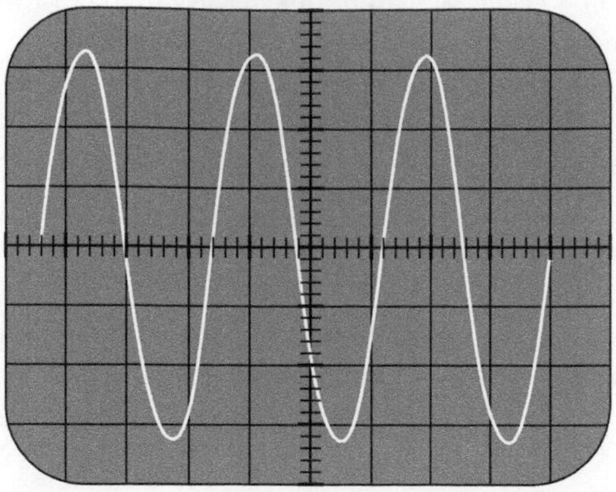

Figure 30

The time base is set at 2 ms per division and the y-gain at 0.5 V per division.

23–1 Determine the peak to peak voltage. [1]

23–2 Determine the peak voltage. [1]

23–3 Determine the RMS voltage. [1]

23–4 Determine the period. [2]

23–5 Determine the frequency. [1]

Total: 6

SB2 p168 | 3.7.5.6
24–1 Draw a diagram of a simple transformer. [2]

SB2 p168 | 3.7.5.6
24–2 Write down the transformer equation that links the number of turns on each coil to the voltage on each coil. [1]

SB2 p168 | 3.7.5.6
24–3 Assuming that the transformer is 100% efficient, show that:

$$\frac{N_p}{N_s} = \frac{I_s}{I_p}$$

[2]

SB2 p168 | 3.7.5.6
24–4 A transformer has a primary of 3600 turns and a secondary of 150 turns. It takes 1.5 amps from the 230 V mains.

1. Calculate the turns ratio. [1]

2. Calculate the output voltage and current. [1]

Total: 7

25 Transformers are very efficient machines, but are not 100% efficient. The best efficiency for a large power station transformer is about 97%.

SB2 p169 | 3.7.5.6
25–1 Outline the reasons for this and discuss whether there is a limit to the power input and output of such a transformer. [3]

SB2 p169 | 3.7.5.6 | MS 2.2, 2.3
25–2 A single phase electricity generating station generates 200 MW at a voltage of 15 000 V_{rms} to supply a large factory.

Show that the RMS current from the generator is about 13 000 A. [1]

SB2 p169 | 3.7.5.6 | MS 2.2, 2.3
25–3 A transmission line takes the output to a factory. Over the whole length of the transmission line, the resistance is 0.75 Ω. Calculate:

1. the energy loss in the power line [1]

2. the energy available to the factory. [1]

SB2 p169 | 3.7.5.6 | MS 2.2, 2.3
25–4 Explain how the energy loss is reduced to an acceptable level. [2]

Total: 8

Topic review: fields and their consequences

1 Uranus is a large gas planet. It has 27 moons that orbit the planet.

   Table 12 shows data for 8 of the largest moons.

Moon	Diameter/km	Orbital radius, r/km	r^3/m³	Orbital period, T/days	T^2/s²
Portia	135	66 100	2.89×10^{23}	0.51	1.94×10^9
Puck	162	86 000	6.36×10^{23}	0.76	4.31×10^9
Miranda	472	129 900	2.19×10^{24}	1.41	1.48×10^{10}
Ariel	1158	190 900	6.96×10^{24}	2.52	4.74×10^{10}
Umbriel	1169	266 000	1.88×10^{25}	4.14	
Titania	1578	436 300	8.31×10^{25}	8.71	5.66×10^{11}
Oberon	1523	583 500		13.46	1.35×10^{12}

Table 12

SB2 p47 3.7.1.4

1–1 State what shape the orbits of the moons have. [1]

SB2 p48 3.7.1.4 MS 0.1

1–2 Copy and complete the rows in **Table 12** for Umbriel and Oberon. [2]

SB2 p46 3.7.1.4 MS 3.1, 3.2

1–3 Plot this data as a graph of T^2 against r^3. [3]

SB2 p48 3.7.1.4 MS 3.1

1–4 Some astronomers think they have found a moon, Moon X, between the orbits of Titania and Oberon. It has an orbital radius of 477 000 km. Use your graph to determine its orbital period in days. [3]

SB2 p48 3.7.1.4 MS 3.4

1–5 Show that the gradient of the graph has a value of about 7×10^{-15}. Give the correct unit. [2]

SB2 p48 3.7.1.4 MS 3.3

1–6 Determine the mass of Uranus using the gradient of the graph. [2]

Total: 13

2 The data in **Table 13** shows how the gravitational field strength varies with distance from the surface of the Earth.

Height/m	Radius, r/m	g/N kg⁻¹	$Log_{10}(r)$	$Log_{10}(g)$
0.00	6.37×10^6	9.81	6.804	0.992
1.00×10^6		7.34		
2.00×10^6		5.69		
3.00×10^6		4.54		
4.00×10^6		3.71		
5.00×10^6		3.09		
6.00×10^6		2.61		
7.00×10^6		2.23		
8.00×10^6		1.93		
9.00×10^6		1.69		
1.00×10^7		1.49		

Table 13

SB2 p39 3.7.1.2

2–1 Define gravitational field strength. [1]

SB2 p39 3.7.1.2 MS 0.5, 2.5

2–2 Copy and complete **Table 13**. One line has been done as an example. [3]

SB2 p39 3.7.1.2 MS 2.2, 2.3, 3.2, 3.4

2–3 Plot a graph of $log_{10}(r)$ against $log_{10}(g)$. [3]

2–4 On your graph, show where the Earth's surface is. [1]

SB2 p39 3.7.1.2 MS 3.3, 3.4, 3.10, 3.11

2–5 1. Calculate the gradient of the graph. [2]

   2. Explain the significance of the gradient of the graph. [2]

Total: 12

3   **Figure 31** shows an alpha particle approaching a uniform electric field.

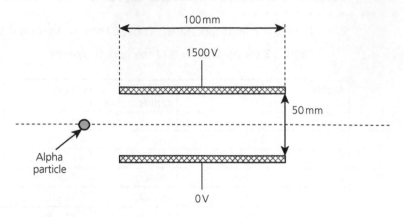

Figure 31

The alpha particle has a mass of 6.64×10^{-27} kg. Its charge is $+2e$. It is travelling at a speed of 1.50×10^5 m s^{-1}.

SB2 p90 | 3.7.2.1 3–1 State what is meant by electric field strength.   [1]

SB2 p90 | 3.7.2.1 | MS 0.1 3–2 Calculate the electric field strength between the plates.   [1]

SB2 p93 | 3.7.2.1 | MS 2.2 3–3 Show that the force acting on the alpha particle is about 10^{-14} N.   [1]

SB2 p92 | 3.7.2.1 3–4 1. Determine the acceleration.   [1]

       2. State the direction of movement and the shape of its path.   [1]

Total: 5

4   **Figure 32** shows an experiment used to investigate the potential around a charged sphere.

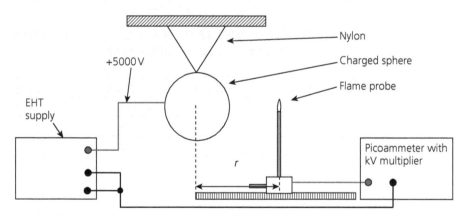

Figure 32

The charged sphere is isolated and at a high potential (+5000 V). The radius of the sphere is 15 cm.

The picoammeter with the kV multiplier acts as a voltmeter. The flame probe is used to detect the potential, V, at distances r.

SB2 p97 | 3.7.2.2 | MS 0.1, 2.2 4–1 Show that the charge on the sphere is about 80 nC.   [1]

6 Fields and their consequences

SB2 p97 3.7.2.2 MS 0.5, 2.5

4–2 **Table 14** shows the results of this experiment. Copy and complete **Table 14**. *[2]*

Radius, r/m	Potential, V/V	$\log_{10}(r/\text{m})$	$\log_{10}(V/\text{V})$
0.15	5000	−0.824	3.70
0.45	1700	−0.347	3.23
0.75	1000	−0.125	3.00
1.05	720	0.021	2.86
1.35	560	0.130	2.75
1.65	445	0.217	2.65
1.95	390		
2.25	320		
2.55	290		
2.85	260		
3.15	240		

Table 14

SB2 p97 3.7.2.2 MS 3.2

4–3 Plot this data on a graph of $\log_{10}$ (potential/V) against $\log_{10}$ (radius/m). *[3]*

SB2 p97 3.7.2.2 MS 3.4

4–4 1. Determine the gradient of the line you have plotted. *[2]*

2. State what the gradient shows. *[1]*

Total: 9

5 The circuit in **Figure 33** was used to study the charge and discharge of a capacitor.

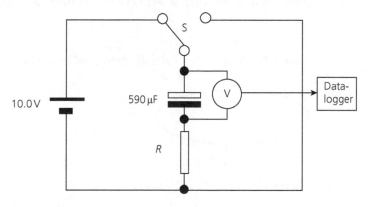

Figure 33

The capacitor is charged up from the battery when the two-way switch, S, is in the position shown. The capacitor is discharged by moving the switch to its second position. It discharges through the resistor of resistance R.

The data is recorded using a data-logger connected to a voltmeter sensor.

Data is recorded every second in **Table 15** (see next page).

Time/s	Voltage/V	ln (V/V)
0.0	10.000	2.303
1.0	6.48	1.869
2.0	4.19	1.433
3.0	2.72	1.001
4.0	1.76	0.565
5.0	1.14	0.131
6.0	0.74	−0.305
7.0	0.48	
8.0	0.31	
9.0	0.20	
10.0	0.13	

Table 15

SB2 p120 3.7.3.4 MS 0.5 **5–1** Copy and complete **Table 15**. *[1]*

SB2 p120 3.7.3.4 MS 3.11 **5–2** Plot the data as a graph of ln (*V*) against time *t*. *[3]*

SB2 p120 3.7.3.4 MS 3.10 **5–3** Use your graph to calculate:

1. the time constant *[3]*

2. the resistance *R* of the resistor. Give your answer to an appropriate
number of significant figures. *[2]*

SB2 p120 3.7.3.4 MS 3.10 **5–4** The battery is now set to provide an output voltage of 5.0 V. Explain what
effect, if any, this change will have on the time constant. *[2]*

Total: 11

SB2 p132 3.7.4.1 **6–1** **Figure 34** shows a thin flat coil being used to produce a magnetic field.

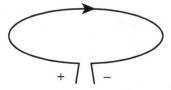

Figure 34

Copy and complete **Figure 34** showing the magnetic field lines. *[1]*

SB2 p139 3.7.4.2 **6–2** **Figure 35** shows a stream of electrons travelling in a straight line before
they enter a force field.

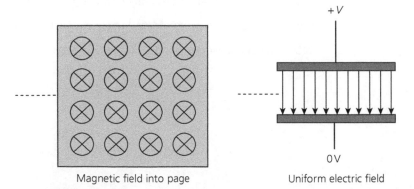

Magnetic field into page　　　　　Uniform electric field

Figure 35

1. Copy and complete the diagram to show the paths of the electrons. *[2]*

2. Explain why the paths have this shape. *[2]*

SB2 p139 | 3.7.4.2 | MS 2.2 | 6–3 A charged particle of mass m and charge q travels at speed v into a magnetic field of flux density B that is perpendicular to the path of the electron.

Show that the radius r of the path is given by:

$$r = \frac{mv}{Bq}$$ [2]

SB2 p139 | 3.7.4.2 | MS 0.2, 1.1, 2.3 | 6–4 An electron travels at $1.5 \times 10^7 \, \mathrm{m\,s^{-1}}$ into a magnetic field of flux density 1.75 mT. Calculate the radius of the path in the magnetic field.

Give your answer to an appropriate number of significant figures. [2]

SB2 p139 | 3.7.4.2 | 6–5 Explain what, if any, would be the difference to the path of the electron if the beam was in the air, rather than in a vacuum. No calculation is needed. [2]

Total: 11

SB2 p167 | 3.7.4.6 | 7–1 1. Draw a diagram to show the structure of a simple transformer. [1]

2. Give its circuit symbol. [1]

SB2 p169 | 3.7.4.6 | 7–2 A physics tutor is demonstrating a transformer in an A-level physics class. The tutor closes the core with a piece of soft iron. The transformer buzzes loudly. The secondary voltage is much lower than expected. The transformer rapidly gets hot, especially the soft iron piece.

1. Explain these observations. [1]

2. State what the tutor has done wrong, and what they should do to correct it. [1]

SB2 p169 | 3.7.4.6 | 7–3 Although transformers are very efficient, they are not quite 100% efficient. Identify the ways that energy is lost in a transformer. [2]

SB2 p169 | 3.7.4.6 | MS 2.2, 2.3 | 7–4 A large industrial transformer steps an input voltage of 132 kV to an output voltage of 1000 V to provide power to an electric arc furnace. The furnace takes a current of 40 000 A. The transformer is 95% efficient. Assume the current is single phase.

1. Calculate the power taken by the furnace. [1]

2. Calculate the input current. [1]

3. Calculate the power lost when the furnace is running. [1]

4. Suggest how the lost power is removed from the transformer to avoid damage due to overheating. [1]

Total: 10

Nuclear physics

Radiation and the nucleus

Quick questions

SB2 p183 | **3.8.1.2**

1   Copy and complete **Table 1** on ionising radiations:

Radiation	Description and charge	Penetration	Ionisation	Effect of E or B field
Alpha (α)				
Beta (β)				
Gamma (γ)				

Table 1

SB2 p183 | **3.8.1.2**

2   The element radium (Ra) has a proton number of 88 and a nucleon number of 226.

a)   What is meant by an isotope?

b)   Write the symbol for radium in nuclide notation.

c)   It decays by alpha radiation with a half-life of 1600 years. Write down the decay equation.

d)   What is meant by half-life?

SB1 p9 | **3.2.1.2, 3.8.1.2**

3a)   Sodium-24 is an unstable isotope of sodium. It decays by beta-minus decay to magnesium. Write its decay equation in nuclide notation.

b)   Sodium-22 is an unstable isotope of sodium. It decays by beta-plus decay to neon. Write its decay equation in nuclide notation.

SB2 p177 | **3.8.1.1**

4   Give **two** points that Rutherford scattering tells us about the structure of the atom.

SB2 p184 | **3.8.1.2**

5   A cobalt-60 source is often used as a gamma source. It also emits beta-minus radiation.

Explain how the source is adapted to emit gamma only.

SB2 p186 | **3.8.1.2**

6   Granite rocks emit a radioactive gas, radon, for which A = 222. Radon decays by alpha emission to Element X by alpha decay with a half-life of 3.82 d.

a)   By writing an appropriate decay equation, identify Element X.

b)   Many houses are built with granite. Explain what the risks are, and how they can be minimised.

SB2 p180 | **3.8.1.1**

7   When electrons strike a magnesium target, they form a diffraction pattern.

Explain why electrons do this.

SB2 p181 | **3.8.1.5** | **MS 1.1**

8   The stable cobalt nucleus is shown in nuclide notation:

$^{59}_{27}\text{Co}$

Show that the nuclear radius is $4.7 \times 10^{-15}\,\text{m}$.

Exam-style questions

SB2 p184 3.8.1.2

9 Cobalt-60 is widely used as a gamma source. Which one of these correctly
describes the emissions from cobalt-60? [1]

A: An alpha particle is emitted, followed by a gamma photon.

B: A gamma photon alone is emitted.

C: A beta-minus particle is emitted, followed by a gamma photon.

D: A neutron is emitted simultaneously with a gamma photon.

Total: 1

SB2 p178 3.8.1.5 MS 2.2, 2.3

10 Alpha particles of kinetic energy 3.0 MeV are used to estimate the diameter
of a silver nucleus. The nuclide notation is $^{107}_{47}$Ag.

Which one of these gives the closest approach of the alpha particles to the
silver nucleus? [1]

A: 9.0×10^{-14} m   **B:** 1.8×10^{-13} m   **C:** 2.0×10^{-13} m   **D:** 6.3×10^{-6} m

Total: 1

SB2 p181 3.8.1.5 MS 0.1, 0.2

11 Which one of these gives the correct diameter in femtometres of a silver nucleus? [1]

The constant $r_0 = 1.2 \times 10^{-15}$ m

A: 1.2 fm   **B:** 5.7 fm   **C:** 11.4 fm   **D:** 12.4 fm

Total: 1

SB2 p187 3.8.1.2 MS 0.3

12 A radiation counter is placed at a distance d from a gamma emitter. The
count rate is A. It is then moved to a new position, $4d$ away from the
original position. What is the count rate in the new position? [1]

A: A **B:** $\dfrac{A}{4}$ **C:** $\dfrac{A}{16}$ **D:** $\dfrac{A}{25}$

Total: 1

SB2 p177 3.8.1.1

13–1 **Figure 1** shows the paths taken by different alpha particles as they
approach different gold nuclei in a Rutherford scattering experiment.

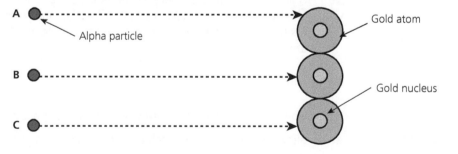

Figure 1

Describe the path of each alpha particle. [2]

SB2 p177 3.8.1.1

13–2 Write down **two** conclusions from this experiment. [2]

SB2 p178 3.8.1.1 MS 2.2, 2.3

13–3 A 5.0 MeV alpha particle approaches a gold nucleus. Calculate:

1. the charges carried by the alpha particle and the gold nucleus [1]

2. the energy in joules of the alpha particle [1]

3. the minimum approach distance. [1]

SB2 p179 3.8.1.1

13–4 Explain why the answer in **Question 13–3–3** is not a reliable measure of
the radius of the nucleus. [2]

Total: 9

14 When we do experiments that count radioactive decay events, we need to carry out background radiation counts.

SB2 p186 3.8.1.2 14–1 Identify **two** different sources of radiation in a school physics laboratory. [2]

SB2 p186 3.8.1.2 14–2 Describe how you would obtain a reliable estimate of the background count in the school physics laboratory. [4]

SB2 p186 3.8.1.2 14–3 A radiation counter is reset to zero, and left in the open laboratory for a period of 15 minutes. After 15 minutes, the count is 867 counts.

Calculate the background radiation count. Give the correct unit. [1]

SB2 p186 3.8.1.2 MS 0.1 14–4 A sample of radioactive material is measured to have an activity of 1800 counts per minute. If the background radiation is 85 counts per minute, calculate:

1. the activity of the material in Bq [1]

2. the number of disintegrations over a 20-hour period. [1]

SB2 p186 3.8.1.2 MS 14–5 The background count in a laboratory is measured as 90 counts per minute. A Geiger–Müller tube is placed 15 cm from a gamma source. The count is 1000 counts per minute. The GM tube is now placed 60 cm from the source. Calculate the expected count for one minute. [3]

Total: 12

SB2 p189 3.8.1.2 15–1 Define the term absorbed dose, and give the SI unit. [2]

SB2 p189 3.8.1.2 15–2 A more common unit is the Gray (Gy). Describe how it is related to the SI unit for absorbed dose. [1]

SB2 p189 3.8.1.2 15–3 1. State the factor other than absorbed dose that determines the biological damage that may be caused. [1]

2. Give the unit for dose equivalent. [1]

SB2 p190 3.8.1.2 15–4 Calculate the absorbed dose when a cancerous tissue of mass 150 g is exposed to 12 J of energy. Give the correct unit. [1]

SB2 p190 3.8.1.2 15–5 Calculate the equivalent dose on the tissue when neutrons are used for the treatment. The weighting factor for neutrons is 3.0. Give the correct unit. [1]

Total: 7

16 A beam of electrons strikes a magnesium target. The electrons have a kinetic energy of 268 MeV. The first diffraction minimum is found at an angle of 55°. The relationship between the angle and the diameter of the atom is given by:

$$\sin\theta = 1.22\frac{\lambda}{d}$$

where λ is the de Broglie wavelength in m and d is the diameter of the nucleus in m.

SB2 p181 3.8.1.5 MS 0.2 16–1 Show that the energy of each electron is about 4×10^{-11} J. [1]

SB2 p181 3.8.1.5 MS 0.2 16–2 Calculate the momentum of each electron. Give the correct unit. [2]

SB2 p181 3.8.1.5 MS 0.2 16–3 Calculate the de Broglie wavelength of the electron. [1]

SB2 p181 3.8.1.5 MS 0.1, 0.2, 2.2, 2.3 16–4 Use the equation above to calculate the radius of the magnesium nucleus. [2]

Total: 6

17 Some students carry out an investigation into the inverse square law for gamma radiation using apparatus like that in **Figure 2**.

The gamma source is cobalt-60.

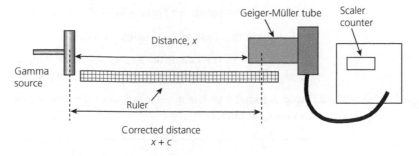

Figure 2

SB2 p185 3.8.1.2 RP 12

17–1 Give **two** reasons why there needs to be a corrected distance. [2]

SB2 p188 3.8.1.2 RP 12

17–2 Explain the need for a corrected count. [1]

The data from the experiment is shown in **Table 2**.

Distance x/m	Count 1/s⁻¹	Count 2/s⁻¹	Corrected count 1/s⁻¹	Corrected count 2/s⁻¹	Average/s⁻¹	1/average$^{0.5}$/ s$^{0.5}$
0.05	8.82	8.93	8.43	8.54		
0.1	3.32	3.1	2.93	2.71		
0.15	1.75	1.76	1.36	1.37		
0.2	1.26	1.23	0.87	0.84		
0.25	0.95	1.1	0.56	0.71		
0.3	0.87	0.82	0.48	0.43		
0.35	0.72	0.71	0.33	0.32		
0.4	0.64	0.62	0.25	0.23		

Table 2

SB2 p188 3.8.1.2 MS 1.2 RP 12

17–3 Copy and complete the last two columns of **Table 2**. [2]

SB2 p188 3.8.1.2 MS 3.1, 3.2

17–4 Plot a graph of the inverse of the square-root of the average count rate against the distance, x. [4]

SB2 p188 3.8.1.2 MS 1.4 RP 12

17–5 Use the graph to estimate the value of the correction factor, c. [2]

Total: 11

18 The **Figure 3** sketch graph shows the electron diffraction pattern for an element.

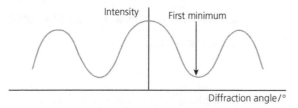

Figure 3

SB2 p182 3.8.1.5

18–1 1. Explain what happens to the diffraction pattern when the energy of the electrons is reduced. [1]

2. Explain what happens to the diffraction pattern when the energy of the electrons is kept the same, but a sample of a lighter element is used. [1]

SB2 p182 3.8.1.5 MS 2.2, 2.3, 4.5

18–2 An electron diffraction experiment is carried out using an imaginary element called tantrum (Tt). In nuclide notation, one isotope of this element is:

$^{243}_{119}$Tt

When electrons of energy 550 MeV are used, the first minimum is observed at 10.6°.

1. Show that the de Broglie wavelength is 2.3×10^{-15} m. [3]

2. Calculate the diameter of the tantrum nucleus. [1]

Total: 6

SB2 p182 3.8.1.5

19–1 The mass of the cobalt-59 nucleus is 58.93 u, where 1 u = 1.661 × 10⁻²⁷ kg.

1. Calculate the mass of the nucleus in kg. [1]

2. The radius of the cobalt nucleus is 4.7 × 10⁻¹⁵ m. Calculate the density of the nuclear material. Assume that the nucleus is a sphere. [2]

SB2 p182 3.8.1.5 MS 0.5, 3.2

19–2 **Table 3** shows how the nuclear radius of the first 20 elements varies with the nucleon number.

Element	Nucleon	$A^{\frac{1}{3}}$	r / fm
H	1	1.00	1.20
He	4	1.59	1.90
Li	7	1.91	2.30
Be	9	2.08	2.50
B	11	2.22	2.67
C	12	2.29	2.75
O	14	2.41	2.89
N	16	2.52	3.02
F	19	2.67	3.20
Ne	20	2.71	3.26
Na	23	2.84	3.41
Mg	24	2.88	3.46
Al	27	3.00	3.60
Si	28	3.04	3.64
P	31	3.14	3.77
S	32	3.17	3.81
Cl	35	3.27	3.93
Ar	38	3.36	4.03
K	39		
Ca	40		

Table 3

1. Write down the data needed to complete **Table 3**. [2]

2. Plot the data on a graph, with radius on the vertical axis and $A^{\frac{1}{3}}$ on the horizontal axis. [3]

3. Sketch a second graph of nuclear density against the nucleon number. [2]

4. State what the second graph shows. [1]

Total: 11

SB2 p183 3.8.1.5 MS 1.1, 2.2, 2.3

20 The formula for working out nuclear radius R is:

$$R = r_0 (A)^{\frac{1}{3}}$$

Nuclide 1 has a radius of R_1 and a nucleon number of A_1, while nuclide 2 has a radius of R_2 and a nucleon number of A_2.

20–1 Derive an expression that allows R_1 to be worked out if R_2 is known. [2]

20–2 The radius of the isotope $^{28}_{14}\text{Si}$ is 3.7 × 10⁻¹⁵ m. Use your derived equation to show that the radius of $^{99m}_{43}\text{Tc}$ is between 5 × 10⁻¹⁵ m and 6 × 10⁻¹⁵ m. Give your answer to an appropriate number of significant figures. [1]

Total: 3

Radioactive decay

Quick questions

SB2 p196 | 3.8.1.3 | MS 2.3

1 A sample of living material contains carbon-14 with an activity of $260\,Bq\,kg^{-1}$. The fraction that is made of carbon-14 is 1.4×10^{-12}.

a) Show that in 1 kg of material there are 6×10^{13} radioactive atoms.

b) Calculate the decay constant.

SB2 p195 | 3.8.1.3 | MS 2.3

2a) Define the term activity for a radioactive element.

b) The decay constant of a radioisotope is $4.29 \times 10^{-12}\,s^{-1}$. A sample contains 3.96×10^{21} particles. Calculate the initial activity.

SB2 p198 | 3.8.1.3 | MS 0.5, 2.2

3 A certain radioisotope has a half-life of 250 years.

a) Show that the decay constant, λ, is about $9 \times 10^{-11}\,s^{-1}$.

b) The initial activity of the sample is 5000 Bq. Calculate the activity after 60 years.

SB2 p199 | 3.8.1.3 | MS 0.5

4 Francium has a half-life of 22 minutes. A sample of francium has an initial activity of 2000 Bq.

a) Calculate the decay constant.

b) Calculate the activity of the sample after 120 minutes.

Exam-style questions

SB2 p195 | 3.8.1.3

5 Which one of these statements is correct about radioactive emissions? *[1]*

A: The emitted particles are radioactive themselves.

B: Radioactive nuclei can split spontaneously.

C: If nuclei have nucleon numbers that are multiples of 4, they can decay by alpha decay.

D: Radioactivity is an entirely random process.

Total: 1

SB2 p201 | 3.8.1.3 | MS 0.5

6 A certain radionuclide has a decay constant of $4.5 \times 10^{-9}\,s$. What is the half-life? *[1]*

A: $8.2 \times 10^7\,s$ **B:** $1.5 \times 10^8\,s$ **C:** $2.2 \times 10^8\,s$ **D:** $6.0 \times 10^8\,s$

Total: 1

SB2 p187 | 3.8.1.3 | MS 2.3

7 A tumour has a mass of 0.20 kg. It is exposed to 6.5×10^{12} protons each of which has an energy of 0.50 MeV. What is the dose of radiation received? W_R for protons is 2. *[1]*

A: $0.52\,Sv$ **B:** $1.04\,Sv$ **C:** $2.6\,Sv$ **D:** $5.2\,Sv$

Total: 1

SB2 p196 | 3.8.1.3 | MS 1.3

8–1 Define the term 'decay constant (λ)'. *[1]*

SB2 p196 | 3.8.1.3 | MS 1.3

8–2 Radioactive decay can be modelled using a large number (e.g. 100) of dice. A 6 represents a nucleus that has decayed. The dice carrying a 6 are removed, and the remainder are thrown again. The procedure is repeated at least 10 times. Work out the decay constant for any one die. *[1]*

SB2 p196 | 3.8.1.3 | MS 2.3

8–3 0.25 kg radium-226 emits alpha particles at a measured rate of $9.0 \times 10^{12}\,s^{-1}$. Calculate the decay constant of radium.

1 mole of radium has a mass of 0.226 kg.

Number of atoms in a mole $= 6.0 \times 10^{23}$. *[3]*

Total: 5

9   **Figure 4** shows the decay of an unstable radionuclide. Initially the sample has 2.70×10^{12} radioactive atoms.

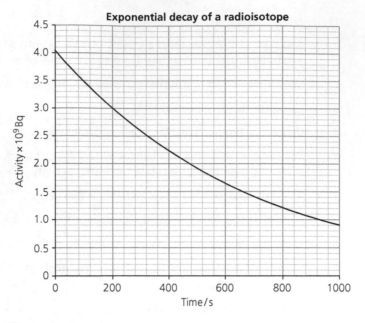

Figure 4

SB2 p198   3.8.1.3   MS 3.6

9–1 Calculate the initial activity using the graph.   [2]

SB2 p198   3.8.1.3   MS 3.1

9–2 Use the graph to determine the half-life of the radioisotope.   [1]

SB2 p198   3.8.1.3   MS 0.5

9–3 Calculate the decay constant of this decay.   [1]

9–4 **Figure 5** shows the area of stability for nuclides.

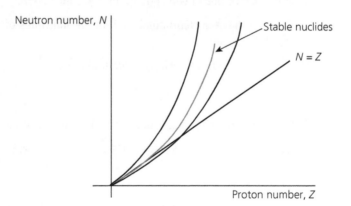

Figure 5

This radioisotope is an alpha emitter.

Explain where this nuclide is most likely to be found.   [3]

Total: 7

10   Rubidium $^{87}_{37}$Rb decays to strontium-87.

The decay has a decay constant of 1.42×10^{-11} year^{-1}.

SB2 p198   3.8.1.3   MS 0.1, 0.5, 2.2

10–1 Give the decay equation in nuclide notation.   [1]

SB2 p198   3.8.1.3   MS 0.1, 0.5, 2.2

10–2 Show that the half-life in years of this decay is 4.9×10^{10} years.   [1]

SB2 p198   3.8.1.3   MS 0.1, 0.5, 2.2

10–3 In a particular ancient rock sample that is known to be about 4.5×10^9 years old, 2.50 mg of rubidium is found. The half-life of rubidium is 4.9×10^{10} years.

1. Determine the mass **in grams** of the rubidium that was in the rock originally.   [3]

2. Calculate the activity of the 2.50 mg of rubidium, giving your answer to the appropriate number of significant figures and with the correct units.   [3]

Total: 8

7 Nuclear physics

SB2 p188 3.8.1.3

11 A source of alpha radiation, X, has an activity 1.7×10^9 Bq. The alpha particles are emitted uniformly from a spherical source that has a radius of 6.74×10^{-3} m.

11–1 Copy and complete the equation:

$$^A_Z X \rightarrow \qquad \qquad [1]$$

11–2 Show that the intensity of the radiation is about 3×10^{12} Bq m^{-2}. [2]

11–3 Calculate the intensity of the radiation at a distance of 5.0 m. Assume that the alpha particles are travelling in a vacuum. [2]

11–4 A detector has a diameter of 1.0 cm. It is connected to a counter. The background count is 1.5 Bq. Calculate the count rate displayed on the detector in one minute. [3]

Total: 7

Nuclear instability

Quick questions

SB2 p206 3.8.1.4

1 Explain how gamma rays are often emitted when a nuclide decays by alpha or beta decay.

SB2 p206 3.8.1.4

2 What is meant by a metastable state in a gamma emitter?

SB2 p206 3.8.1.4

3 What is meant by a metastable nucleus? Explain how you would expect a sample consisting of a large number of metastable nuclei to behave.

SB2 p205 3.8.1.4

4 Use a grid like **Figure 6** to show these two chains on the same decay chain diagram. Polonium-216 (start) and lead-208 (end) are shown.

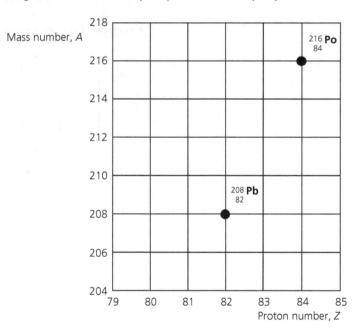

Figure 6

SB2 p213 3.8.1.4 MS 0.1, 0.5

5 Thorium decays by alpha decay to radium with the following equation:

$$^{228}_{90}\text{Th} \rightarrow {}^a_b\text{Ra} + \alpha$$

a) Determine a and b.

b) Show that the energy of the emitted alpha particle is about 5.5 MeV. The data to use is given below:

- mass of thorium nucleus = 227.97932 u

- mass of radium nucleus = 223.971888 u

- mass of alpha particle = 4.00150 u.

Exam-style questions

SB2 p215 3.8.1.4

6 Which one of the graphs of binding energy per nucleon against nucleon number in **Figure 7** is correct? [1]

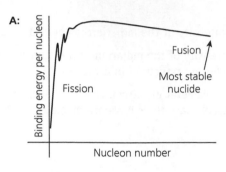

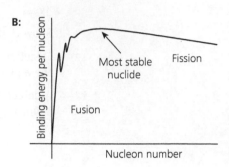

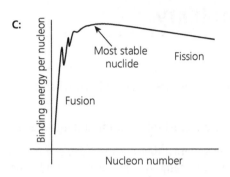

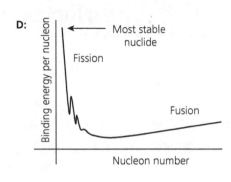

Figure 7

Total: 1

SB2 p205 3.8.1.4

7 **Figure 8** shows the decay of a nuclide. What decay could it represent? [1]

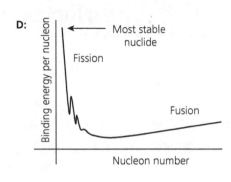

Figure 8

A: Alpha

B: Neutron capture

C: Beta-minus

D: Beta-plus

Total: 1

SB2 p205 3.8.1.4

8–1 Thorium-228 decays by alpha decay to radium-224. The equation is shown below:

$$^{228}_{90}\text{Th} \rightarrow {}^{224}_{88}\text{Ra} + {}^{4}_{2}\text{He}$$

Draw this decay on a graph of nucleon number against proton number. [3]

SB2 p205 **3.8.1.4**

8–2 **Figure 9** shows neutron number, *N*, against proton number, *Z*.

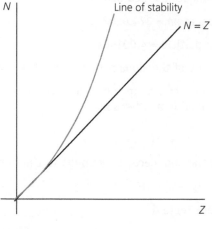

Figure 9

On a copy of **Figure 9**, mark on it where you would expect to see:

1. Alpha decay [1]

2. Beta-minus decay [1]

3. Beta-plus decay [1]

4. The limit below which alpha decay does not happen. [1]

SB2 p205 **3.8.1.4**

8–3 The isotope chromium-57 decays to an isotope of vanadium by K-capture, with a half-life of 27.7 days.

1. Describe how this happens. [1]

2. Write an equation to show the process. [1]

Total: 9

SB2 p205 **3.8.1.4**

9 $^{64}_{29}$Cu is an isotope of copper. It can decay by positron emission or electron capture (or K-capture) to an isotope of nickel (Ni). **Figure 10** shows nucleon number, *A*, against proton number, *Z*.

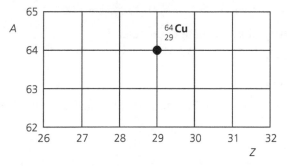

Figure 10

9–1 On a copy of **Figure 10**, show this decay. [2]

9–2 Write equations for both the positron emission and the electron capture. [2]

9–3 Explain why this decay happens. [1]

9–4 Describe what happens in electron capture. [1]

Total: 6

SB2 p213 **3.8.1.4**

10–1 In the examinations, the data that is used in the nuclear energy calculations is **nuclear** masses. However, many sources give **atomic** masses. Explain the effect on calculations of mass defect that would result from such a mistake. [2]

SB2 p213 3.8.1.4 | MS 0.1, 0.5

10–2 A student finds this data for the alpha decay of thorium:

- mass of thorium atom = 228.02873 u
- mass of radium atom = 224.02020 u
- mass of alpha particle = 4.00150 u.

Calculate the energy of the alpha particle using this data. [2]

SB2 p213 3.8.1.4

10–3 Explain why the mass of the alpha particle was quoted as 4.00150 u instead of 4.00260 u as found in another source. [1]

Total: 5

11 The equation for the beta decay of actinium to thorium is:

$$^{228}_{89}\text{Ac} \rightarrow {}^{228}_{90}\text{Th} + {}^{0}_{-1}e^- + \bar{v}_e$$

SB2 p213 3.8.1.4

11–1 Copy and complete **Table 4**. [3]

Component	Atomic mass/u	Number of electrons	Mass of electrons/u	Nuclear mass/u
Actinium	228.03310			
Thorium	228.02873			
Beta-minus	0.000549	1	0.000549	0.000549
Electron antineutrino	0			

Table 4

SB2 p213 3.8.1.4 | MS 2.1

11–2 Calculate the energy of this decay:

$$^{228}_{89}\text{Ac} \rightarrow {}^{228}_{90}\text{Th} + {}^{0}_{-1}e^- + \bar{v}_e$$ [2]

SB2 p205 3.8.1.4

11–3 Some isotopes can decay by different pathways. Polonium-216 decays to lead-208 (stable) by these two decay chains:

Chain 1:

$$^{216}_{84}\text{Po} \rightarrow {}^{212}_{82}\text{Pb} \rightarrow {}^{212}_{83}\text{Bi} \rightarrow {}^{212}_{84}\text{Po} \rightarrow {}^{208}_{82}\text{Pb}$$

Chain 2:

$$^{216}_{84}\text{Po} \rightarrow {}^{212}_{82}\text{Pb} \rightarrow {}^{212}_{83}\text{Bi} \rightarrow {}^{208}_{81}\text{Tl} \rightarrow {}^{208}_{82}\text{Pb}$$

Identify which of these decays are:

1. Alpha [2]

2. Beta. [2]

(Po – polonium; Bi – bismuth; Tl – thallium; Pb – lead.)

Total: 9

Mass and energy

Quick questions

SB2 p213 3.8.1.6

1a) Define the atomic mass unit.

b) Give its mass in kg, and energy in eV, MeV and J.

SB2 p214 3.8.1.6 | MS 2.3

2 The most common isotope of calcium is $^{40}_{20}\text{Ca}$

It has a nuclear mass of 39.9626 u.

a) Calculate the mass defect of this isotope.

b) Calculate the binding energy in MeV.

- mass of a proton = 1.00728 u
- mass of a neutron = 1.00867 u

SB2 p215 | 3.8.1.6 | MS 2.3 3a) Show that the mass defect in a helium nucleus (alpha particle) is 0.03039 u.

b) Calculate the binding energy per nucleon in MeV.

SB2 p216 | 3.8.1.6 4a) Write down an equation to show how helium is formed from hydrogen by fusion.

b) Explain the conditions needed for the reaction to occur.

SB2 p213 | 3.8.1.6 5 Calculate the mass of material that is converted to 5.0×10^{13} J.

SB2 p180 | 3.8.1.7 | MS 2.2, 2.3 6 A thermal neutron has a kinetic energy that is in the infrared region, about 1.0 eV.

a) Calculate the wavelength of a photon that has energy of 1.0 eV.

b) Calculate the speed of a thermal neutron.

7 A typical alpha decay is shown by the decay of radon-220:

$$^{220}_{86}\text{Rn} \rightarrow\ ^{A}_{Z}\text{X} + \alpha + \text{energy}$$

SB2 p213 | 3.8.1.6 | MS 0.1, 0.2 a) Give the values for A and Z for both X and α.

b) Explain which of these products is stable, and which is unstable.

c) Calculate the energy released in MeV and J by 1 nucleus of radon.

Data:

- mass of Rn-220 nucleus = 219.916886 u
- mass of nucleus of isotope X = 213.902874 u
- mass of alpha = 4.00150 u.

SB2 p213 | 3.8.1.6 | MS 1.1, 2.1 8 In this reaction, a proton and a deuterium nucleus fuse to form a helium nucleus, releasing energy:

$$^{1}_{1}\text{p} +\ ^{2}_{1}\text{H} \rightarrow\ ^{A}_{2}\text{He} + \text{energy}$$

a) Write down the isotope of helium formed, in isotope notation.

b) Calculate the energy given out by this interaction.

Data:

- mass of proton = 1.00728 u
- mass of deuterium = 2.01355 u
- mass of helium = 3.01493 u.

SB2 p213 | 3.8.1.6 | MS 0.1, 0.2 9 Beryllium decays by electron capture to lithium with a half-life of 53.28 days. The equation is:

$$^{7}_{4}\text{Be} +\ ^{0}_{-1}e^{-} \rightarrow\ ^{7}_{3}\text{Li} + \nu_{e}$$

Calculate the energy change in this event.

Data:

- mass of beryllium nucleus = 7.014727 u
- mass of lithium nucleus = 7.014356 u
- mass of electron = 0.000549 u
- mass of electron neutrino is negligible.

Exam-style questions

SB2 p215 | 3.8.1.6 | MS 2.3 10–1 Beryllium (Be) has a proton number of 4. Its nuclear mass is 9.01218 u.

1. Calculate the number of neutrons beryllium has. [1]

2. Calculate the mass defect of the beryllium nucleus. [2]

SB2 p217 3.8.1.6 10–2 In very old stars, helium can fuse together to form beryllium-8. However, beryllium-8 is never found in the remains of old stars. Suggest why. [1]

SB2 p215 3.8.1.6 MS 2.3 10–3 Beryllium-8 has a nuclear mass of 8.00531 u.

1. Calculate the mass defect of a beryllium-8 nucleus. [2]

2 Show that the energy from the mass defect in a beryllium-8 nucleus is 54.48 MeV. [1]

3. Calculate the binding energy per nucleon. [1]

SB2 p217 3.8.1.6 MS 2.3 10–4 The binding energy worked out from the mass defects of each helium nucleus is 28.31 MeV. The two alpha particles fuse as shown by the equation:

$$^4_2\text{He} + ^4_2\text{He} \rightarrow ^8_4\text{Be}$$

1. Calculate the total binding energy of the two alpha particles. [1]

2. Calculate the energy change as the reaction proceeds. [1]

Total: 10

11 Use the data for the nuclear masses to answer this question.

- proton mass = 1.00782 u
- neutron mass = 1.00867 u
- deuterium mass = 2.01355 u
- tritium mass = 3.01550 u
- helium-3 mass = 3.01493 u
- helium-4 mass = 4.00151 u

SB2 p217 3.8.1.6 MS 2.3 11–1 This reaction involves fusion of two deuterium nuclei:

$$^2_1\text{H} + ^2_1\text{H} \rightarrow ^3_2\text{He} + ^1_0\text{n}$$

1. Show that the change in mass (mass defect) in this reaction is about 3.5×10^{-3} u. [1]

2. Determine the energy in MeV that is emitted. [1]

SB2 p217 3.8.1.6 MS 2.3 11–2 An alternative fusion reaction for deuterium nuclei is:

$$^2_1\text{H} + ^2_1\text{H} \rightarrow ^3_1\text{H} + ^1_1\text{H}$$

1. Show that the mass defect in this reaction is about 4.4×10^{-3} u. [1]

2. Determine the energy in MeV that is emitted. [1]

SB2 p217 3.8.1.6 MS 2.3 11–3 Deuterium and tritium fuse to give a helium-4 nucleus and a neutron:

$$^2_1\text{H} + ^3_1\text{H} \rightarrow ^4_2\text{He} + ^1_0\text{n}$$

1. Show that the mass defect in this reaction is about 1.9×10^{-2} u. [1]

2. Determine the energy in MeV that is emitted. [1]

SB2 p217 3.8.1.6 MS 2.3 11–4 Deuterium and helium-3 react to form helium-4 and a proton:

$$^2_1\text{H} + ^3_2\text{He} \rightarrow ^4_2\text{He} + ^1_1\text{p}$$

1. Show that the mass defect is about 0.02 u. [1]

2. Determine the energy in MeV. [1]

SB2 p217 3.8.1.6 MS 2.2, 2.3 11–5 The total energy released in these reactions is about 43 MeV. 6.0×10^{23} particles react.

1. Show that the total energy is about 43 MeV. [1]

2. Calculate the total energy released in J. [1]

3. Calculate the total mass that has been turned into energy. [1]

Total: 11

SB2 p218   3.8.1.7

12–1 A student says that 'nuclear fission can happen with all elements that are alpha emitters'. Discuss this statement. [3]

12–2 Describe the process of fission of a single uranium nucleus. [3]

12–3 Describe what conditions are needed for a chain reaction and how it occurs. [4]

Total: 10

SB2 p213   3.8.1.6   MS 1.1, 2.1

13–1 In a certain fission reaction from U-235, the mass change is 3.1×10^{-28} kg for each nucleus.

    1. Determine the energy equivalence to this mass change. [1]

    2. Calculate how many nuclei would be needed to give 1.0 J. Give your answer to an appropriate number of significant figures. [1]

SB2 p214   3.8.1.6

13–2 A nucleus written in isotope notation is $^{A}_{Z}X$. It has a mass M. If the mass of a proton is m_p and the mass of a neutron is m_n:

    1. Write an expression to determine the mass defect Δm of the nucleus. [1]

    2. Hence, write an expression to determine the binding energy per nucleon, E. [1]

Total: 4

Topic review: nuclear physics

1 A research facility has an underground gamma-bunker for irradiating materials with gamma rays from a cobalt-60 source. The diagram shown in **Figure 11** is a simplified version of the arrangement.

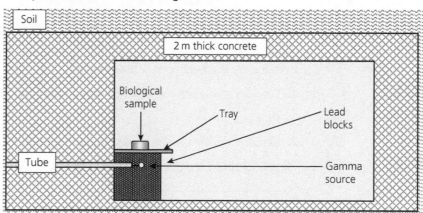

Figure 11

The gamma source is normally stored in a tube that is surrounded by thick concrete. When it's needed, it is pushed forward by an electric motor through the tube into a box made of lead blocks and open at the top. The biological sample being irradiated is placed on a tray to be exposed to the source. Radiation detectors and timers are used to ensure that the dose of radiation to which the sample is exposed is measured accurately.

There are locking systems in place to ensure that people are outside the bunker before the source can be exposed. Nor can they go back in until the source is withdrawn.

SB2 p183   3.8.1.2

1–1 Cobalt-60 decays by beta-minus decay to nickel. Copy and complete the equation:

$$^{60}_{27}Co \rightarrow$$ [1]

SB2 p183   3.8.1.2

1–2 1. Explain how a pure gamma ray is achieved from this source. [1]

    2. Explain how the gamma rays are produced. [2]

SB2 p204 | 3.8.1.4 | MS 2.3

1–3 **Figure 12** shows some of the energy levels of the cobalt as it emits gamma rays.

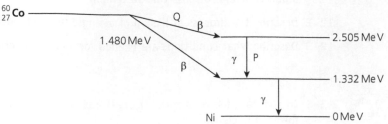

Figure 12

Calculate the transitions P (gamma) and Q. [2]

SB2 p188 | 3.8.1.3 | MS 0.1, 2.2

1–4 The source has a total activity of 1.0×10^9 Bq. It is spherical with a radius of 5.0 cm.

1. Show that the intensity of the radiation is about 3×10^{10} Bq m^{-2}. [1]

2. The irradiated sample is 60 cm from the centre of the source. Determine the intensity of the radiation at this point. [1]

3. The irradiated biological sample is in the form of a thin square of side 5.0 cm. Calculate the radiation received per second by the sample. [1]

SB2 p188 | 3.8.1.3 | MS 0.1

1–5 The average photon energy of a gamma photon is considered to be 1.25 MeV.

1. Calculate the energy per second with which the sample is irradiated. [1]

2. The biological sample has a mass of 1.5×10^{-4} kg. Calculate the dose received in 1 hour. [1]

3. Give the dose equivalent of 1 hour. Give the correct unit. [2]

Total: 13

2 Cobalt-60 is an artificial isotope of cobalt. It is made from cobalt-59, which is stable, by bombarding it with neutrons.

SB2 p204 | 3.8.1.4

2–1 1. Copy and complete the equation in nuclide notation: [1]

$^{59}_{27}\text{Co} +$

2. Describe the conditions needed for cobalt-60 to be formed. [2]

SB2 p205 | 3.8.1.4

2–2 **Figure 13** shows N against Z.

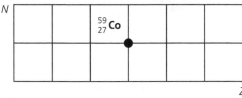

Figure 13

1. Copy and label **Figure 13** with appropriate values. [1]

2. Draw an arrow to show the new isotope that is formed. [1]

3. On a copy of **Figure 13**, show what happens when cobalt decays to nickel. [1]

SB2 p199 | 3.8.1.3 | MS 0.5, 1.3

2–3 The half-life of cobalt-60 is 5.27 years.

Calculate the probability of a single nucleus decaying in 1 hour. [2]

SB2 p199 | 3.8.1.3 | MS 0.5, 2.3

2–4 A cobalt-60 source contains about 60 µg of the radioactive isotope.

1. Calculate how many atoms of cobalt-60 there are in the source. [1]

2. Calculate the activity of the source when freshly prepared. [1]

3. Calculate the activity when the source is 15 years old. [1]

Total: 11

3   There are many kinds of fission nuclear reactors. Most use the fissionable isotope uranium-235 as a fuel.

SB2 p218 3.8.1.7

3-1   1. Define critical mass.   [1]

2. Describe how fission occurs in a fissile nucleus.   [3]

SB2 p219 3.8.1.7

3-2   In a nuclear power station, all reactors have a moderator, control rods and coolant.

Give the function of each of the components.   [3]

SB2 p215 3.8.1.6   MS 0.2

3-3   Uranium-235 has the following nuclide notation: $^{235}_{92}\text{U}$

Use this data to calculate the binding energy per nucleon: [2]

• mass of U-235 nucleus = 235.044 u

• mass of proton = 1.00728 u

• mass of neutron = 1.00867 u.

SB2 p215 3.8.1.6

3-4   **Figure 14** shows the variation of binding energy per nucleon with the nucleon number, *A*.

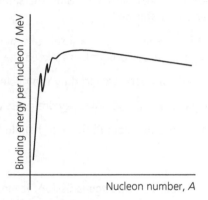

Figure 14

1. State where uranium-235 and the fission fragments are on the graph. (No values are needed.)   [1]

2. Explain why fission energy is released.   [3]

3-5   List **two** advantages and **two** disadvantages of using nuclear power.   [2]

Total: 15

4   **Figure 15** shows the binding energy per nucleon against the number of nucleons.

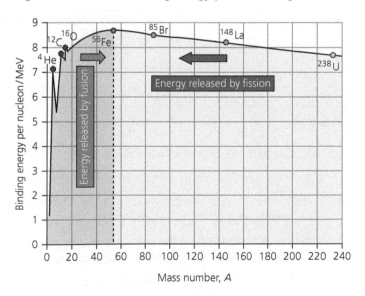

Figure 15

SB2 p218 3.8.1.6 MS 2.2, 2.3

4–1 Use **Figure 15** to determine the binding energy per nucleon for:

1. U-235 *[1]*

2. Sr-90 *[1]*

3. Xe-144. *[1]*

SB2 p218 3.8.1.6 MS 2.2, 2.3

4–2 Use the data to calculate the binding energy released for the reaction:

$$^{235}_{92}U + \ ^{1}_{0}n \rightarrow \ ^{144}_{56}Xe + \ ^{90}_{38}SR + 2^{1}_{0}n$$ *[2]*

SB2 p219 3.8.1.7

4–3 Explain the role of these key components of a nuclear power station:

1. the moderator *[2]*

2. the control rods *[3]*

3. the coolant. *[2]*

Total: 12

SB2 p220 3.8.1.8

5–1 Describe **three** safety measures taken to ensure the safety of personnel working in a nuclear power station. *[3]*

SB2 p223 3.8.1.7

5–2 A reactor is stopped in an emergency as a result of a power cut. Suggest **two** further measures that need to be taken to keep the reactor safe. *[2]*

SB2 p223 3.8.1.8

5–3 1. State the **three** kinds of waste produced by the nuclear industry. *[1]*

2. Explain what they consist of and how they are dealt with. *[6]*

SB2 p224 3.8.1.8

5–4 Explain why spent fuel rods are more difficult to handle than fresh fuel rods. *[4]*

Total: 16

6 When a nucleus of uranium-235 undergoes fission, it mostly splits into two fragments. The fragments can be a range of different elements. The most commonly occurring fragments have nucleon numbers of about 95 and 137.

SB2 p218 3.8.1.6

6–1 One such reaction gives:

$$^{235}_{92}U + \ ^{1}_{0}n \rightarrow \left[\ ^{236}_{92}U \right] \rightarrow \ ^{144}_{56}Ba + \ ^{A}_{Z}X + 3^{1}_{0}$$

Determine the proton number and the nucleon number of the isotope of the element X. *[1]*

6–2 This reaction gives out energy:

$$\left[\ ^{236}_{92}U \right] \rightarrow \ ^{144}_{56}Ba + \ ^{A}_{Z}X + 3^{1}_{0}n$$

The nuclear masses are shown in **Table 5**:

Isotope	Nuclear mass / u
Uranium 236	235.995061
Barium 144	143.892211
X	88.898071
Neutron	1.00867

Table 5

SB2 p218 3.8.1.6 MS 2.2, 2.3

1. Calculate the change in mass given by the reaction. *[1]*

2. Show that the energy emitted is about 170 MeV. *[1]*

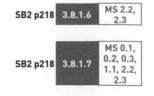

6–3 The critical mass of uranium is about 13 kg. The molar mass of the fissionable uranium is 0.235 kg mol⁻¹.

1. Assuming that 5% of the uranium splits by fission, calculate the energy released. Give your answer to an appropriate number of significant figures. *[2]*

2. Diesel oil has a calorific value of 43 MJ kg⁻¹. Determine how much diesel oil would release the same energy as in **Question 6–1**. *[1]*

SB2 p218 | 3.8.1.6 | MS 0.1, 0.2, 0.3, 1.1, 2.2, 2.3

6–4 A nuclear power station has an output power of 1500 MW. The power station is about 40% efficient.

    1. Show that the amount of fissionable material used in a day is less than 5 kg per day, assuming that 1 kg of the uranium fuel contains 6.8×10^{13} J. [2]

    2. If the fuel contains 5% of fissionable material, determine the total amount of fuel that is needed each day. [1]

Total: 9

7 The isotope $^{238}_{92}$U decays by a combination of alpha and beta-minus decays to $^{234}_{92}$U. The grid shows the neutron number N plotted against the proton number Z. The positions of the two isotopes are shown in **Figure 16**.

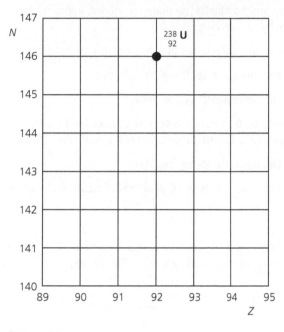

Figure 16

SB2 p205 | 3.8.1.4

7–1 On a copy of the grid, mark the position of the U-234 isotope. [1]

SB2 p205 | 3.8.1.4

7–2 On your grid, draw **three** arrows that show the three decays. Mark each one with α or β as appropriate. [2]

SB2 p205 | 3.8.1.4

7–3 The intermediate isotopes are thorium (Th) and protactinium (Pa). Write the equations that show the decays. [2]

SB2 p198 | 3.8.1.3 | MS 0.1, 0.5, 1.1, 3.9

7–4 Uranium-238 has a half-life of 4.468×10^9 years.

    1. Show that the decay constant is about 5×10^{-18} s^{-1}. [2]

    2. Calculate the activity of a sample of uranium-238 of mass 1.0 kg. State the correct unit and give your answer to an appropriate number of significant figures. [3]

    Molar mass of U-238 = 0.238 kg mol^{-1}.

Total: 10

SB2 p206 | 3.8.1.4

8–1 Explain how gamma radiation is emitted from a radioactive isotope. [3]

SB2 p206 | 3.8.1.4 | MS 0.1, 0.2

8–2 Cobalt-60 is a commonly used gamma source. The initial decay is beta-minus emission to an isotope of nickel.

    1. Write an equation for this decay. [1]

2. The energy level diagram for the decay is given in **Figure 17**.

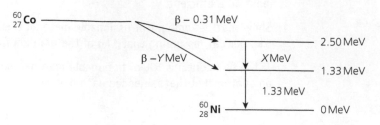

Figure 17

State the kind of emission that is represented by *X*. [1]

SB2 p206 | 3.8.1.4 | MS 0.1, 0.2 8–3 Copy and complete the diagram, giving appropriate values for *X* and *Y*.
Show your working. [2]

SB2 p206 | 3.8.1.4 | MS 0.1, 0.2 8–4 One of the gamma ray photons has an energy of 1.33 MeV. Calculate:

1. the energy in joules of the photon [1]

2. the wavelength of the photon. [2]

SB2 p199 | 3.8.1.3 | MS 0.5 8–5 A cobalt-60 source in a school physics lab had an activity of 1000 Bq when it was new. Cobalt-60 decays with a half-life of 5.272 years.

1. Calculate the decay constant. [2]

2. The source is now 25 years old. Calculate the activity now. [2]

Total: 14

9 Technetium-99m has a metastable nucleus. The energy levels in the metastable nuclei are shown in **Figure 18**.

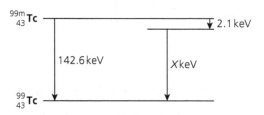

Figure 18

SB2 p206 | 3.8.1.4 | MS 2.2, 2.3 9–1 1. Calculate the energy of the transition *X*. [1]

2. Calculate the wavelength of the 2.1 keV photon. [1]

3. Give the part of the electromagnetic spectrum it is in. [1]

SB2 p206 | 3.8.1.4 | MS 2.2, 2.3 9–2 A patient has a medical diagnostic test that requires that she is injected with a sample of technetium-99m. She is advised that she should not see her grandchildren for 48 hours.

Technetium-99m has a half-life of 6 h. Calculate what fraction of the dose is left after 48 h. [1]

SB2 p198 | 3.8.1.5 | MS 1.3 9–3 Once technetium-99m has emitted a gamma photon, the nucleus goes to its ground state. It decays by beta-minus decay to ruthenium-99, which is a stable isotope. The half-life of the decay is 211 000 years.

1. Give the proton number and number of neutrons ruthenium has. [1]

2. State the probability of a given technetium nucleus decaying in any one second. [2]

3. A dose of 10^{10} atoms of technetium are given to a patient. Discuss whether there would be long-term health implications for a patient who has undergone diagnostic tests with technetium. [3]

Total: 10

Astrophysics

Telescopes

Quick questions

SB2 p229 3.9.1.1

1   Construct a ray diagram using the principal axis and two parallel rays to show how a thin convex (converging) lens refracts light onto the principal focus (focal point).

SB2 p230 3.9.1.1

2   What is meant by the following terms in optics that describe the image?

a)   Real

b)   Virtual

c)   Magnified

d)   Diminished

e)   Upright

f)   Inverted

SB2 p230 3.9.1.1

3   A convex lens has a focal length of 4.0 cm. Construct ray diagrams for the following object distances from the lens axis and state the properties of each image:

a)   object at 10 cm

b)   object at 8 cm

c)   object at 6 cm

d)   object at 4 cm

e)   object at 2 cm.

SB2 p231 3.9.1.1

4   In a refracting telescope, state the image properties of:

a)   objective lens

b)   eyepiece lens.

SB2 p231 3.9.1.1

5   Explain why an astronomical telescope would not be useful for birdwatching.

SB2 p232 3.9.1.1 MS 4.1, 4.5, 4.6

6   Define the term 'angular magnification'.

SB2 p234 3.9.1.2

7   What advantages and disadvantages are there in a reflecting telescope compared with a refracting telescope?

SB2 p234 3.9.1.2

8   What is the relationship between the diameter of a telescope and the collecting power (the intensity of light)?

SB2 p234 3.9.1.4 MS 0.1

9   A small refracting telescope has a lens diameter of 5.0 cm. A large telescope has a mirror diameter of 35 cm. What is the ratio of the light collected by the large telescope compared with the small telescope?

Does it matter what type of telescope it is?

Exam-style questions

10   A refracting telescope has an objective focal length of 45 cm, and an eyepiece focal length of 2.5 cm.

SB2 p231 3.9.1.1 MS 2.1, 2.3

10–1 Calculate its magnification.

[1]

SB2 p231 | 3.9.1.1 | MS 2.1, 2.3

10–2 Calculate the distance between the lens axis of the objective and the lens axis of the eyepiece. [1]

SB2 p232 | 3.9.1.1 | MS 2.1, 2.3

10–3 Another refracting telescope has an objective focal length of 85 cm, and an eyepiece focal length of 2.5 cm. The image in the eyepiece subtends an angle of 0.050 rad. Calculate the angle of the object that subtends to the unaided eye. [2]

SB2 p232 | 3.9.1.1 | MS 4.1, 4.5, 4.6

10–4 The diameter of the Moon is about 3500 km. Its distance is about 410 000 km. It is viewed with a telescope in which the distance between the objective lens and the eyepiece lens is 1.00 m. The focal length of the eyepiece is 0.050 m.

1. Calculate the angle subtended to the naked eye. [1]

2. Calculate the angle made when the Moon is observed through the eyepiece. [1]

3. Determine the apparent diameter of the Moon as observed through the eyepiece. [1]

Total: 7

SB2 p232 | 3.9.1.1 | MS 2.2, 4.1, 4.5, 4.6

11–1 The focal length of the concave mirror of a reflecting telescope is 2.50 m. The eyepiece lens has a focal length of 0.050 m. It is used to observe an asteroid that has a diameter of 150 km and is a distance of 20 million km from the Earth.

1. Calculate the magnification. [1]

2. Calculate the angle α for the asteroid. [1]

3. Calculate the angle β as seen in the eyepiece. [1]

SB2 p236 | 3.9.1.4 | MS 2.1, 4.1, 4.5, 4.6

11–2 By considering the aperture of a telescope of diameter D as a single slit, derive an equation that allows us to determine the minimum angle the telescope will resolve. [2]

SB2 p236 | 3.9.1.4 | MS 2.1, 4.1, 4.5, 4.6

11–3 State the circumstances in which objects are:

1. not able to be resolved [1]

2. just able to be resolved [1]

3. able to be resolved easily. [1]

SB2 p236 | 3.9.1.4 | MS 0.1, 1.1, 2.3, 4.1, 4.5, 4.6

11–4 The pupil in the human eye is on average 4 mm in diameter.

1. Determine the angular resolution of the human eye. Assume that the average wavelength of light is 550 nm. [1]

2. Two distant rocks on the skyline of a mountain ridge are known to be 2.0 m apart. They can just be resolved by an observer. Calculate the distance from the rocks to the observer. Give your answer in km and to an appropriate number of significant figures. [1]

Total: 10

SB2 p236 | 3.9.1.4 | MS 0.1, 1.1, 2.3, 4.1, 4.5, 4.6

12 A telescope has an aperture of 20 cm. It is used to observe two asteroids that are 30 000 km apart, and are 500×10^6 km from the Earth. Assume the average wavelength of visible light is 550 nm.

12–1 Show that the angular separation between the two asteroids is 6.0×10^{-5} rad. [1]

12–2 Calculate the angular resolution of the telescope. [1]

12–3 State whether the telescope will resolve the two asteroids. Support your answer with an appropriate calculation. [1]

Total: 3

13 The Lovell Radio Telescope is located at Jodrell Bank in Cheshire. It has a parabolic dish that is 76 m across, which focuses radio waves onto an antenna that is mounted at the principal focus. It is used to investigate sources of wavelength 21 cm.

SB2 p236 | 3.9.1.3 | MS 0.1, 1.1, 2.3, 4.1, 4.5, 4.6

13–1 Calculate the frequency of the radio waves. [1]

13–2 Calculate its minimum resolution of sources of wavelength 21 cm. [1]

13–3 Two 21 cm radio sources can just be resolved with the Lovell telescope. They are known to be 1200 light years away. Determine the distance between the sources. [1]

Several radio telescopes can be arranged in an **array**, which means that the effective diameter is much larger than that of a single instrument. It is said that the Very Large Array will have an effective diameter of 36 km.

13–4 Determine the collecting power compared with the Lovell Telescope. [1]

13–5 The Very Large Array is used to investigate sources of wavelength 21 cm. Calculate its minimum resolution of such sources. [1]

13–6 Two 21 cm radio sources can just be resolved with the Very Large Array. They are known to be 1200 light years away. Calculate how far they are apart. [1]

Total: 6

14 Up to about 30 years ago, many astronomers used photographic film to record their images. In the 1960s, the first charge-coupled devices (CCDs) were developed. Nowadays, CCDs are used by almost all astronomers. Discuss the advantages and disadvantages of film; and the advantages and disadvantages of CCDs. [6]

Total: 6

15–1 Draw a ray diagram for a refracting telescope in normal adjustment.

Your diagram should show the paths of **two** axial rays and **three** non-axial rays. You should show the focal lengths of both lenses, the angle subtended by the naked eye (α) and the angle subtended by the image (β). [3]

15–2 The objective lens has a focal length of 65 cm. The eyepiece has a focal length of 3.5 cm.

1. Determine the distance between the lenses when the telescope is in normal adjustment. [1]

2. Show that the magnification is about 19 times. [1]

15–3 The objective lens has a diameter of 10.0 cm. The telescope is used to observe two objects that are 1.5×10^8 km away. The two objects are 3.0×10^3 km apart.

1. Determine the angle at the objective lens. [1]

2. Using a wavelength of 570 nm, and an appropriate calculation, determine whether the telescope can resolve the two objects. [2]

Total: 8

16–1 1. State what is meant by spherical aberration. [1]

2. State what is meant by chromatic aberration. [1]

16–2 Describe how spherical aberration can be reduced. [3]

16–3 Give **two** other disadvantages of refracting telescopes. [2]

16–4 Give **two** advantages of the refracting telescope. [2]

16–5 A small telescope has a lens diameter of 5.0 cm, while a more expensive instrument has a lens diameter of 15 cm. Determine the ratios of the collecting powers. [1]

Total: 10

17 Professional quality astronomical telescopes are usually Cassegrain reflecting telescopes.

SB2 p234 | 3.9.1.2
17–1 Draw a ray diagram of a Cassegrain telescope in normal adjustment. [2]

SB2 p234 | 3.9.1.4 | MS 2.3, 4.6, 4.7
17–2 The focal length of the concave mirror is 2.25 m, while the eyepiece has a focal length of 3.5 cm. It is used to observe the Moon, which has a diameter of 3500 km. The distance from the Moon to the Earth is 384 000 km.

1. Show that the magnification of the telescope is about 64 times. [1]

2. Calculate the angle of the Moon's disc when observed by the naked eye. [1]

3. Calculate the angle of the Moon's disc when observed through the telescope. [1]

4. State your answer to **Question 17–2–3** in degrees. [1]

SB2 p236 | 3.9.1.4 | MS 2.3, 4.6
17–3 The telescope is 25 cm in diameter. Calculate:

1. the resolution of the telescope, using the wavelength of light as 570 nm [1]

2. the diameter of the smallest crater that can be seen on the Moon's surface. [1]

SB2 p238 | 3.9.1.4
17–4 Telescopes of this kind are sent into orbit.

1. Give the equipment used to observe optical images. [1]

2. Give **one** advantage and **one** disadvantage of using a space telescope. [2]

Total: 11

18 The University of Manchester uses the Lovell Telescope (Mark 1 telescope) for radio astronomy. It has a parabolic dish 76 m in diameter to pick up electromagnetic radiation from objects in the universe. It is steered by electric motors so that it can track the objects of interest.

The nearby Mark 2 telescope has an elliptical dish 38 m across.

SB2 p234 | 3.9.1.3 | MS 0.3
18–1 Determine the collecting power of the Mark 1 compared with Mark 2 telescope. Express your answer as a ratio. [1]

SB2 p234 | 3.9.1.3 | MS 2.2
18–2 A pulsar is emitting radio waves of frequency 400 MHz, which are observed by the Mark 2 telescope. It emits pulses of radio waves every 1.8 ms.

1. Show that the wavelength of the radio waves is 0.75 m. [1]

2. Describe a pulsar. [2]

3. Determine the frequency of these pulses. [1]

SB2 p234 | 3.9.1.3 | MS 2.2
18–3 The Mark 1 telescope is used to observe a pair of pulsars that are thought to be in the same star system, 15 ly apart. The frequency of the radio waves is 400 MHz. The two sources can just be resolved by the Mark 1 telescope.

1. Determine the angle that separates the two pulsars. [1]

2. Determine whether the Mark 2 telescope could resolve the two pulsars. Use an appropriate calculation. [1]

SB2 p239 | 3.9.1.3
18–4 State **one** similarity and **one** difference between the radio telescope and the reflecting optical telescope. Give reasons for your answer. [2]

18–5 Suggest a way that the resolution of radio telescopes could be improved. [1]

Total: 10

Classification of stars

Quick questions

SB2 p242 | 3.9.2.1 | MS 0.3
1 Star A has an apparent magnitude of 2 and Star B has an apparent magnitude of 4. How much brighter is Star A than Star B?

SB2 p242 | 3.9.2.1

2   What is the main problem with the apparent magnitude scale?

SB2 p244 | 3.9.2.2 | MS 0.5, 0.6

3   An imaginary star, PLC 2.44, has an apparent magnitude of +5.32. It is 8.7 pc from the Earth. Calculate its absolute magnitude.

SB2 p244 | 3.9.2.2 | MS 0.1, 0.2

4   The apparent magnitude of the Sun is −26.7.

a)   How many parsecs is 1 AU?

b)   Determine the absolute magnitude of the Sun.

SB2 p246 | 3.9.2.3 | MS 0.1, 0.2

5   A star has a surface temperature of 6500 K. Determine its peak wavelength.

SB2 p246 | 3.9.2.3

6   The formula for Stefan's law is:

$$P = \sigma A T^4$$

Copy and complete **Table 1**, filling in the missing information about the terms, meanings and units used in the formula.

Term	Meaning	Unit
	Power	Watt (W)
σ		$5.67 \times 10^{-8}\,\mathrm{W\,m^{-2}\,K^{-4}}$
A		$\mathrm{m^2}$
T	Temperature	

Table 1

SB2 p249 | 3.9.2.4

7   What is meant by the Balmer series in a hydrogen atom?

SB2 p247 | 3.9.2.3 | MS 0.3

8   A red giant star, R, and a main sequence star, M, both have an absolute magnitude of 0. Their surface temperatures are 4000 K and 10 000 K, respectively. How many times larger is the diameter of the red giant than the main sequence star?

SB2 p250 | 3.9.2.5

9   Discuss how the lifetime of stars varies with their size.

SB2 p251 | 3.9.2.5 | MS 2.3

10   When a Sun-like star collapses at the end of its life, it forms a very small body that is intensely hot. This body is the same size as the Earth.

a)   What is this body called?

b)   Determine the density of such a body if the star has a mass of 0.5 solar masses.

   • mass of the Sun = 2.0×10^{30} kg

   • radius of the Earth = 6.4×10^{6} m

SB2 p252 | 3.9.2.5

11   At the very end of their lives, stars that are slightly larger than the Sun can fuse helium with other elements to form elements such as calcium. What element fuses with helium to form calcium? Write down the equation.

SB2 p255 | 3.9.2.6

12   Give reasons for supernova explosions being accompanied by bursts of neutrinos and gamma rays.

Exam-style questions

13   The star Arcturus has an apparent magnitude of −0.04, while the star Deneb has a magnitude of 1.25.

SB2 p242 | 3.9.2.1 | MS 0.3

13–1 Show that Arcturus is about 3 times brighter than Deneb.   *[1]*

At a certain point in the Universe, the intensity of light from Arcturus is found to be $1.0 \times 10^{-15}\,\mathrm{W\,m^{-2}}$.

SB2 p242 | 3.9.2.1 | MS 0.5, 2.2, 2.3, 2.5

13–2 Determine the intensity of the light coming from Deneb, using this relationship:

$$\Delta m = 2.5 \log\left(\frac{I_\mathrm{A}}{I_\mathrm{B}}\right)$$

   [2]

Total: 3

SB2 p243 | 3.9.2.2 | MS 0.1, 4.6, 4.7

14–1 1 arc second = $\dfrac{1}{3600}$ degree. Determine what this is in radians. [1]

14–2 A distant star is a distance d from the Earth. The Earth's orbital radius is R. The distant star subtends an angle θ. Derive an expression to enable the distance in parsecs to a distant star to be worked out. [2]

14–3 Calculate how many light years is 1 parsec. [1]

Total: 4

SB2 p246 | 3.9.2.3 | MS 2.3

15–1 The Sun has a radius of 6.96×10^8 m and a surface temperature of 5800 K. Show that the luminosity of the Sun is about 4×10^{26} W. [1]

SB2 p247 | 3.9.2.3 | MS 2.2, 2.3

15–2 Dubhe is a large and bright star in the constellation of *The Great Bear*. Its surface temperature is 4660 K. Its luminosity is 1.23×10^{29} W.

1. Calculate the peak wavelength and give its colour. [1]

2. Calculate the radius of Dubhe. [2]

3. Compare this to the radius of the Sun. [1]

SB2 p248 | 3.9.2.4 | MS 0.1, 0.2, 2.2, 2.3

15–3 The diagram in **Figure 1** shows energy levels of the hydrogen atom:

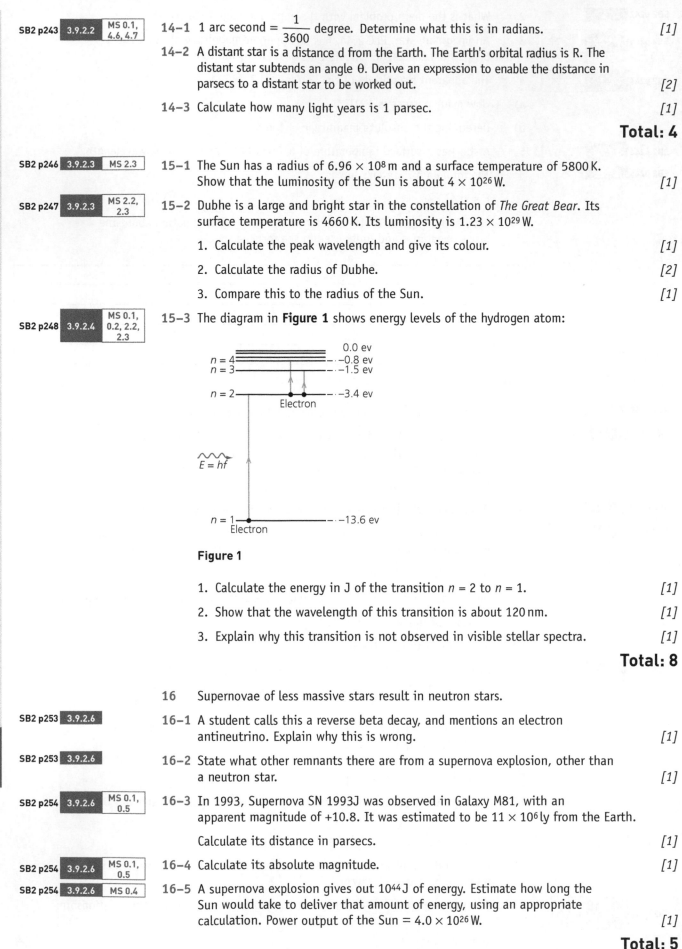

Figure 1

1. Calculate the energy in J of the transition $n = 2$ to $n = 1$. [1]

2. Show that the wavelength of this transition is about 120 nm. [1]

3. Explain why this transition is not observed in visible stellar spectra. [1]

Total: 8

16 Supernovae of less massive stars result in neutron stars.

SB2 p253 | 3.9.2.6

16–1 A student calls this a reverse beta decay, and mentions an electron antineutrino. Explain why this is wrong. [1]

SB2 p253 | 3.9.2.6

16–2 State what other remnants there are from a supernova explosion, other than a neutron star. [1]

SB2 p254 | 3.9.2.6 | MS 0.1, 0.5

16–3 In 1993, Supernova SN 1993J was observed in Galaxy M81, with an apparent magnitude of +10.8. It was estimated to be 11×10^6 ly from the Earth.

Calculate its distance in parsecs. [1]

SB2 p254 | 3.9.2.6 | MS 0.1, 0.5

16–4 Calculate its absolute magnitude. [1]

SB2 p254 | 3.9.2.6 | MS 0.4

16–5 A supernova explosion gives out 10^{44} J of energy. Estimate how long the Sun would take to deliver that amount of energy, using an appropriate calculation. Power output of the Sun = 4.0×10^{26} W. [1]

Total: 5

SB2 p254 | 3.9.2.6 | MS 0.4, 0.5, 0.6

17–1 A neutron star has a mass about 1.5 times the mass of the Sun. It has a diameter of about 25 km. Use an appropriate calculation to estimate its density. *[1]*

When a star of 20 or more solar masses collapses as a supernova, a black hole is formed.

SB2 p254 | 3.9.2.6

17–2 State what is meant by a black hole. *[2]*

SB2 p254 | 3.9.2.6

17–3 Give another circumstance in which a black hole can be formed. *[1]*

SB2 p255 | 3.9.2.6 | MS 2.3

17–4 A black hole has a mass of 3.5 solar masses. Show that the radius of the event horizon is about 10 km. *[1]*

SB2 p255 | 3.9.2.6 | MS 2.2

17–5 Another black hole has a photon sphere of 60 km in diameter.

    1. Determine the Schwarzschild radius. *[1]*

    2. Determine the mass of the star in solar masses of the star that made the black hole. *[2]*

Total: 8

SB2 p239 | 3.9.2.1

18–1 State **two** factors that affect the brightness of a star. *[1]*

SB2 p241 | 3.9.2.1 | MS 2.5

18–2 The early astronomer Hipparchus produced the first method of classifying stars in a scientific manner.

    1. State how he did this. *[2]*

    2. Derive an expression to give a ratio of brightness between stars that are one magnitude apart. *[2]*

SB2 p241 | 3.9.2.1 | MS 0.3

18–3 The star Betelgeuse has an apparent magnitude of +0.4, and the star Antares has an apparent magnitude of +1.1.

    1. Explain which one of these stars is brighter. *[1]*

    2. Determine how much brighter the brighter of these two stars is than the dimmer. *[1]*

Total: 7

19 In astrophysics, distances are measured in astronomical units (AU), light years (ly) and parsecs (pc).

SB2 p243 | 3.9.2.2

19–1 Define astronomical unit, light year and parsec. *[3]*

SB2 p243 | 3.9.2.2 | MS 4.5, 4.6, 4.7

19–2 Show that 1 pc is 3.26 ly. *[2]*

SB2 p245 | 3.9.2.2 | MS 2.5

19–3 Two stars, A and B, have intensities I_A and I_B. They have apparent magnitude m_A and m_B, and are at distances d_A and d_B. Show that the difference in their apparent magnitudes Δm is related to their distances by:

$$\Delta m = 5 \log_{10} \left(\frac{d_B}{d_A} \right)$$ *[4]*

Total: 9

SB2 p245 | 3.9.2.2

20–1 Explain what is meant by the absolute magnitude of a star. *[1]*

SB2 p245 | 3.9.2.2 | MS 0.5

20–2 The apparent magnitude of a star at a distance of 10 light years is +5.6.

    1. Calculate the absolute magnitude. *[2]*

    2. State, with a reason, whether the star is brighter or dimmer than its magnitude suggests. *[1]*

SB2 p245 | 3.9.2.2 | MS 0.5, 2.2

20–3 Two stars, A and B, are found to have the same absolute magnitude of −4.0. Star A is known to be 300 light years from the Earth.

    1. Determine the distance in parsecs to Star A. *[1]*

    2. Determine the apparent magnitude of Star A. *[2]*

    3. Star B is at a distance of 58 light years. Discuss whether it would have a higher or lower apparent magnitude than A. No calculation is needed. *[1]*

Total: 8

SB2 p246 3.9.2.3

21–1 1. State Wien's law. [1]

2. Give **two** features of a black body. [2]

SB2 p246 3.9.2.3 MS 1.1

21–2 A star has a surface temperature of 5600 K. Calculate the peak wavelength.
Give your answer to an appropriate number of significant figures. [2]

SB2 p246 3.9.2.3

21–3 **Figure 2** shows that the intensity of peak wavelength varies with the
wavelength when the surface temperature is 6000 K.

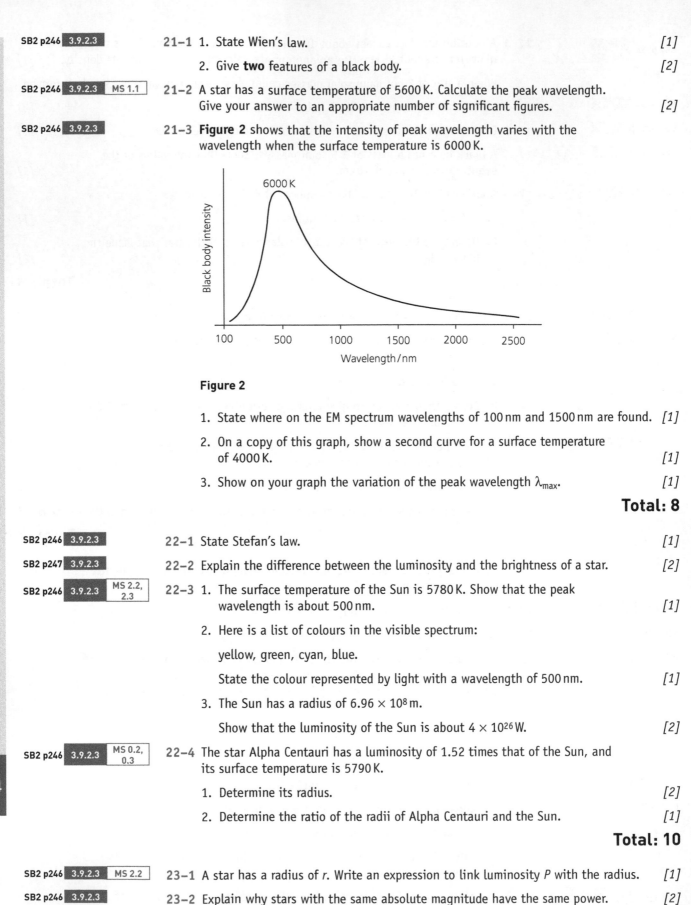

Figure 2

1. State where on the EM spectrum wavelengths of 100 nm and 1500 nm are found. [1]

2. On a copy of this graph, show a second curve for a surface temperature
of 4000 K. [1]

3. Show on your graph the variation of the peak wavelength λ_{max}. [1]

Total: 8

SB2 p246 3.9.2.3

22–1 State Stefan's law. [1]

SB2 p247 3.9.2.3

22–2 Explain the difference between the luminosity and the brightness of a star. [2]

SB2 p246 3.9.2.3 MS 2.2, 2.3

22–3 1. The surface temperature of the Sun is 5780 K. Show that the peak
wavelength is about 500 nm. [1]

2. Here is a list of colours in the visible spectrum:

yellow, green, cyan, blue.

State the colour represented by light with a wavelength of 500 nm. [1]

3. The Sun has a radius of 6.96×10^8 m.

Show that the luminosity of the Sun is about 4×10^{26} W. [2]

SB2 p246 3.9.2.3 MS 0.2, 0.3

22–4 The star Alpha Centauri has a luminosity of 1.52 times that of the Sun, and
its surface temperature is 5790 K.

1. Determine its radius. [2]

2. Determine the ratio of the radii of Alpha Centauri and the Sun. [1]

Total: 10

SB2 p246 3.9.2.3 MS 2.2

23–1 A star has a radius of r. Write an expression to link luminosity P with the radius. [1]

SB2 p246 3.9.2.3

23–2 Explain why stars with the same absolute magnitude have the same power. [2]

SB2 p247 3.9.2.3

23–3 Two stars, P and Q, have the same absolute magnitude and temperature.
Show that the stars both have the same radius. [2]

Total: 5

24 Astronomers classify stars according to their spectral class. Significant data is show in the table below.

24–1 Copy and complete **Table 2**. [3]

Spectral class	Surface temp/K	Colour	H Balmer series	Other elements
	40000	Blue	Weak	Ionised He
	20000		Medium	He atoms
	10000	Blue–white	Strong	Ionised metals
	7500	White	Medium	Ionised metals
			Weak	Medium ionised and neutral metals
	4500		Weaker	
	3000	Red	Very weak	Neutral atoms, strong TiO

Table 2

24–2 1. Explain why a star with a surface temperature of 4500 K has weak Balmer lines. [1]

2. Explain why the Balmer lines are weak in hot stars. [2]

24–3 In stars like the Sun, molecules like water (H_2O) have been found. In hotter stars, no molecules are found.

1. Explain why molecules can be found in cooler stars. [2]

2. Suggest what kind of molecules are found. [2]

24–4 Green light has a wavelength of about 530 nm.

1. Calculate the temperature of a star whose peak wavelength is 530 nm. [1]

2. Determine which class of stars includes green stars. [1]

3. Explain why green stars are never seen as green. [2]

Total: 14

25 The Hertzsprung–Russell diagram is a useful tool for astrophysicists. **Figure 3** shows an incomplete sketch.

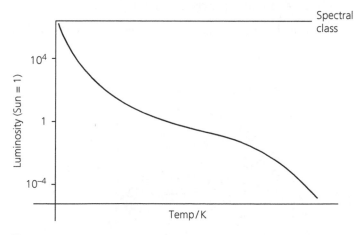

Figure 3

The diagram shows the main sequence.

25–1 Copy the diagram and mark on it:

1. these temperatures: 50000 K, 6000 K, and 2500 K [1]

2. the spectral classes corresponding to these temperatures [1]

3. the position of the Sun on the main sequence [1]

4. a white dwarf and a red supergiant. [1]

SB2 p251 3.9.2.4

25-2 1. Explain what is meant by a main sequence star. [1]

2. Explain how a main sequence star maintains its size. [1]

SB2 p246 3.9.2.3 MS 2.2

25-3 1. Show that the intensity of a star of temperature T is given by: [1]

$I = \sigma T^4$

2. Show that the intensity of the Sun is about $6 \times 10^7 \, W\,m^{-2}$. [2]

SB2 p246 3.9.2.3 MS 2.3

25-4 A main sequence star has a surface temperature of 25 000 K and has an absolute magnitude of −5.1. The Sun has an absolute magnitude of +4.8.

1. Show that the star has an intensity of about 9000 times that of the Sun. [2]

2. Determine the intensity of the star. [1]

3. Determine the peak wavelength. [1]

Total: 13

26 Stars are formed in nebulae.

SB2 p250 3.9.2.4

26-1 1. State what a nebula is. [1]

2. Describe how a star is formed in a nebula. [3]

SB2 p250 3.9.2.4

26-2 Some astronomers regard Jupiter as a failed star. Give **two** reasons why Jupiter did not become a star. [2]

SB2 p251 3.9.2.4

26-3 The Sun is a middle-aged star that is about 4.6×10^9 years old.

Draw the evolution of the Sun from when it starts as a nebula to when it ends up as a white dwarf on a simple Hertzsprung–Russell diagram. Mark each significant point on your diagram. [3]

Total: 9

SB2 p251 3.9.2.5

27-1 Explain what will happen when the hydrogen fuel runs out in a star like the Sun. [6]

SB2 p103 MS 0.1, 0.2

27-2 1. The Sun has a mass of 1.99×10^{30} kg. The Earth has a radius of 6.37×10^6 m. Calculate the density of a white dwarf. [1]

2. Determine the gravitational field constant, g, at the surface of the white dwarf. [1]

3. Determine the escape speed from the white dwarf. Express your answer as a fraction of the speed of light, c. [3]

SB2 p250 3.9.2.5

27-3 Explain why very large main sequence stars are rarely seen. [2]

Total: 13

28 **Figure 4** shows a Sun-like star that is coming to the end of its life.

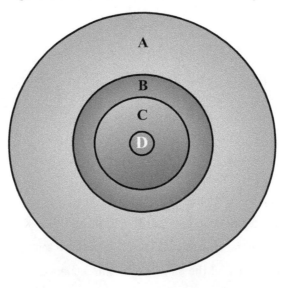

Figure 4

SB2 p251 3.9.2.5

28–1 Copy **Table 3** and match the layers with the letters in **Figure 4**. [1]

Name of layer	Layer
Helium fusion shell	
Low density envelope	
Heavier elements	
Hydrogen fusion shell	

Table 3

SB2 p246 3.9.2.3 MS 0.3, 2.3

28–2 A red giant has a radius of 5.2 AU. Its surface temperature is 3500 K.

1. Determine its radius in SI units. [1]

2. Calculate the luminosity. Give your answer to an appropriate number of significant figures. [2]

3. Compare this to the luminosity of the Sun (3.83×10^{26} W). [1]

SB2 p250 3.9.2.5

28–3 During this phase of the star's life, helium fuses to form bigger elements.

1. Explain the conditions that are needed for this to occur. [2]

2. Give an equation to show how beryllium is formed by helium fusion. [1]

3. A second reaction involving an alpha particle makes a stable element. Give the equation for this reaction. [1]

4. Explain why beryllium is never found in the core of stars. [1]

Total: 10

29 In stars bigger than 8 solar masses, the final fusion reactions take place over a short period of time (in the order of days). A chain of reactions occurs, part of which is shown:

$$^{28}_{14}\text{Si} \rightarrow {}^{4}_{2}\text{He} \rightarrow {}^{32}_{16}\text{S} \;\; \rightarrow\rightarrow {}^{32}_{16}\text{S} + {}^{4}_{2}\text{He} \rightarrow {}^{36}_{18}\text{Ar} \rightarrow\rightarrow \text{X}$$

SB2 p252 3.9.2.6

29–1 1. Give equation X. [1]

2. Explain why it is much more likely for a sulfur nucleus to fuse with a helium nucleus than with a second sulfur nucleus. [2]

SB2 p253 3.9.2.6

29–2 Draw a diagram to show the structure of a large star in the last few days of its existence. [2]

SB2 p253 3.9.2.6

29–3 1. Explain what the heaviest element produced by fusion in large stars is. [1]

2. Describe what happens when all the fusion fuel is used up. [5]

SB2 p253 3.9.2.6

29–4 When a supernova explodes, a burst of neutrinos is observed.

1. State what the source of the neutrinos is. [1]

2. Give the reaction that causes the neutrinos. [1]

Total: 13

Cosmology

Quick questions

SB2 p258 3.9.3.1 MS 2.1

1. A locomotive whistle sounds at 512 Hz. The locomotive is travelling at 40 m s^{-1} along a straight track. Calculate the frequency as the engine:

a) approaches

b) moves away.

Speed of sound = 340 m s^{-1}

SB2 p258 3.9.3.1 MS 2.1

2. A star is moving away from the Earth at 5000 km s^{-1}. A certain wavelength has been detected in its spectrum which corresponds to a line of wavelength 350 nm as measured in a laboratory. What is the wavelength of this line?

SB2 p258 3.9.3.1 MS 2.1

3. A spectral line of a particular element is known to be 547 nm in the lab. On analysis of the spectrum of a star, it is found to differ by −0.264 nm.

a) State whether the star is moving towards or away from the Earth.

b) Calculate its speed relative to the Earth.

SB2 p258 3.9.3.2 MS 2.1, 2.3

4. The wavelength of a spectral line in the spectrum of light from a distant galaxy was measured at 428.6 nm. The same line measured in the laboratory has a wavelength of 422.3 nm. Calculate:

a) the speed of recession of the galaxy

b) the distance to the galaxy. ($c = 3.0 \times 10^8$ m s^{-1}, $H = 65$ km s^{-1} Mpc^{-1})

SB2 p261 3.9.3.2 MS 2.2

5a) What is the SI unit for the Hubble constant?

b) Show that the age, t, of the Universe can be worked out by $t = \dfrac{1}{H}$.

SB2 p261 3.9.3.2 MS 2.3

6a) Show that the value of H in SI units is about 2×10^{-18} s^{-1}.

b) What is the age of the Universe?

SB2 p260 3.9.3.2

7. Outline the evidence that suggests that the Universe started with the Big Bang.

SB2 p260 3.9.3.2 MS 2.3

8. The cosmic background radiation consists of microwaves of wavelength 1.8 mm.

Calculate the photon energy of the microwaves.

SB2 p262 3.9.3.3

9. Quasars were discovered in the early 1960s. They are intense sources of radio waves. What are the main features of a quasar (or quasi-stellar object)?

SB2 p264 3.9.3.4

10a) What is meant by an exoplanet?

b) What kind of exoplanets have been directly observed?

SB2 p264 3.9.3.4

11. Many exoplanets have been discovered. Explain why such planets have not been directly observed.

SB2 p264 3.9.3.4

12. Explain how the existence of an exoplanet can be shown by Doppler shift. What is the limitation of this method?

Exam-style questions

13 Astrophysicists are observing a star which has the characteristic absorption spectrum of a particular element. In the laboratory there is a certain line that has a wavelength of 434.2 nm. While observing the star, the same line is seen to be at 439.7 nm.

SB2 p258 | 3.9.3.1

13–1 1. State the term used for this observation. [1]

2. Explain how this arises. [1]

SB2 p258 | 3.9.3.1 | MS 0.5

13–2 Calculate the speed of the star and state whether it is coming towards the Earth or moving away from the Earth. [2]

SB2 p259 | 3.9.3.2 | MS 0.5

13–3 Determine how far this star is away from the Earth. [1]

SB2 p259 | 3.9.3.2

13–4 Explain what would be seen if a similar star were closer to the Earth. [2]

Total: 7

14 While Hubble was making his observations in the early part of the twentieth century, most astrophysicists believed in the Steady State model of the Universe.

SB2 p260 | 3.9.3.2

14–1 Hubble's work suggested that the Universe was expanding.

Compare the two theories. [2]

SB2 p261 | 3.9.3.2

14–2 The study of red shift suggested that the Universe is expanding. Explain **two** other pieces of evidence that lead present-day astronomers to consider that the Universe is expanding. [4]

SB2 p261 | 3.9.3.2 | MS 0.1

14–3 The value of the Hubble constant is given in the data sheet as $65 \, \text{km s}^{-1} \, \text{Mpc}^{-1}$.

1. Show that the correct unit for the Hubble constant is s^{-1}. [1]

2. Calculate the value for the Hubble constant in SI units. [1]

3. Determine the age of the Universe in years. [2]

4. State the assumption you made. [1]

Total: 11

15 Quasars were first discovered in the 1960s. They are intense radio sources. Observation of the night sky revealed what seemed to be very distant stars which were more luminous than entire galaxies. Some revealed spectral patterns indicating the presence of unknown elements.

SB2 p262 | 3.9.3.3

15–1 Astronomers thought that they were very distant. Give **one** piece of evidence that supports this theory. [1]

SB2 p262 | 3.9.3.3 | MS 2.3

15–2 Astronomers concluded that a particular quasar was travelling at 15% of the speed of light. Determine its distance in megaparsecs. [2]

SB2 p262 | 3.9.3.3 | MS 2.3

15–3 Astronomers working on the quasar identified the spectral pattern of a particular element which had a particularly strong absorption line at 425 nm when observed in the laboratory. Colleagues working at a radio telescope knew that radio emissions of wavelength 21 cm were a feature of this type of quasar.

1. Determine the wavelength at which the strong absorption line appeared. [1]

2. In order to identify the radio waves, the radio telescope had to be tuned to a particular frequency. Determine the frequency to which the instrument had to be tuned. [2]

Total: 6

16 Astronomers have estimated a particular quasar to be 780 Mpc from the Earth. It has an apparent magnitude of +14.6.

SB2 p262 3.9.3.3

16–1 1. State what astronomers think a quasar is. [1]

2. Explain the significance of quasars to astronomers. [1]

SB2 p263 3.9.3.3

16–2 Describe the way quasars behave. [3]

SB2 p263 3.9.3.3 | MS 0.3, 0.5, 2.2, 2.3

16–3 The quasar is thought to have a mass of 5.0×10^8 times the mass of the Sun. (Solar mass = 1.99×10^{30} kg.)

1. Determine the absolute magnitude of the quasar. [2]

2. The intensity of the Sun is 7.3×10^7 W m^{-2}.

   Determine the intensity of the quasar. [2]

3. Determine the radius of the event horizon of the quasar. [2]

4. Show that the luminosity of the quasar is about 4×10^{18} times the luminosity of the Sun. [2]

Total: 13

17 Astrophysicists agree that the Universe originated from a single point, known as a singularity.

SB2 p260 3.9.3.2

17–1 Suggest **two** properties of such a singularity. [2]

SB2 p260 3.9.3.2

17–2 1. Name the event that was the result of the singularity. [1]

2. Describe the events that happened immediately afterwards. [3]

SB2 p260 3.9.3.2

17–3 Draw a diagram to illustrate the idea that the Universe is expanding. [2]

SB2 p261 3.9.3.2

17–4 Astrophysicists are considering what the fate of the Universe will be in thousands of millions of years' time. **Figure 5** shows the relative size of the Universe against time.

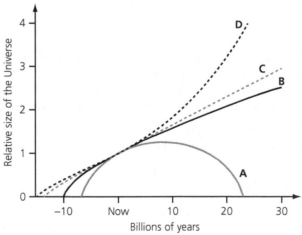

Figure 5

It is thought that there is a critical density of the Universe that will determine the fate of the Universe. Each of the four lines in **Figure 5** show the different fates of the Universe according to its density.

For each of the curves **A–D**, explain the possible fates of the Universe. [4]

Total: 12

Topic review: astrophysics

1   A star is being observed from a high mountain using a telescope of diameter 1.50 m.

The star has a radius of 6.96×10^8 m, and a luminosity of 3.83×10^{26} W. It is 50 light years (ly) from the Earth.

SB2 p243 | 3.9.2.2
1–1   Define the light year.   *[1]*

SB2 p244 | 3.9.2.3 | MS 2.2, 2.3
1–2   1. Show that the intensity at the star's surface is about 6×10^7 W m^{-2}.   *[2]*

2. Calculate the intensity of the light at the Earth's surface.   *[1]*

SB2 p238 | 3.9.2.4 | MS 2.2, 2.3
1–3   The star gives out light with a peak wavelength of 483 nm.

1. Show that the photon energy is about 4×10^{-19} J.   *[1]*

2. Calculate the number of photons entering the telescope every second.   *[1]*

SB2 p238 | 3.9.1.4 | MS 2.2, 2.3
1–4   This telescope uses a charge-coupled device (CCD), which has a high 'quantum efficiency'.

1. Define quantum efficiency.   *[1]*

2. The CCD has a quantum efficiency of 87%. Determine how many electrons come out of the CCD every second.   *[1]*

SB2 p240 | 3.9.1.3 | MS 2.2, 2.3
1–5   X-rays cannot be observed using a telescope like this one.

1. Explain why not.   *[1]*

2. Describe how X-ray sources can be observed.   *[1]*

SB2 p258 | 3.9.3.1 | MS 0.1
1–6   A second star is observed using the same telescope. Using accurate spectroscopic techniques, the star is found to have a spectrum which has a distinctive line at 488.0293 nm. In the laboratory, the same line is found at 488.0265 nm.

1. Show that the speed of recession is about 1.7 km s^{-1}.   *[2]*

2. Find the distance of this star from the Earth in parsecs. Give your answer to an appropriate number of significant figures.   *[2]*

SB2 p245 | 3.9.2.2 | MS 0.5
1–7   The apparent magnitude of the star is +10.3. Using the distance you worked out in **Question 1–6–2**, calculate the absolute magnitude of the star.   *[2]*

SB2 p260 | 3.9.3.2 | MS 0.5
1–8   Discuss the uncertainty in the value for your answer to **Question 1–7**. No calculation is needed.   *[1]*

Total: 17

2   The data in **Table 4** was used by the astronomer E P Hubble.

Distance, d/Mpc	Velocity, v/km s⁻¹
15	1100
20	1200
32	2400
34	1800
46	3600
50	3100
58	4300
61	3300
64	5300
91	5300
93	7500
97	6700
103	5100
120	9000
122	9900
145	10700
145	9600
158	8900
185	9500

Table 4

SB2 p259   3.9.3.2   MS 3.2

2–1   Plot the data on a graph of velocity against distance.

Draw a line of best fit.   *[4]*

SB2 p259   3.9.3.2   MS 3.4

2–2   Determine the gradient of your line of best fit.   *[2]*

SB2 p259   3.9.3.2   MS 1.5

2–3   The value for the constant given in the datasheet is 65 km s⁻¹ Mpc⁻¹.

1. Show on your graph how you would estimate the uncertainty in your line of best fit.   *[1]*

2. Give the highest value and the lowest value.   *[2]*

3. Determine the value of the uncertainty.   *[1]*

SB2 p260   3.9.3.2

2–4   The Hubble constant has a wide range in values from 45 km s⁻¹ Mpc⁻¹ to 90 km s⁻¹ Mpc⁻¹. Suggest a reason for this.   *[1]*

Total: 11

3   It has been suspected for many years that there are planets outside the Solar System. These are called exoplanets. Evidence for such bodies was first found in 1992.

SB2 p264   3.9.3.4

3–1   Explain why exoplanets are not easy to observe directly.   *[2]*

SB2 p260, p264   3.9.3.2, 4   MS 2.1

3–2   One method of detecting the presence of a planet is by variation of Doppler shift.

1. Explain how an exoplanet orbiting a star causes variation in Doppler shift.   *[2]*

2. A star is being moved by a giant planet so that its recession velocity is 140 m s⁻¹. Calculate the change in wavelength of a spectral line of known wavelength 436.3 nm.   *[1]*

3. Estimate how far the star is away from the Earth if the star system is receding from the Earth at 20 m s⁻¹. Give your answer in parsecs (pc).   *[2]*

3–3   Explain, with the help of a diagram, how the transit of a planet can be detected.   *[3]*

3–4   Some astronomers are searching for signs of life on exoplanets. They talk of the 'Goldilocks zone'. Explain what is meant by this.   *[2]*

Total: 12

4   Absorption spectra are an important tool in astrophysics.

SB2 p248   3.9.2.4

4–1   1. State what an absorption spectrum is.   [1]

       2. Explain what absorption spectra show.   [2]

       3. Describe how they are useful to astronomers.   [1]

SB2 p248   3.9.2.4

4–2   Balmer lines for hydrogen are often used in the study of stars.

       1. State what the Balmer lines are.   [1]

       2. Explain how the Balmer lines are formed.   [3]

       3. Explain where in the star the Balmer lines have their origin.   [2]

SB2 p248   3.9.2.4   MS 2.3

4–3   1. Calculate the wavelength for the transition $n = 3$ ($-1.51\,eV$) to $n = 2$ ($-3.41\,eV$).   [2]

       2. State the colour of the photons.   [1]

Total: 13

5   A white dwarf is the remains of a star that is about the same mass as the Sun.

SB2 p252   3.9.2.5

5–1   Which astronomical feature is often formed at the same time as a white dwarf?

       Copy the list in **Table 5** and put a tick (✓) by the correct answer.   [1]

A comet	
A spiral arm	
Planetary nebula	
A number of small black holes the same mass as a planet	
A new planet consisting of iron and carbon	

Table 5

SB2 p252   3.9.2.5

5–2   A white dwarf is very dense, and has a low luminosity.

       Write down **two** other features of a white dwarf.   [2]

SB2 p252   3.9.2.5   MS 0.4, 4.3

5–3   Assume that a white dwarf is the size of the Earth and the mass of the Sun.

       Determine the luminosity compared with the Sun using an appropriate calculation.   [3]

SB2 p252   3.9.2.5

5–4   A white dwarf lasts a long time. Explain why this is and what will happen to the white dwarf in the distant future.   [3]

Total: 9

6   A supernova is observed with a magnitude of +3.25. It is known to be about 250 000 ly from the Earth.

SB2 p254   3.9.2.6   MS 2.3

6–1   1. Show that the distance in parsecs is about 80 kpc.   [1]

       2. Calculate the absolute magnitude.   [2]

SB2 p254   3.9.2.6

6–2   In a star of about 1.5 solar masses, a neutron star is formed. It has a radius of about 12 km.

       Describe how the neutron star is formed.   [2]

SB2 p255   3.9.2.6   MS 0.4

6–3   Estimate the gravitational field strength around a neutron star.   [2]

SB2 p254 3.9.2.6 6–4 **Figure 6** shows a neutron star. It spins around its axis of rotation.

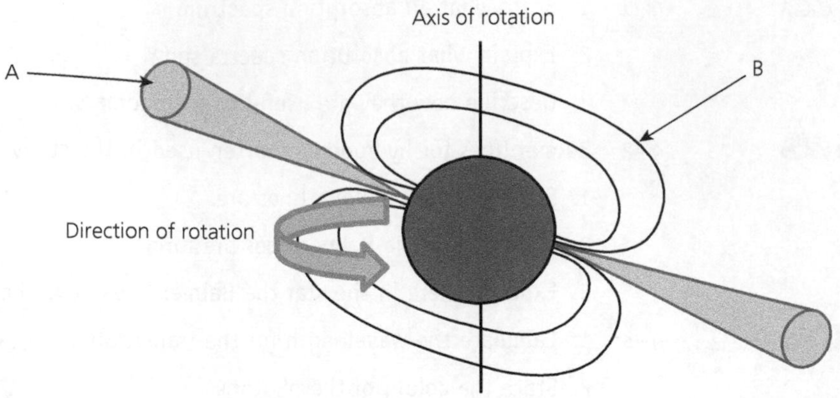

Figure 6

1. State what A and B are. *[1]*
2. Give the name of this type of neutron star. *[1]*

Total: 9

7 A star of about 20 solar masses collapses into a black hole.

SB2 p254 3.9.2.6 7–1 State the main feature of a black hole. *[1]*

SB2 p255 3.9.2.6 MS 0.3, 2.2 7–2 Black holes have an event horizon, which has the Schwarzschild radius, R_S.

1. Explain what the event horizon is. *[1]*
2. Show that the Schwarzschild radius is given by the equation: *[2]*

$$R_S = \frac{2GM}{c^2}$$

3. The Schwarzschild radius of a particular black hole is 30 km. Calculate the mass of the black hole. *[2]*
4. Determine how much more massive than the Sun this black hole is. *[1]*

SB2 p255 3.9.2.6 7–3 Black holes cannot be observed directly. State **one** piece of evidence that black holes are present in a star system. *[1]*

SB2 p255 3.9.2.6 7–4 Supergiant stars collapse into large black holes. As they do so, they give out gamma bursts.

Explain what they tell astrophysicists about the nature of a supernova explosion. *[2]*

Total: 10

8 Type 1a supernovae are very interesting objects to astrophysicists. They always have an absolute magnitude of −19.5, and are used as standard candles.

SB2 p254 3.9.2.6 8–1 State what is meant by:

1. absolute magnitude *[1]*
2. standard candle. *[1]*

SB2 p256 3.9.2.6 8–2 Describe the processes that occur during the formation of a Type 1a supernova. *[5]*

8–3 The explosion gives a characteristic intensity curve as shown in **Figure 7**.

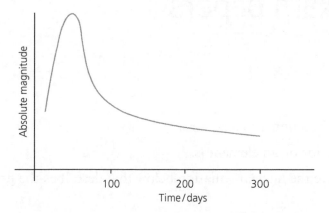

Figure 7

1. Copy the diagram and mark where cobalt and nickel decay occur. [1]

2. Explain how the nickel was formed. [1]

3. Explain how iron-56 is the end result of these decays. [1]

8–4 A Type 1a supernova has an apparent magnitude of +2.1 when observed. Calculate the distance it is from the Earth. [2]

Total: 12

Practice exam papers

Paper 1

Section A

1   An isotope of carbon undergoes β⁺ decay.

1–1   State what an isotope of an element is.   *[1 mark]*

1–2   Copy and complete the nuclear equation below that describes the β⁺ decay of carbon-10.

$$^{10}_{6}C \rightarrow \ \underline{\ \ }B + \ \ ^{\ }_{-1}e^{+} + \underline{\ \ \ \ \ }$$ *[2 marks]*

1–3   Draw a diagram of this nuclear decay showing the correct change of quarks in the nucleon involved in the interaction.   *[3 marks]*

1–4   State which type of nuclear interaction the β⁺ decay of carbon-10 is.   *[1 mark]*

Total: 7

2   A thin block of potassium is connected to the metal plate of an electroscope charged positively. When green light hits the potassium plate, the electroscope discharges quickly.

2–1   Explain why this phenomenon shows particle-like behaviour of light.   *[2 marks]*

2–2   The wavelength of the light hitting the potassium block is 520 nm and the electrons escaping the metal show kinetic energy ranging from 0 to 0.078 eV.

Calculate the work function of potassium.

Work function = __ J   *[4 marks]*

2–3   Explain why not all photoelectrons escape the potassium block with an energy of 0.078 eV.   *[2 marks]*

2–4   The potassium block is now connected to the circuit in **Figure 1**. Suggest what could be done to increase the current through the resistor.   *[1 mark]*

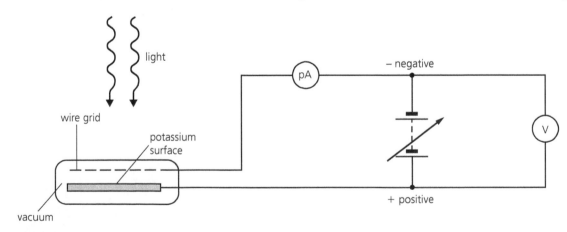

Figure 1

2–5   The metal of the block in the circuit from **Question 2–4** is changed to a zinc plate. Suggest why a current is not detected by the ammeter.   *[2 marks]*

Total: 11

3 A red and violet beam of light are incident on a block of borosilicate glass. Both beams of light hit the surface of the glass at the same point and at an angle of 55° to the normal, as shown in **Figure 2**.

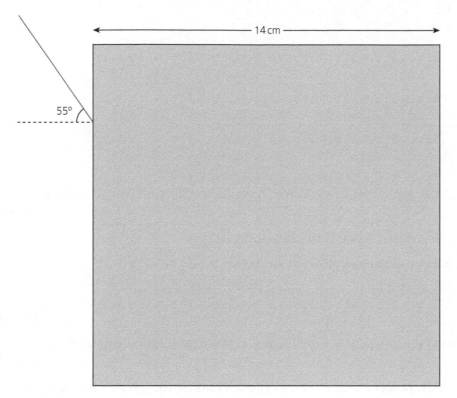

Figure 2

3–1 The borosilicate glass manufacturer gives the information in **Table 1**.

Colour of light	Wavelength / nm	Refractive index
Red	640	1.50917
Yellow	589	1.51124
Green	509	1.51534
Blue	486	1.51690
Violet	434	1.52136

Table 1

Deduce the angles of refraction for the red and violet beams of light to 4 s.f. using the information in **Table 1** and in **Figure 2**. *[3 marks]*

3–2 Calculate separation between the red and violet beam when they come out of the borosilicate block. *[2 marks]*

3–3 On a copy of **Figure 2**, draw a **single** line to show the path of the light beams inside the borosilicate glass. *[1 mark]*

3–4 Explain the potential effects of using pulses of light of different wavelengths in optical fibres. *[2 marks]*

3–5 Describe an experiment that could be used to find a suitable material to use for cladding to make an optical fibre that has borosilicate glass at its core. *[4 marks]*

Total: 12

4 A mining cart of mass 233 kg is pushed downhill for a distance of 5 m with a force of 140 N by a miner. The miner stops pushing the cart at point X in **Figure 3** and the cart continues to travel downhill. The distance XY is 23 m. The cart is initially at rest.

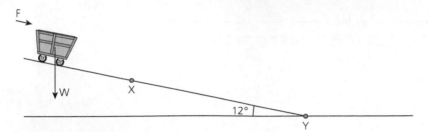

Figure 3

4–1 Calculate the resultant force parallel to the slope. Assume the friction on the cart to be negligible. *[2 marks]*

4–2 Calculate the velocity of the cart when it reaches point X. *[2 marks]*

4–3 At point Y, the brakes are applied and the cart stops in a time t = 3.2 s.

Calculate the distance travelled by the cart from point Y until it stops.

Distance travelled = __ m *[3 marks]*

4–4 Suggest **two** assumptions that were made in arriving at your calculation of stopping distance. *[2 marks]*

4–5 Explain why it would require less effort to empty the content of the cart by tilting it on the slope rather than on the flat surface. *[2 marks]*

Total: 11

5 The main cables supporting the Golden Gate suspension bridge in San Francisco are composed of 27 572 steel wires grouped together and tightly compressed.

The length of one main cable is 2332 m and its mass is approximately 8.15×10^7 kg.

The density of steel is about 7.7×10^3 kg m^{-3}.

Figure 4 shows a cross-sectional fragment of one of the wires.

Figure 4

5–1 Determine the cross-sectional area of one of the wires composing the cable. *[3 marks]*

5–2 The Young modulus of steel is approximately 195 GPa. At one point during the day, 156 cars were crossing the bridge, adding an additional tension to one of the main cables of approximately 1.39×10^6 N.

Show that the extension of the cable under this tension is about 3.7×10^{-3} m. *[3 marks]*

5–3 Suggest an advantage of using multiple wires to build the main cables of the Golden Bridge. *[1 mark]*

Total: 7

6   A student suggests that it is possible to use a set of wires of the same length, material and cross-sectional area to determine the resistivity of the material of the wires.

Describe a method that could be used to investigate the student's claim. Include a circuit diagram for your investigation. *[6 marks]*

Total: 6

7   **Figure 5** shows the graph of resistance against the temperature of a thermistor (T_1).

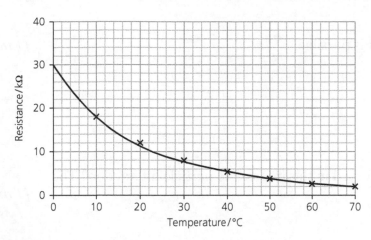

Figure 5

The thermistor in **Figure 5** is used in a potential divider circuit connected to the circuits of a washing machine. The current through the Load is used to regulate the temperature of the water in the drum. T_1 is immersed in the water of the washing machine.

Figure 6 shows the potential divider and the Load.

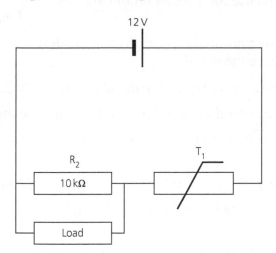

Figure 6

7–1   The washing machine heats cold water up to certain temperatures for different washing programs.

One program requires water to be heated to 30 °C. At this temperature, the potential difference (p.d.) across T_1 is 7.5 V.

Calculate the resistance of the Load. *[5 marks]*

7–2   For another washing program, the resistance of T_1 decreases by 2.5 kΩ from its value in **Question 7–1**. Deduce the temperature of the water in this situation. *[1 mark]*

Total: 6

Section B

8 What exchange particle mediates the repulsive force between two protons in the nucleus of an atom? *[1 mark]*

    A: W^-

    B: W^+

    C: gluon

    D: gamma photon

9 Choose the **only** interaction that is possible. *[1 mark]*

    A: $n \rightarrow p + e^- + v_e$

    B: $\mu^- \rightarrow e^- + v_e + v_\mu$

    C: $K^0 \rightarrow \pi^+ + \pi^-$

    D: $p \rightarrow n + e^+ + \bar{v}_e$

10 Which conservation law is **not** obeyed in this impossible particle interaction? The quark composition of Σ^+ is uus. *[1 mark]*

$$p + \pi^0 \rightarrow \Sigma^+ + \pi^0$$

    A: Conservation of charge

    B: Conservation of baryon number

    C: Conservation of strangeness

    D: Conservation of lepton number

11 Which is the correct definition of threshold frequency for a metal illuminated by electromagnetic radiation? *[1 mark]*

    A: The minimum frequency needed for a photon absorbed by an electron to shift enough energy to allow the electron to escape the metal.

    B: The minimum energy needed for photoelectrons to escape the metal surface.

    C: The frequency needed for a photon to be absorbed by an electron and escape the metal.

    D: The minimum energy the photon needs to shift enough energy to the electron to allow it to escape the metal.

12 Which is the correct definition of ionisation energy for an atom? *[1 mark]*

    A: The energy needed to remove an electron from its ground state to the highest energy level.

    B: The maximum energy needed to remove an electron from its ground state to the highest energy level.

    C: The minimum energy needed to remove an electron from its atom completely.

    D: The minimum frequency of a photon absorbed by an electron to transfer enough energy for the electron to escape the atom completely.

13 An electron gun shoots a beam of electrons through a metal plate with a narrow gap 2.3 μm wide. What is the velocity the electrons should have to show diffraction through the gap? *[1 mark]*

  A: $3.16 \times 10^{-3}\,\text{m s}^{-1}$

  B: $316\,\text{m s}^{-1}$

  C: $3.0 \times 10^{8}\,\text{m s}^{-1}$

  D: $3.16 \times 10^{-4}\,\text{m s}^{-1}$

14 What is the correct outcome after the polaroid filter at 45° from the horizontal filter in **Figure 7**? *[1 mark]*

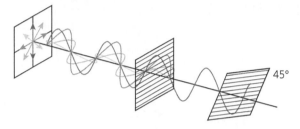

Figure 7

  A: Half of the light will go through, but it will no longer be polarised.

  B: All the polarised light will go through.

  C: None of the polarised light will go through.

  D: Half of the polarised light will go through.

15 What are the units of μ in the first harmonic equation? *[1 mark]*

  A: kg m^{-1}

  B: s^{-1}

  C: s

  D: kg m

16 Which statement about coherent waves is correct? *[1 mark]*

  A: Two waves are coherent when they interfere constructively.

  B: Two waves are coherent when they are in phase.

  C: Two waves are coherent when they have a fixed phase difference and have the same frequency.

  D: Two waves are coherent when they have the same frequency.

17 Which safety statement applies to the dangers associated with the reflections of a laser beam? *[1 mark]*

  A: Do not exceed the operating potential difference across a laser diode.

  B: Do not use laser pointers of power greater than 1 mW.

  C: Always point a laser beam away from people's eyes.

  D: Use a dark matt surface for the screen in a Young double slit experiment.

18 A laser beam goes through a diffraction grating of 180 slits per millimetre. A diffraction pattern is shown on a screen 1.5 m from the diffraction grating and the 3rd order maxima are 80 cm from each other. What is the wavelength of the laser light? *[1 mark]*

A: $\lambda = 9.88 \times 10^{-7}$ m

B: $\lambda = 2.96 \times 10^{-6}$ m

C: $\lambda = 4.76 \times 10^{-7}$ m

D: $\lambda = 1.48 \times 10^{-6}$ m

19 A light ray is shone through a block of diamond joined to a block of glass. The refractive index of diamond is 2.42 and the refractive index of glass is 1.52. *[1 mark]*

What is the critical angle for the diamond–glass boundary?

A: 0.6°

B: 1.7°

C: 52°

D: 38.9°

20 What does Newton's 3rd law of motion state? *[1 mark]*

A: When one body exerts a force on a second body, the second body exerts an equal and opposite force on the first body.

B: When two objects interact with each other, they exert an equal and opposite force on each other.

C: Every force on a body has a balanced force.

D: When two objects interact with each other, the forces on each body are balanced.

21 A cyclist accelerates from 10 ms⁻¹ to 15 ms⁻¹ over a distance of 35 m. A student wants to find the acceleration of the cyclist. *[1 mark]*

Which is the correct equation of uniform linear motion for this situation?

A: $v = u + at$

B: $s = \dfrac{u+v}{2}t$

C: $s = ut + \dfrac{at^2}{2}$

D: $v^2 = u^2 + 2as$

22 Choose the image in **Figure 8** that best describes the forces acting on a tennis ball flying over a net. *[1 mark]*

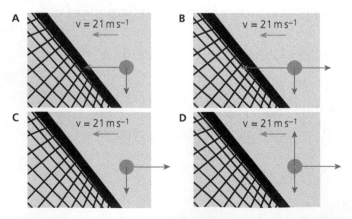

Figure 8

23 **Figure 9** shows a baby mobile with toys hanging on different arms. The mobile is in equilibrium.

What is the mass of the ring? *[1 mark]*

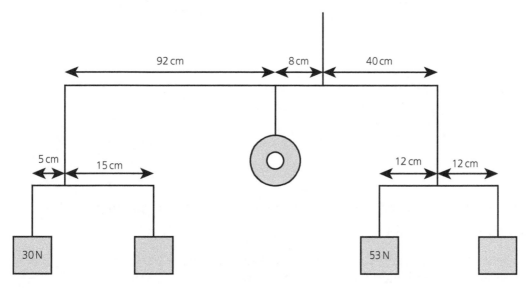

Figure 9

A: 51 g

B: 5 N

C: 5 kg

D: 3.05 kg

24 Why do boxers use boxing gloves? *[1 mark]*

A: Because the soft material reduces the impulse of the force, hence reducing the risk of injuries to the boxer's knuckles.

B: Because the soft material increases the impulse of the force and reduces the risk of injuries to the boxer's knuckles.

C: Because the glove increases the time of contact with the fist, hence increasing the force of the punch.

D: Because the glove decreases the time to change the momentum of the fist, hence increasing the force of the punch.

25 **Figure 10** shows force–extension graphs for a rubber band being stretched (loading) and unstretched (unloading).

Why are the two graphs different?   [1 mark]

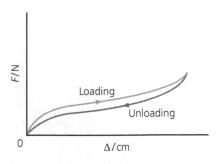

Figure 10

   A: Because some of the energy transferred to the rubber band in loading was lost.

   B: Because the rubber band has been permanently deformed.

   C: Because some of the energy stored in stretching the rubber band has increased the internal energy of the rubber band.

   D: Because some of the bonds in the elastic band where broken during loading and the band needs less force to be stretched after loading.

26 An edible strawberry flavoured string, 30 cm long and with a 2 mm diameter, is stretched by a force of 2 N. Under this force, the string extends by 2.5 mm.

What is the Young modulus of the string?   [1 mark]

   A: $5.3 \times 10^3\,\mathrm{Nm^{-2}}$

   B: $7.6 \times 10^7\,\mathrm{Nm^{-2}}$

   C: $1.9 \times 10^7\,\mathrm{Nm^{-2}}$

   D: $1.3 \times 10^3\,\mathrm{Nm^{-2}}$

27 Which statement best describes an electric current through a circuit?   [1 mark]

   A: Electric current is the amount of charge passing through a point of the circuit.

   B: Electric current is the rate of flow of electricity through a circuit.

   C: Electric current is the charge passing through a point of the circuit multiplied by the time taken.

   D: Electric current is the amount of charge passing through a point of the circuit in the unit of time.

28 Which statement about a superconductor is correct?   [1 mark]

   A: A current set up in a loop of superconducting material will eventually decrease to zero.

   B: All known superconducting materials exhibit superconductor properties only at very low temperatures.

   C: Below the critical temperature, a semiconductor has very low electrical resistance.

   D: Superconductors include magnetic fields inside them.

29   What change would increase the current through R_2 to 300 mA in **Figure 11**?   *[1 mark]*

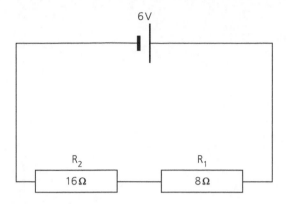

Figure 11

  A: Add a 3.2 V cell in series with the 6 V cell.

  B: Add a 120 Ω resistor in parallel with R_1 and R_2.

  C: Add an 8 Ω resistor in parallel with R_1.

  D: Add a 32 Ω resistor in parallel with R_2.

30   Which statement about the resistance of the component represented by the I–V characteristic in the sketch in **Figure 12** is correct?   *[1 mark]*

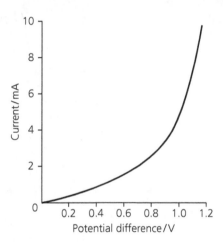

Figure 12

  A: As the potential difference across the component increases, the resistance decreases by larger and larger increments.

  B: As the potential difference across the component increases, the resistance increases by larger and larger increments.

  C: As the potential difference across the component increases, the resistance increases by smaller and smaller increments.

  D: As the potential difference across the component increases, the resistance decreases by smaller and smaller increments.

31 What is the correct value for V_{out} in the potential divider circuit in **Figure 13**? *[1 mark]*

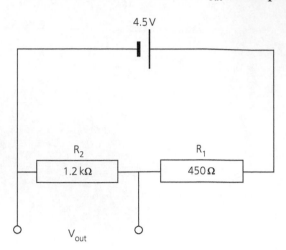

Figure 13

A: 1.69 V

B: 3.27 V

C: 1.22 V

D: 2.25 V

32 Which statement about the circuit in **Figure 14** is correct? *[1 mark]*

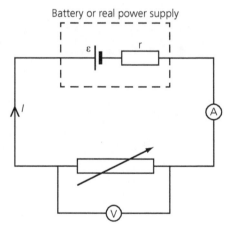

Figure 14

A: The ammeter affects the potential difference across the variable resistor.

B: The potential difference across the internal resistor is equal to Ir.

C: The potential difference across r must be the same as the potential difference across the variable resistor.

D: The e.m.f. e is the potential difference across the variable resistor.

Total: 25

Paper 1 total: 85

Paper 2

Section A

1 **Figure 1** shows a diagram of an outdoor swimming pool.

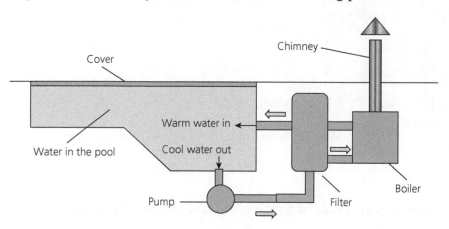

Figure 1

Cold water is pumped through a filter. It passes into a boiler, where it is heated before going back into the pool.

A cover is placed over the pool when it is not in use, to prevent heat escaping and things falling into the pool.

The owner has just filled the pool with clean water which has a temperature of 10 °C.

1–1 Define the 'specific heat capacity' of a material.   *[1 mark]*

1–2 The boiler has a power of 35 kW. The owner has just switched it on, and leaves it running.

The cold water is pumped through the boiler at a rate of 5.0 kg s⁻¹.

Calculate the temperature rise of the water going into the pool.

(Specific heat capacity of water = 4200 J kg⁻¹ K⁻¹)

Temperature rise = __ K   *[1 mark]*

1–3 The pool holds 9.0×10^4 kg of water. The sides of the pool are well-insulated and the cover is a very good insulator.

Calculate the temperature in °C of the pool after 1 day.

Temperature after 1 day = __ °C   *[2 marks]*

1–4 When the cover is removed from the surface, the water in the pool starts to evaporate. Evaporation removes energy from the pool.

The data suggests that, for a pool of this size, the rate of evaporation is 9.0 kg per hour for a certain temperature.

Calculate the power needed to maintain the temperature at this value, assuming the cover is left off.

(Specific latent heat of fusion of water = 2.26×10^6 J kg⁻¹)

Power = __ W   *[1 mark]*

1–5 In reality, the rate of evaporation is governed by factors other than the specific latent heat.

Suggest **one** such factor.   *[1 mark]*

Total: 6

2   **Figure 2** shows argon gas contained in a cylinder.

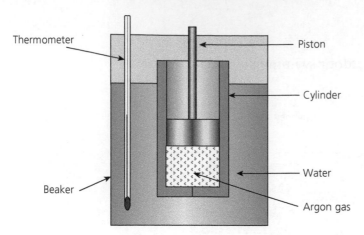

Figure 2

The argon gas in the cylinder is at standard temperature and pressure (0 °C and 1.01×10^5 Pa).

The volume of gas in the cylinder is 1.5×10^{-4} m³.

The water is then heated and the temperature and volume are recorded. The data is plotted on a graph, which is used to calculate absolute zero.

2–1   Sketch a graph to show the variation in volume and temperature that will allow you to determine absolute zero.   *[1 mark]*

2–2   Calculate the number of moles, *n*, in the cylinder at a temperature of 45 °C.

Number of moles = __ mol   *[1 mark]*

2–3   Calculate the number of molecules in the cylinder.

Number of molecules = __   *[1 mark]*

2–4   Calculate the work done on the cylinder as the temperature rises from 0 °C to 45 °C.

Work done = __ J   *[2 marks]*

2–5   Argon is an ideal gas because its molecules are single atoms, so that all the energy in the gas is kinetic. It has an atomic mass of 40 u.

Give **one** assumption used in the kinetic theory of molecules.   *[1 mark]*

2–6   Calculate the root mean square speed of the argon molecules at a temperature of 45 °C.

Give your answer to an appropriate number of significant figures.

Root mean square speed = __ m s⁻¹   *[3 marks]*

2–7   State what the translational kinetic energy of a gas molecule depends on.   *[1 mark]*

Total: 10

3   A satellite is orbiting a planet in a circular orbit.

The planet has a radius of 6.5×10^6 m and a mass of 5.51×10^{24} kg.

The satellite has a mass of 5500 kg. The orbital height of the satellite is 1500 km above the surface of the planet.

3–1   Define 'gravitational field strength'.   *[1 mark]*

3–2 Calculate the gravitational field strength acting on the satellite. *[1 mark]*

3–3 The orbital period of a satellite depends on its radius from the centre of a planet.

Show that the orbital period of the satellite is given by:

$$T^2 = \frac{4\pi^2 r^3}{GM}$$

[2 marks]

3–4 Calculate the orbital period of this satellite.

Orbital period = ___ s *[2 marks]*

3–5 It can be shown that the escape velocity of the satellite is:

$v^2 = 2gr$

Calculate the escape velocity of the satellite from its orbit.

Escape velocity = ___ m s⁻¹ *[1 mark]*

Total: 7

4 Students are investigating the electric field between two point sources, one of which is positive and the other is negative. They use the apparatus shown in **Figure 3**.

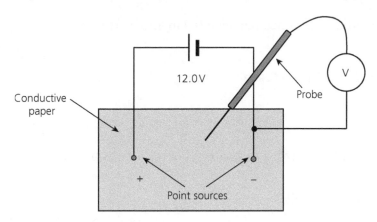

Figure 3

The probe is placed at many points around the conductive paper to get a trace of the equipotentials. The potential differences are displayed on the voltmeter.

From the equipotentials, the students can determine the shape of the electric field.

4–1 On a copy of **Figure 4**, draw the electric field around and between the two point sources on your diagram. *[1 mark]*

Figure 4

4–2 The students repeat the experiment for a uniform field. Copy **Figure 5** and draw the equipotentials on your diagram and show the potentials. *[1 mark]*

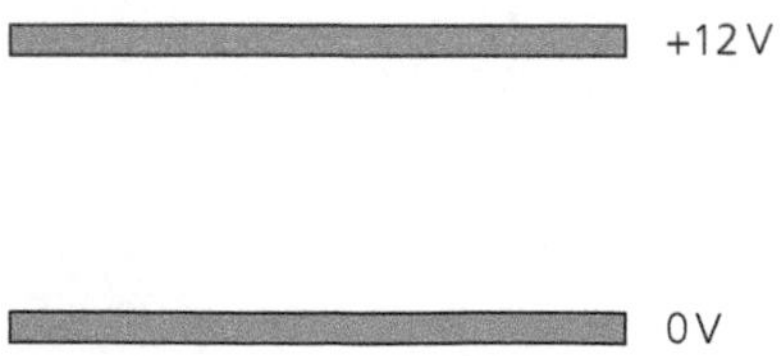

▓▓▓▓▓▓▓▓▓▓ +12 V

▓▓▓▓▓▓▓▓▓▓ 0 V

Figure 5

4–3 In another experiment, the electric field strength is measured about a point charge of +80 nC. The data is logged and plotted in **Figure 6**.

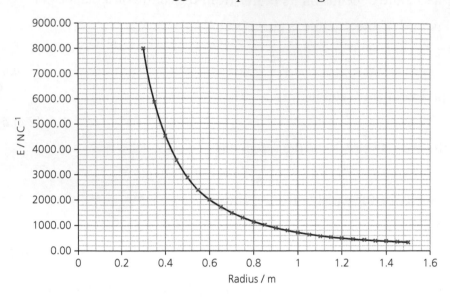

Figure 6

Use the graph to calculate the potential difference between 0.4 m and 1.0 m.

Potential difference = __ J C⁻¹   *[2 marks]*

4–4 A positive test charge of +12 nC is moved from 0.4 m to 1.4 m.

Calculate the energy transferred.

Energy transferred = __ J   *[2 marks]*

4–5 Wood dust in a joiner's workshop can be a hazard to health. Therefore as much of it as possible should be filtered out.

A traditional filter gets clogged up after a while.

One possible method is to use an electrostatic precipitator, shown in **Figure 7**. It gives dust particles a small negative charge, so that they are attracted to the positive plate and stick there.

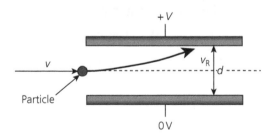

Figure 7

A dust particle of mass m has a negative charge of Q. It approaches the plates with a velocity v. The path of the particle is at the midpoint between the two plates. The plates have a potential difference V and are separated by a distance d.

The particle accelerates along a parabolic path, and strikes the positive plate with a resultant velocity of v_R.

Show that the resultant velocity v_R is given by:

$$v_R = \left(v^2 + \frac{QV}{m} \right)^{0.5}$$

[3 marks]

Total: 9

5 Figure 8 shows electrons of mass m and charge q travelling in a vacuum at a speed of v into a magnetic field of flux density B. The field lines are perpendicular to the page.

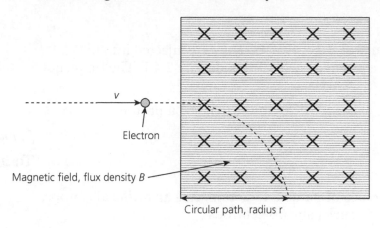

Figure 8

5–1 Define 'magnetic flux density'. *[1 mark]*

5–2 Describe the movement of the electrons when they enter the magnetic field. *[1 mark]*

5–3 Show that the radius, r, of the circular path is given by the expression:

$$r = \frac{mv}{Bq}$$

[2 marks]

5–4 An electron moves into a magnetic field of flux density $0.15\,mT$ at a speed of $3.5 \times 10^6\,m\,s^{-1}$.

Calculate the radius of the circular path. Give your answer to an appropriate number of significant figures.

Radius = __ m *[2 marks]*

5–5 A positively charged particle of mass m and charge Q passes into a magnetic field of flux density B at a speed of v. It follows a circular path of radius r.

The magnetic field is now reduced to $0.25B$. A second particle of mass $2m$ and charge $2Q$ passes into the field at a speed of $2v$.

Calculate the radius of the circular path of the second particle. *[1 mark]*

5–6 Figure 9 shows a cyclotron, a machine used to accelerate charged particles. It consists of two D-shaped electrodes (the dees) which are placed in a magnetic field. The arrangement is viewed from above.

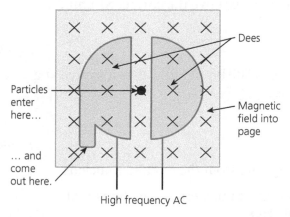

Figure 9

Copy Figure 9 and draw the path the particles take on your diagram. *[1 mark]*

5–7 For a cyclotron, it can be shown that:

$$f = \frac{Bq}{2\pi m}$$

Particles have a charge magnitude of $3.2 \times 10^{-19}\,C$. They are injected into a cyclotron. The dees are within a magnetic field of flux density 2.4 T. The frequency of the alternating electric field is 3.6 MHz.

Calculate the ratio of the mass of the particles to the mass of a proton.

Ratio = __

[2 marks]

Total: 10

6 Cobalt-60 is widely used as a source of gamma radiation. It is an artificial isotope, made in a reactor when cobalt-59 nuclei absorb neutrons.

It decays by beta-minus decay to nickel-60 with a half-life of 5.27 years. It also emits gamma rays. The proton number of cobalt is 27.

6–1 Copy and complete the decay equation for cobalt.

$$^{60}_{27}\text{Co} \rightarrow {}^{60}_{28}\text{Ni}$$

[1 mark]

6–2 The apparatus in **Figure 10** can be used to investigate the inverse square law for gamma rays.

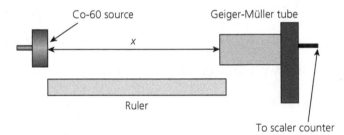

Figure 10

The data is recorded and plotted on a graph.

The graph is a straight line.

State the variable that should be plotted on each axis.

[2 marks]

6–3 The half-life of cobalt-60 is 5.27 y. Sketch a graph to show the way that the activity decays with time.

[1 mark]

6–4 Calculate the decay constant of the cobalt-60 nucleus. Give the correct SI unit for the decay constant.

Decay constant = __ unit = __

[2 marks]

6–5 A brand new cobalt-60 source has an activity of 2500 Bq. A tutor uses a source at a college and finds that its activity is only 150 Bq.

Calculate the age of the source in years.

Age = __ y

[2 marks]

6–6 The gamma photons in a cobalt-60 source originate from the daughter nucleus.

The simplified **Figure 11** shows the two most important energy transitions that release gamma photons from the daughter nucleus.

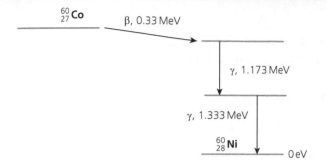

Figure 11

Calculate the mass defect between the cobalt and nickel nuclei.

Mass defect = __ u

[2 marks]

Total: 10

7   Electricity has been generated in nuclear power stations for about sixty years. However, it is a controversial process due to the dangers associated with the handling of nuclear waste materials.

7–1

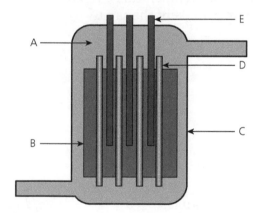

Figure 12

Figure 12 shows a diagram of a nuclear reactor. Identify these components:

- pressure vessel

- coolant

- control rods

- moderator

- fuel rods.   *[1 mark]*

7–2 There are three levels of nuclear waste:

- high level

- intermediate level

- low level.

For **one** of these levels, explain how **one** example of a waste item is disposed of safely.   *[1 mark]*

7–3 It has been suggested that nuclear energy can help a great deal in the fight against human-made global warming.

Explain the advantages and disadvantages of this approach.   *[6 marks]*

Total: 8

Section B

8 A tutor demonstrates Brownian motion using a smoke cell under a microscope. The students make observations. Which one statement is the correct interpretation of these observations? *[1 mark]*

A: That collisions between air molecules can be seen.

B: That large smoke particles are bombarded randomly by invisible air particles.

C: That invisible air particles are bombarded randomly by large smoke particles.

D: That there are collisions between large particles.

9 $3.0 \times 10^{-3}\,m^3$ of an ideal gas is at a pressure of $3.03 \times 10^5\,Pa$ at a temperature of $10\,°C$. The pressure is reduced to $1.01 \times 10^5\,Pa$ and the temperature is raised to $20\,°C$. What is the volume of the gas? *[1 mark]*

A: $1.0 \times 10^{-3}\,m^3$

B: $2.0 \times 10^{-3}\,m^3$

C: $9.3 \times 10^{-3}\,m^3$

D: $18 \times 10^{-3}\,m^3$

10 The speed distributions of molecules in the same gas at two temperatures, $100\,K$ and $1000\,K$, are shown in **Figure 13**. Pressure and volume are kept the same.

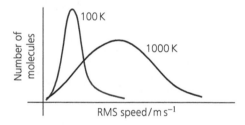

Figure 13

Which statement gives the correct interpretation of these graphs? *[1 mark]*

A: The molecules travel fastest at $100\,K$.

B: The fastest molecules are to be found at the peak of the $1000\,K$ plot.

C: Molecules at $100\,K$ are at a higher pressure than those at $1000\,K$.

D: It is possible for molecules at $100\,K$ to travel as fast as those at $1000\,K$.

11 A satellite is orbiting a planet of radius r at a height of h_1 above the surface of the planet. The height h_1 is equal to the planetary radius, r. It moves to a new orbital height of $2r$ above the surface of the planet.

The gravitational field strength is g at orbital height h_1. What is the gravitational field strength at the new orbital height of h_2? *[1 mark]*

A: $\dfrac{g}{9}$ **B:** $\dfrac{g}{4}$ **C:** $\dfrac{4g}{9}$ **D:** $\dfrac{9g}{4}$

12 A satellite is orbiting a planet of radius r at a height of h_1 above the surface of the planet. The height h_1 is equal to the planetary radius, r. It moves to a new orbital height of $2r$ above the surface of the planet.

The sketch graph in **Figure 14** shows force against distance.

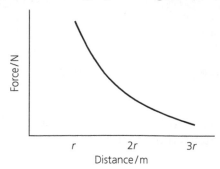

Figure 14

How can you determine, from the graph, the work done on the satellite?   *[1 mark]*

A: The area under the graph between 2*r* and 3*r*.

B: The gradient of the graph between 2*r* and 3*r*.

C: The area under the graph between *r* and 2*r*.

D: The gradient of the graph between *r* and 2*r*.

13 What is the work done, *W*, as a satellite orbits at a height *h* above the surface of a planet, radius *r*?   *[1 mark]*

A: $W = 0$

B: $W = \dfrac{GMm}{r}$

C: $W = \dfrac{2\pi GMm}{r}$

D: $W = \dfrac{mv^2}{2\pi GMr}$

14 A graph is plotted of orbital period, *T*, against radius, *r*. The graph is a straight line.

Which line in **Figure 15** shows the relationship between orbital period, *T*, and orbital radius, *r*? Choose an answer from **Table 1**.   *[1 mark]*

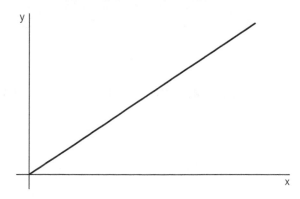

Figure 15

	x	y
A	r	T
B	r	T^2
C	r^2	T^3
D	r^3	T^2

Table 1

15 What is the similarity between gravitational and electrical fields? *[1 mark]*

A: Both can be attractive or repulsive.

B: Both have an infinite range.

C: The constant of proportionality for each can be changed by adding a material.

D: Both are strong at short range.

16 A parallel plate capacitor of capacitance C has plates separated by air by a distance d. It has a charge Q on its plates, and the plates have a potential difference V across them. The electric field is E.

The capacitor is disconnected so that the charge on the plates remains the same. The plates are moved so that their separation is $2d$.

What is the new value of the electric field between the plates? *[1 mark]*

A: $\dfrac{E}{4}$ **B:** $\dfrac{E}{2}$ **C:** E **D:** $2E$

17 A capacitor of capacitance C is charged to a potential difference V. Which graph in **Figure 16** shows the energy held on the plates, E, against the potential difference, V? *[1 mark]*

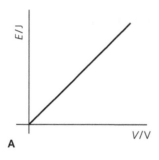

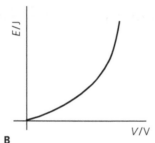

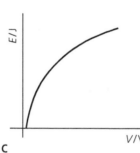

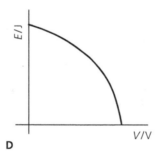

Figure 16

18 What is the relationship between the terms RC and $T_{\frac{1}{2}}$ as a capacitor charges up? *[1 mark]*

A: $T_{\frac{1}{2}} = 0.257\ RC$

B: $T_{\frac{1}{2}} = 0.368\ RC$

C: $T_{\frac{1}{2}} = 0.500\ RC$

D: $T_{\frac{1}{2}} = 0.693\ RC$

19 A capacitor has capacitance 220 µF. It is charged up from 0 V to 12 V through a 15 kΩ resistor.

What is the voltage after 5.0 s?   *[1 mark]*

A: 0 V

B: 2.6 V

C: 9.4 V

D: 12 V

20 A wire carries a current of 6.5 A in a magnetic field of 0.663 T. The length of the magnetic field is 5.0 cm. The wire is placed at an angle of 75° to the field lines.

What is the magnitude of the force on the wire?   *[1 mark]*

A: 0.056 N

B: 0.21 N

C: 0.22 N

D: 21 N

21 A 500 turn coil of wire, of area 0.085 m², is placed in a magnetic field of flux density 0.663 T so the coil is perpendicular to the field lines. This is shown in **Figure 17**.

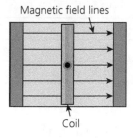

Magnetic field lines

Coil

Figure 17

The coil is now moved so that it makes an angle of 60° with the field lines. What is the change in flux?   *[1 mark]*

A: 3.8 Wb turns

B: 14 Wb turns

C: 24 Wb turns

D: 28 Wb turns

22 **Figure 18** shows a sinusoidal AC waveform displayed on a CRO screen.

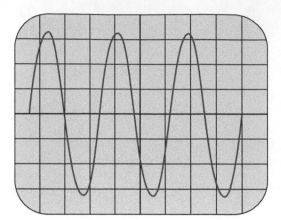

Figure 18

The time-base is set to 2.5 ms per division, while the voltage gain is set to 0.5 V per division.

Which line gives the correct peak voltage, RMS voltage and frequency? Choose an answer from **Table 2**. *[1 mark]*

	V_0/V	V_{RMS}/V	f/Hz
A	1.1	1.6	0.14
B	1.6	1.1	0.14
C	1.1	1.6	140
D	1.6	1.1	140

Table 2

23 A transformer is connected to a $240\,V_{RMS}$ supply and takes $0.5\,A_{RMS}$. Its primary consists of 12 000 turns, while its secondary consists of 600 turns.

What are the secondary voltage and current? Assume the transformer is 100% efficient. Choose an answer from **Table 3**. *[1 mark]*

	V_{Sec}/V	I_{Sec}/A
A	12	0.025
B	12	10
C	4 800	0.025
D	4 800	10

Table 3

24 Rutherford scattering and electron scattering are two methods of determining the radius of a nucleus.

Which statement is correct? *[1 mark]*

A: Rutherford scattering does not depend on the closest approach of an alpha particle.

B: Rutherford scattering suggests a nuclear radius of about 45 nm.

C: Electron scattering uses the kinetic energy of the electron to calculate the de Broglie wavelength.

D: Electron scattering gives a nuclear radius of about 2 fm.

25   A radiation detector is placed $0.12\,\text{m}$ from gamma source which has a diameter of $0.10\,\text{m}$. The background count is $2\,\text{Bq}$. The recorded count on the radiation detector is $100\,\text{Bq}$. What is the count on the surface of the gamma source?   *[1 mark]*

   A: $576\,\text{Bq}$

   B: $564\,\text{Bq}$

   C: $235\,\text{Bq}$

   D: $141\,\text{Bq}$

26   The ruthenium isotope $^{107}_{44}\text{Ru}$ is an unstable radionuclide which has a half-life of 3.75 days.

   The sample contains 2.8×10^{16} atoms.

   What is the decay constant λ, the initial activity A_0, and the activity A after 3.00 days? Choose an answer from **Table 4**.   *[1 mark]*

	λ/s^{-1}	A_0/Bq	A/Bq
A	4.34×10^{-6}	1.2×10^{11}	4.0×10^{10}
B	2.14×10^{-6}	6.0×10^{10}	3.4×10^{10}
C	2.14×10^{-6}	6.0×10^{10}	1.0×10^{11}
D	0.267	1.2×10^{11}	4.4×10^{10}

Table 4

27   Which equation shows a K-capture event?   *[1 mark]*

   A: $^{26}_{13}\text{Al} + \,^{0}_{-1}e^- \rightarrow \,^{26}_{12}\text{Mg} + v_e$

   B: $^{26}_{13}\text{Al} + \,^{0}_{-1}e^- \rightarrow \,^{26}_{12}\text{Mg} + \bar{v}_e$

   C: $^{26}_{13}\text{Al} + \,^{0}_{+1}e^+ \rightarrow \,^{26}_{14}\text{N} + \bar{v}_e$

   D: $^{26}_{13}\text{Al} + \,^{0}_{-1}e^+ \rightarrow \,^{26}_{14}\text{N} + v_e$

28   Which one of these statements is NOT correct about a metastable nucleus?   *[1 mark]*

   A: The nucleus is excited after a decay event and remains in an excited state for a much longer period than normal.

   B: Metastable nuclei lose their energy by emitting gamma photons.

   C: The rate of gamma photon emission from metastable nuclei can be regulated by changing physical conditions such as temperature.

   D: Metastable nuclei decay exponentially.

29   Lead $^{208}_{82}\text{Pb}$ is the largest stable nuclide. What is the nuclear radius?   *[1 mark]*

   $(R_0 = 1.2 \times 10^{-15}\,\text{m})$

   A: $1.7 \times 10^{-14}\,\text{m}$

   B: $7.1 \times 10^{-15}\,\text{m}$

   C: $6.0 \times 10^{-15}\,\text{m}$

   D: $5.2 \times 10^{-15}\,\text{m}$

30 A sketch graph of binding energy per nucleon against nucleon number is shown in **Figure 19**.

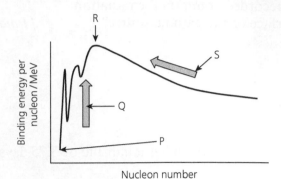

Figure 19

On this sketch graph, there are four features, P, Q, R and S. Which line is the one that describes them correctly? Choose an answer from **Table 5**. *[1 mark]*

	P	Q	R	S
A	Hydrogen nucleus	Fission	Iron nucleus	Fusion
B	Hydrogen nucleus	Fusion	Iron nucleus	Fission
C	Deuterium nucleus	Fission	Iron nucleus	Fusion
D	Deuterium nucleus	Fusion	Iron nucleus	Fission

Table 5

31 The bismuth isotope $^{212}_{83}$Bi decays by alpha decay to form the stable isotope of thallium, $^{208}_{81}$Tl. This data can be used to calculate the energy released:

- mass of bismuth nucleus = 211.94562 u
- mass of thallium nucleus = 207.93746 u
- mass of alpha particle = 4.00150 u. *[1 mark]*

A: 5.9×10^{-10} J

B: 6.9×10^{-11} J

C: 1.0×10^{-12} J

D: 3.0×10^{-21} J

32 High level nuclear waste is potentially very harmful to the environment. Which one of these is NOT an acceptable way of disposing of nuclear waste? *[1 mark]*

A: Throwing drums overboard from a ship so that they sink into a deep ocean trench.

B: Storing the waste in vitreous form in dry repositories deep underground.

C: Storing the waste in deep ponds for many years at a secure site in a reprocessing site, before storing the waste deep underground or in specially constructed silos.

D: Storing the waste in sealed drums in specially designed ventilated silos with thick concrete walls, floors and roofs.

Total: 25

Paper 2 total: 85

Paper 3

Part A

1 A group of students wants to investigate the simple harmonic motion (SHM) of a jumping toy's spring between the ground and the highest point of its jump. The height of the compressed toy is 3.0 cm.

Figure 1 shows the toy in its starting position and at the highest point it reaches.

The students use a 1200 frames per second (fps) slow-motion camera to record the motion of the jumping toy.

2.5 cm

55 cm

1–1 The students notice that the toy reaches its highest point at frame 383 from the start of the jump.

Calculate the initial velocity of the toy. *[2 marks]*

1–2 Describe a method the students could use to record a table of results for the displacement of the base of the toy from its centre of oscillation and study the base SHM. The students can use information from their slow-motion camera. *[3 marks]*

Figure 1

1–3 The students measured the mass of the toy with a digital balance and found it to be 23 g.

Figure 2 shows a portion of the displacement–time graph of the base of the toy.

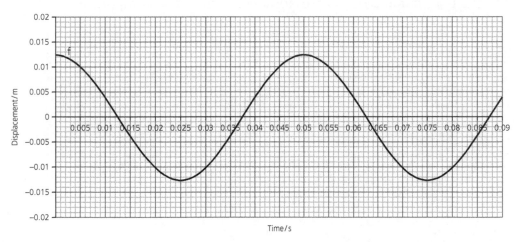

Figure 2

Use the information from **Figure 2** to deduce the spring constant of the spring in the jumping toy. *[3 marks]*

1–4 Explain why the students used the mass of the whole toy rather than just the mass of the base to calculate the spring constant. *[3 marks]*

Total: 11

2   A group of students is given the mystery box in **Figure 3** to investigate.

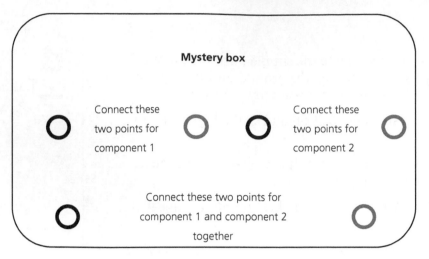

Figure 3

The students need to identify what component 1 and component 2 are and how they are connected together in the two bottom sockets.

The students decide to plot the I–V characteristic curves of component 1 and component 2 individually and then of components 1 and 2 together (bottom sockets).

2–1   Draw a circuit diagram that the students could use to investigate the I–V characteristics of the components in the mystery box.   *[3 marks]*

2–2   **Figure 4** shows the I–V characteristic curves for component 1 and component 2.

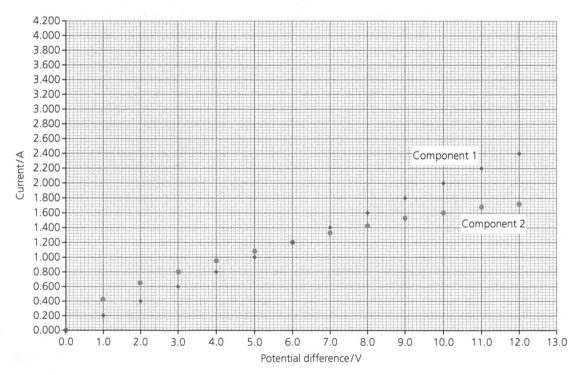

Figure 4

Plot the I–V characteristic curve of components 1 and 2 together using the data in **Table 1**.   *[3 marks]*

Components 1 and 2 together	
V / V	I / A
0.0	0.00
1.0	0.63
2.0	1.05
3.0	1.40
4.0	1.75
5.0	2.08
6.0	2.40
7.0	2.73
8.0	3.03
9.0	3.33
10.0	3.60
11.0	3.88
12.0	4.13

Table 1

2–3 Explain why having the current on the *x*-axis and the p.d. on the *y*-axis would not be appropriate. *[1 mark]*

2–4 Describe the resistance shown by each graph line. *[2 marks]*

2–5 Suggest what components 1 and 2 could be. *[2 marks]*

2–6 Draw a circuit diagram to show how components 1 and 2 might be connected between the two bottom sockets in the mystery box. *[2 marks]*

Total: 13

3 A group of students wants to estimate the stopping force on a toy car running down a ramp when started from different heights. The car is travelling on a flat carpet once it reaches the end of the ramp.

Figure 5 shows the apparatus they used.

Figure 5

3–1 Describe a method for taking accurate measurements of the stopping distance and of the velocity of the car at the end of the ramp. *[4 marks]*

3–2 The students decide to plot a graph of the kinetic energy at the end of the ramp against the stopping distance and calculate the gradient of the line to find the stopping force.

State whether this is the correct method and explain your answer. *[2 marks]*

3–3 In their investigation, the students started the toy car by positioning its front in line with the height mark. However, in their calculations, they found that the gravitational potential energy they had calculated for each height was consistently lower than the kinetic energy calculated using the mean velocity found using the light gate system at the bottom of the ramp.

Explain why this might have happened. *[1 mark]*

3–4 Suggest **one** way the students could improve their results. *[1 mark]*

3–5 **Table 2** shows the students' results. The mass of the toy car was measured with a digital scale and was 24 g.

Calculate the gravitational potential energy (GPE) for each height. Copy and complete **Table 2**. *[2 marks]*

Height of drop/m	v_1/ $m s^{-1}$	v_2/ $m s^{-1}$	v_3/ $m s^{-1}$	v_{mean}/ $m s^{-1}$	s_1/m	s_2/m	s_3/m	s_{mean}/m	GPE/J	KE/J
0.10	1.98	2.12	2.23	**2.11**	0.390	0.375	0.375	**0.380**		0.053
0.15	1.98	2.40	2.44	**2.27**	0.635	0.530	0.515	**0.560**		0.062
0.20	2.67	2.48	2.71	**2.62**	0.725	0.635	0.725	**0.695**		0.082
0.25	2.68	2.86	2.89	**2.81**	0.745	0.760	0.670	**0.725**		0.095
0.30	3.17	2.94	3.08	**3.06**	0.740	0.990	1.020	**0.917**		0.113
0.35	2.91	3.08	3.25	**3.08**	0.985	0.825	0.950	**0.920**		0.114

Table 2

3–6 Draw a graph of the kinetic energy (KE) against the mean stopping distance (s_{mean}). *[3 marks]*

3–7 Estimate the stopping force on the toy car using information from the graph. *[3 marks]*

3–8 Suggest what factors might have affected the stopping distance of the toy car in this experiment. *[2 marks]*

3–9 Suggest a method that the students could use to calculate the instantaneous deceleration of the toy car along the whole distance travelled from the end of the ramp. *[3 marks]*

Total: 21

Paper 3A total: 45

Paper 3

Part B: astrophysics

1–1 **Figure 1** shows a simplified diagram of a reflecting telescope.

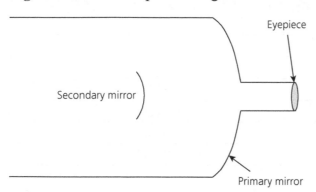

Figure 1

Copy **Figure 1** and draw a ray diagram to show the passage of two parallel rays though the telescope.

Mark on it the principal focus.   *[1 mark]*

1–2 The eyepiece has a focal length of 5.0 cm. The combination of the primary and secondary mirrors has a focal length of 1.25 m.

Calculate the magnification of the telescope.

Magnification = __   *[1 mark]*

1–3 The telescope has a diameter of 0.20 m.

Calculate the smallest angle that this telescope can resolve for light of wavelength 550 nm.

Angle = __ rad   *[1 mark]*

1–4 Reflecting telescopes are used by most astronomers.

Give **two** advantages of reflecting telescopes over refracting telescopes.

Give a reason for each advantage.   *[2 marks]*

1–5 Light from a Sun-like star is picked up by a reflecting telescope of which the area of the aperture is 3.5×10^{-2} m². The light has a wavelength of 550 nm.

The light from the star has an intensity of 2.2×10^{-10} W m².

Calculate the number of photons arriving each second at the light detector, a charge-coupled device (CCD).

Number of photons = __ s⁻¹   *[2 marks]*

1–6 The quantum efficiency of the CCD is 70%. Calculate the number of electrons emitted every second by the CCD.

Number of electrons = __ s⁻¹   *[1 mark]*

1–7 Some kinds of radiation, for example ultraviolet light, cannot be observed from the ground. Explain why and how astrophysicists can observe sources of these radiations.   *[1 mark]*

Total: 9

2 Alpha Centauri A is a bright star. It is the largest star in a system that consists of it, Alpha Centauri B, a smaller star, and a red dwarf called Proxima Centauri.

The A and B stars orbit each other every 82 years, while Proxima Centauri is much further away from the A and B system, orbiting every 550 000 years.

Table 1 shows some data for Alpha Centauri A and B.

Property	A	B
Apparent magnitude	−0.01	+1.33
Absolute magnitude	+4.4	+5.71
Surface temperature	5 790 K	5260 K
Luminosity (Sun = 1.0)	1.519	0.445
Mass (Sun = 1.0)	1.1	0.907

Table 1

2–1 Calculate the distance of the star Alpha Centauri from the Earth and state the unit.

Distance = __ unit = __ *[2 marks]*

2–2 Calculate the peak wavelength of spectrum of Alpha Centauri A.

Wavelength = __ m *[1 mark]*

2–3 Determine the diameter of Alpha Centauri A compared to that of the Sun.

The surface temperature of the Sun is 5778 K.

Diameter of Alpha Centauri = __ times the diameter of the Sun *[2 marks]*

2–4 **Figure 2** shows a simplified Hertzsprung–Russell diagram.

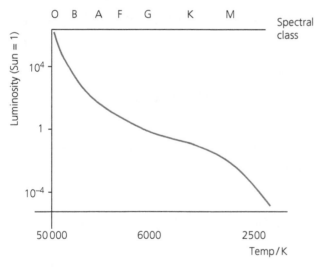

Figure 2

State which spectral class Alpha Centauri A belongs to. *[1 mark]*

2–5 Copy **Figure 2**, and draw on your diagram to show the place of Alpha Centauri A on the main sequence. *[1 mark]*

2–6 It is possible that, at the end of their lives, Alpha Centauri A and B could form a Type 1a supernova.

Describe the formation of a Type 1a supernova. *[2 marks]*

2–7 Explain how astrophysicists use a Type 1a supernova. *[1 mark]*

Total: 10

3 Most stars occur in binary systems, meaning that two stars are orbiting each other about a central point.

This is observed using Doppler shift.

The idea is shown in **Figure 3**.

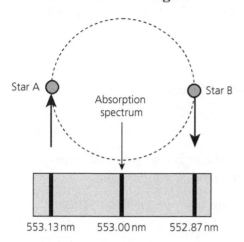

553.13 nm   553.00 nm   552.87 nm

Figure 3

A certain element gives a strong absorption line at a wavelength of 553.00 nm. The light from Star A gives the equivalent absorption line at 553.13 nm, while Star B has the same absorption line at 552.87 nm.

Studies on both stars suggest that they have the same mass.

3–1 Determine the recession speed of Star A.

Recession speed = __ m s⁻¹ *[2 marks]*

3–2 It is noticed that Star B appears to move very slightly from its predicted orbit in a periodic fashion.

Explain what is most likely to be causing this. *[2 marks]*

3–3 Observations of another star show that the intensity of the light is slightly reduced in a periodic fashion as shown in **Figure 4**.

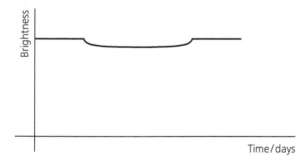

Figure 4

Explain what is causing this observation. *[2 marks]*

3–4 Planets are thought to be very common throughout the Universe, but are not easy to observe directly. However, it is hoped to find life on exoplanets.

Describe the evidence that would satisfy scientists that life is most likely to be found on a planet. *[3 marks]*

Total: 9

4 Quasars (quasi-stellar objects) were first discovered in 1960. They are some of the most distant objects known. They are smaller than galaxies.

4–1 Give **two** features of a quasar. *[2 marks]*

4–2 The visible light spectrum of a quasar is being studied. A strong absorption line for a particular element is found at 550 nm in the laboratory, but is identified at 660 nm.

Calculate the speed of recession of the quasar.

Recession speed = __ km s^{-1} *[1 mark]*

4–3 Calculate the distance of the quasar to the Earth. Give your answer to an appropriate number of significant figures.

Distance = __ Mpc *[2 marks]*

4–4 Astrophysicists believe that the speed of recession of quasars is one piece of strong evidence that the Universe is expanding in all directions from a single point.

Explain **two** other pieces of evidence that suggest that the Universe is expanding in all directions from a single point. *[2 marks]*

4–5 Determine the age of the Universe with a suitable calculation.

Age = __ *[1 mark]*

Total: 8

Paper 3B total: 36

Paper 3 total: 81

Answers

Particles (p.1)

Quick questions

1   Two atoms with the same proton number but different nucleon number, i.e. different numbers of neutrons in their nuclei.

2   Gravitational, electromagnetic, strong force, weak force

3   They annihilate and the energy released can be transferred as high energy photons and the formation of other particle/antiparticle pairs.

4   γ-photons

5   Baryons made up of three quarks (*or antiquarks for their antiparticles*) and mesons made up of a quark–antiquark pair.

6   An up (u) quark in a proton changes character to become a down (d) quark, so the proton changes into a neutron.

Exam-style questions

7–1   X is a proton [1] and Y a neutron [1].

7–2   W+ is an exchange particle [1] and this decay is an example of weak interaction [1].

7–3   The question mark (?) in **Figure 1** is showing an electron neutrino (v_e) [1].

7–4   Charge is conserved because a u quark changes into a d quark, so change of charge $\Delta q = \left(-\frac{1}{3}\right) - \left(+\frac{2}{3}\right) = -1$, but a positron was formed, which carries a +1 charge. So the net charge in this interaction is 0 [1]. The baryon number is conserved because both u and d quarks have a baryon number of $+\frac{1}{3}$ [1].

7–5   An electron neutrino must be formed in this interaction because the positron e+ has lepton number −1 [1], so the v_e conserves the lepton number, as it has lepton number +1 [1].

7–6   All baryons will eventually decay into a proton [1].

8–1   The total energy of the electron–positron pair is
3.512 MeV × 2 = 7.024 MeV
= 7.024 × 1.6 × 10⁻¹³ J = 1.124 × 10⁻¹² J [1]
$E = hc/v \rightarrow v = hc/E$
$$= \frac{6.63 \times 10^{-34}\,m^2\,kg\,s^{-1} \times 3 \times 10^8\,m\,s^{-1}}{1.124 \times 10^{-12}\,kg\,m^2\,s^{-2}}\ [1]$$
$= 1.77 \times 10^{-13}$ m [1]

8–2   Energy [1] and charge [1]

8–3   Because the rest energy of a single muon is 105.659 MeV [1], but the total energy available was only 7.024 MeV [1] (*Allow ecf from **Answer 8–1***)

8–4   Minimum energy needed $E = 105.659$ MeV × 2
= 211.318 MeV
= 211.318 × 1.6 × 10⁻¹³ J
= 3.381 × 10⁻¹¹ J [1]
$E = hf \rightarrow f = E/h$
$$= \frac{3.381 \times 10^{-11}\,kg\,m^2\,s^{-2}}{6.63 \times 10^{-34}\,m^2\,kg\,s^{-1}}\ [1] = 5.11 \times 10^{22}\ Hz\ [1]$$

9–1   It means that carbon atoms can have different numbers of neutrons in their nuclei [1], but they will always have 6 protons [1].

9–2   14 [1]

9–3   8 [1]

9–4 $^{14}_{6}C \rightarrow ^{14}_{7}X + ^{0}_{-1}e + \bar{v}_e$ [3]

9–5 $\bar{v}_e$ is an anti-electron-neutrino [1] and it needs to be emitted in a beta decay to conserve the lepton number [1], because the β-particle e− has lepton number +1 and the antineutrino −1 [1].

10–1   W− [1]

10–2   Weak interaction/force [1]

10–3   Strange particles that are produced through the strong interaction [1].

10–4   A β-particle is produced in this nuclear decay, so the W− carries the negative charge required to conserve the total charge of the system [1].

10–5   10⁻¹⁸ m [1]

11–1   The overall charge is conserved [1], the baryon number is also conserved [1] and strangeness is also conserved in the pair production of K− and K+, with strangeness S = −1 and S = +1, respectively [1], so this interaction is possible.

11–2   K− has charge −1 and strangeness S = −1, π^- and π^+ have strangeness S = 0, but this is a weak interaction, so the strangeness can change by +1 [1], the charge of π^- and π^+ cancel each other leaving an overall charge of −1, as in the K−, i.e. charge is conserved [1], so the interaction is possible [1].

11–3   The baryon number is conserved [1], charge is not conserved because a β-particle is produced which carries a −1 charge [1], also the v_e should be an antineutrino to conserve the lepton number which was 0 before the interaction and it is +2 after [1]. The interaction is not possible.

11–4   The charge is conserved [1], strangeness is conserved because Σ^+ has strangeness −1 and K+ has strangeness +1 and because this interaction shows the formation of strange particles, strangeness needs to be conserved [1]. However, the lepton number is not conserved, as it is −1 before the interaction in the e+ and +1 after in the e− [1], so the interaction is not possible.

12–1   Particle X is a proton [1] and Y is a π^- [1].

12–2   Conservation of charge → charge before = 0 and charge after = +1 − 1 = 0 [1], baryon number is conserved as $+\frac{1}{3} + \frac{1}{3} + \frac{1}{3} = 1$ before and after the interaction [1], strangeness changes from −1 to 0 which is acceptable for a weak interaction like this one [1].

12–3   The proton is held together because the strong nuclear force acts at a short range of 10⁻¹⁴ to 10⁻¹⁵ m [1], and it is 137 times larger than the electromagnetic repulsion of the charges carried by the quarks [1], and because the quark forming a proton is within the range of action of the strong nuclear force, this overcomes the electromagnetic force between them and holds the proton together [1].

13–1   • Overlap a laminated graph paper over the cloud chamber.
 • Accurately measure the range of a number of traces (at least 10) from the source position to the end of the trail.
 • Calculate the mean range of the alpha particles in the chamber.
 • Calculate the absolute uncertainty on the trail length by dividing the difference of the longest trace and the shortest trace by 2.
 • Calculate the percentage uncertainty by dividing the absolute uncertainty by the mean value and multiplying by 100.

149

13–2 Using Fleming's Left Hand Rule the alpha particles can be seen as a current moving up and the magnetic field enters the page [1], so the alpha particles would bend to the left [1] with decreasing curvature as the particles slow down to a stop [1].

13–3 $^{226}_{88}Ra \rightarrow\ ^{222}_{86}Rn +\ ^{4}_{2}He$ [2]

14–1 Annihilation [1]

14–2 An electron and a positron meet each other and annihilate each other [1]. The mass of the particle–antiparticle pair is converted into two gamma photons travelling in opposite directions [1].

14–3 The rest energy of an electron/positron is 0.510999 MeV
= $0.510999 \times 1.6 \times 10^{-13}$ J = 8.176×10^{-14} J [1].
$E = hc/\nu \rightarrow \nu = hc/E$
$$= \frac{6.63 \times 10^{-34} m^2 kg s^{-1} \times 3 \times 10^8 ms^{-1}}{8.176 \times 10^{-14} kg m^2 s^{-2}}:$$
$$= 2.43 \times 10^{-12} m \quad [2]$$
$$f = c/\nu = \frac{3 \times 10^8 ms^{-1}}{2.43 \times 10^{-12} m} = 1.23 \times 10^{20} Hz \quad [1]$$

14–4 Gamma photons are ionising radiation [1], so they can cause mutations in healthy cells in the patient's body and lead to cancer [1].

15–1 The β-particle is a fast-moving electron [1], from within the nucleus [1] (not an electron from the electron shells of the atom).

15–2 Weak interaction/force [1]

15–3 *Any six from:*
• The W– particle transfers <u>momentum</u> from the decaying quark to the electron/beta-particle and neutrino [1].
• Energy is also transferred in this <u>mediation</u> [1].
• The W– particle acts over a <u>very short range</u> and quickly <u>decays</u> into a β-particle and an antineutrino [1].
• Because the mass of the decaying nucleus is much larger than the mass of the β-particle and antineutrino [1] …
• … the decaying nucleus will only experience a <u>small acceleration</u> [1] …
• … while the β-particle and antineutrino will experience a <u>very large acceleration</u> [1] …
• … the force mediated by the W– particle is equal in <u>magnitude</u> for the nucleus, the β-particle and antineutrino [1].

16–1 The baryon number is conserved because both Σ+ and the proton have three quarks [1], the charge is conserved because both the proton and Σ+ are positively charged and the π0 and K0 are neutral [1], strangeness is conserved because K0 contains an anti-strange quark (S = +1) and Σ+ a strange quark (S = −1) [1].

16–2 Strong interaction [1], because strange particles are produced, so strangeness must be conserved [1].

16–3 uus [1], because $+\frac{2}{3} + \frac{2}{3} - \frac{1}{3}$ [1] = +1 [1]

16–4 In this interaction, the strange particle Σ+ is produced [1], but π0 does not contain an anti-strange quark, so this reaction would not occur [1].

16–5 The strange quark in the Σ+ particle has strangeness S = −1 [1], but the strangeness changes to S = 0 in the products, so this is a weak nuclear decay [1].

Electromagnetic radiation and quantum phenomena (p.5)

Quick questions

1 The minimum energy of a photon needed to release a photoelectron from a metal.

2 A discrete packet of electromagnetic energy.

3 The energy needed to move an electron through a potential difference of one volt.

4 The minimum energy needed to remove an electron from its atomic electron clouds completely.

5 Emission spectra are formed when electrons 'fall' from higher to lower energy states within the atom. The photons emitted in this energy drop will have the same energy as the energy gaps between these energy states, so only photons of the equivalent frequencies will be emitted, showing as well-defined lines in the spectrum formed.

6 When a cathode is heated up to high temperatures, the free electrons in the metal conductor gain more and more energy, until some have enough energy to escape the metal.

Exam-style questions

7–1 Convert 940 nm to 9.4×10^{-7} m [1],
$f = c/\lambda = 3 \times 10^8$ m s^{-1}/$(9.4 \times 10^{-7}$ m) = 3.19×10^{14} s^{-1} [1],
the $E_{k(max)}$ for that frequency is greater than 0, so this metal is suitable [1].

7–2 Calculate the gradient of the graph: choose large triangle [1] to calculate $\Delta y/\Delta t$ [1], value of Planck constant between 6.63 and 6.8×10^{-34} J s [1].

7–3 3.00×10^{14} s^{-1} [1]

7–4 $\phi = hf_0$ [1] = 6.63×10^{-34} J s $\times 3.00 \times 10^{14}$ s^{-1}
= 1.99×10^{-19} J [1]

7–5 Straight line with same gradient [1], but to the right of the original line [1].

8–1 Each photon colliding with the metal plate can transfer its energy to a free electron on the metal surface [1]; if the frequency of the incident photon is above the threshold frequency of the metal, the electrons will have enough energy to escape the metal surface [1]. The electrons that have enough kinetic energy to reach the wire grid will be registered as a small current by the picoammeter [1].

8–2 The photoelectrons emitted with maximum kinetic energy $E_{k(max)}$ are the electrons that were on the surface of the potassium plate when they absorbed a photon [1]. However, some electrons deeper inside the metal plate could absorb photons [1], so some of the energy transferred to them by the photons is needed to reach the surface before they can be emitted [1].

8–3 The stopping potential is p.d. needed to reduce the current in the circuit to zero [1].

8–4 *Any six from:*
• Increase the p.d. across the potassium plate and wire grid until no current is registered by the picoammeter [1].
• Reduce the p.d. in small amounts until a current is recorded [1].
• Make a note of the p.d. when the current was first recorded [1].
• This will be the highest frequency emitted by the gas lamp [1].
• Continue to decrease the p.d. until the current increases a second time, and so on [1].
• Make a note of the frequency at each increment of the current [1].
• Each current change/increase matches an emission frequency [1].

8–5 410, 434 and 486 nm wavelengths can be detected, but not 656 nm [1]. This is because $f = c/\lambda$
$$= \frac{3 \times 10^8 ms^{-1}}{656 \times 10^{-9} m} = 4.57 \times 10^{14} Hz \quad [1]$$
which is below 0 p.d. in **Figure 6**. While $f = c/\lambda$
$$= \frac{3 \times 10^8 ms^{-1}}{486 \times 10^{-9} m} = 6.17 \times 10^{14} Hz \quad [1]$$
which is found at a p.d. of about 0.2 V [1].

8–6 To be able to detect the red wavelength of the hydrogen emission spectrum, we could reverse the p.d. across the potassium plate and wire grid [1] and change the p.d. until a further increase in the current in the circuit is measured by the picoammeter [1].

9-1 Convert 410 nm to 4.1×10^{-7} m [1]
Violet photons energy $E = hc/\lambda$

$$= \frac{6.63 \times 10^{-34}\,\text{Js} \times 3 \times 10^8\,\text{ms}^{-1}}{4.1 \times 10^{-7}\,\text{m}} = 4.9 \times 10^{-19}\,\text{J} \ [1]$$

$\Phi = hf_0$ [1] $= 6.63 \times 10^{-34}\,\text{Js} \times 6.5 \times 10^{14}\,\text{s}^{-1}$

$\quad = 4.3 \times 10^{-19}\,\text{J}$ [1]

$(4.9 - 4.3) \times 10^{-19}\,\text{J} = 0.6 \times 10^{-19}\,\text{J}$ [1] $\rightarrow$

$\dfrac{0.6 \times 10^{-19}}{1.6 \times 10^{-19}} = 0.38\,\text{eV}$ [1]

9-2 $E_{k(max)} = 0.6 \times 10^{-19}\,\text{J} = \dfrac{1}{2} m_e v_e^2$ [1] $\rightarrow v_e =$

$$\sqrt{\frac{2 \times 0.6 \times 10^{-19}\,\text{kg m}^2\,\text{s}^{-2}}{9.11 \times 10^{-31}\,\text{kg}}} \ [1] = 362937\,\text{ms}^{-1} \ [1]$$

10-1 $E_3 - E_1 = -1.51\,\text{eV} - (-13.60)\,\text{eV} = 12.45\,\text{eV}$ [1]
$= 12.45 \times 1.6 \times 10^{-19} = 1.99 \times 10^{-18}\,\text{J}$ [1]
$E_3 - E_1 = hc/\lambda \rightarrow$

$$\lambda = \frac{hc}{E_3 - E_1} = \frac{6.63 \times 10^{-34}\,\text{Js} \times 3 \times 10^8\,\text{ms}^{-1}}{1.99 \times 10^{-18}\,\text{J}} \ [1]$$

$\quad = 9.99 \times 10^{-8}\,\text{m}$ [1]

10-2 $E_3 - E_2 = -1.51\,\text{eV} - (-3.41)\,\text{eV} = 1.90\,\text{eV} \rightarrow$
$1.90 \times 1.6 \times 10^{-19} = 3.04 \times 10^{-19}\,\text{J}$ [1]
$E_3 - E_2 = hf \rightarrow$

$f = \dfrac{E_3 - E_2}{h} = \dfrac{3.04 \times 10^{-19}\,\text{J}}{6.63 \times 10^{-34}\,\text{Js}} = 4.59 \times 10^{14}\,\text{s}^{-1}$ [1]

$E_2 - E_1 = -3.41\,\text{eV} - (-13.60)\,\text{eV} = 10.19\,\text{eV} \rightarrow$
$10.19 \times 1.6 \times 10^{-19} = 1.63 \times 10^{-18}\,\text{J}$ [1]
$E_2 - E_1 = hf \rightarrow$

$f = \dfrac{E_2 - E_1}{h} = \dfrac{1.63 \times 10^{-18}\,\text{J}}{6.63 \times 10^{-34}\,\text{Js}} = 2.46 \times 10^{15}\,\text{s}^{-1}$ [1]

The photon emitted in the drop from level 3 to 2 is in the visible spectrum and from 2 to 1 in the UV spectrum [1].

10-3 Ionisation energy is the minimum [1] energy needed to remove an electron from the atom completely [1].

10-4 $13.60\,\text{eV}$ [1] $\rightarrow 13.60 \times 1.6 \times 10^{-19} = 2.18 \times 10^{-18}\,\text{J}$ [1]

11-1
- When the sodium gas is heated, electrons absorb energy and reach higher energy states [1] ...
- ... but after a short time, these excited electrons will fall back to lower energy levels [1].
- When these drops to lower energy levels occur, photons of discrete energy are emitted [1].
- Only two lines are visible in the sodium emission spectrum [1] ...
- ... this supports the model that the electrons in the sodium atoms can only occupy a small number of discrete energy levels [1].
- The energy difference between these energy levels matches the energy of the photons emitted [1].

11-2 $-5.14\,\text{eV}$ is the energy at energy level 1 and converting to joules $\rightarrow -5.14 \times 1.6 \times 10^{-19}\,\text{J} = 8.22 \times 10^{-19}\,\text{J}$ [1]
$E_3 - E_1 = -3.10 \times 10^{-19}\,\text{J} - (-8.22 \times 10^{-19})$ J
$= 5.12 \times 10^{-19}\,\text{J}$ [1]

$E = \dfrac{hc}{\lambda} \rightarrow \lambda = \dfrac{hc}{E} = \dfrac{6.63 \times 10^{-34}\,\text{Js} \times 3 \times 10^8\,\text{ms}^{-1}}{5.12 \times 10^{-19}\,\text{J}}$ [1]

$\quad = 3.88 \times 10^{-7}\,\text{m}$ [1]

11-3 Ultraviolet [1] (*Allow ecf from* **Answer 11-2**)

11-4

level 2 ————————— −1.94 eV

level 1 ————————— −3.03 eV

ground state ————————— −5.14 eV

Diagram includes:
- level 1 as −5.14 eV [1]
- energy of two photons calculated correctly [2]
- energies converted to eV [1]
- level 2 and 3 drawn with −3.03 and −1.94 eV [1].

12-1 The free electrons between the anode and the cathode are accelerated to very high velocity inside the tube [1]. These fast-moving free electrons can collide inelastically with the atoms in the mercury vapour [1], and if they transfer enough energy to these atoms, they can cause electrons in their electron shells to jump to higher energy levels, or even cause ionisation of the atoms [1].

12-2 After a mercury atom is excited, the electron that has moved to a higher energy level will eventually drop down to lower energy levels [1] emitting an ultraviolet photon [1].

12-3 After ionisation, the mercury atom becomes a positive ion [1] and can absorb a free electron inside the fluorescent tube [1]. This electron will drop to lower energy levels emitting an ultraviolet photon [1].

12-4 The UV photons emitted by mercury atoms after excitation and/or ionisation will be absorbed by the phosphor molecules [1] and cause their electrons to move to higher energy levels [1]. These electrons eventually drop down to lower energy levels and the energy gaps of these drops are in the visible range of the electromagnetic spectrum [1].

13-1 $\lambda = 0.025\,\mu\text{m} = 2.5 \times 10^{-8}\,\text{m}$ [1], from de Broglie's equation

$$v = \frac{h}{m\lambda} \ [1] = \frac{6.63 \times 10^{-34}\,\text{Js}}{9.11 \times 10^{-31}\,\text{kg} \times 2.5 \times 10^{-8}\,\text{m}}$$

$\quad = 29110\,\text{ms}^{-1}$ [1]

13-2 Because, at that velocity, the de Broglie's wavelength of the incident electrons would be of the same magnitude as the distance between the atoms in the crystal [1] and diffraction is greatest when the wavelength is roughly equal in size to the gaps/separations [1].

13-3 $E_k = \dfrac{1}{2} mv^2 = \dfrac{1}{2} 9.11 \times 10^{-31}\,\text{kg} \times 29110^2\,\text{ms}^{-1}$

$\quad = 3.86 \times 10^{-22}\,\text{J}$ [1] $= 2.41 \times 10^{-3}\,\text{eV}$ [1]

(*Allow ecf from* **Answer 13-1**)

13-4 To halve the E_k the velocity of the electrons should be $\dfrac{v}{\sqrt{2}}$, because E_k depends on the square of the velocity [1].

So, $\dfrac{E_k}{2} = \dfrac{1}{2} m\left(\dfrac{v}{\sqrt{2}}\right)^2 = \dfrac{1}{2} m \dfrac{v^2}{2} = \dfrac{1}{4} mv^2$ [1]. A reduction in velocity reduces momentum and so would increase the wavelength. [1]

14-1 $5\,\text{MeV} \times 1.6 \times 10^{-19}\,\text{J} = 8 \times 10^{-19}\,\text{J}$ [1],
$m_\alpha = 2(1.673 \times 10^{-27} + 1.675 \times 10^{-27})$ kg
$\quad = 6.656 \times 10^{-27}\,\text{kg}$ [1]

$E_k = \dfrac{1}{2} mv^2 \rightarrow v = \sqrt{\dfrac{2E_k}{m}}$ [1] $= \sqrt{\dfrac{2 \times 8 \times 10^{-19}\,\text{J}}{6.656 \times 10^{-27}\,\text{kg}}}$

$\quad = 15504\,\text{ms}^{-1}$ [1]

14-2 $\lambda = \dfrac{h}{mv} = \dfrac{6.63 \times 10^{-34}\,\text{Js}}{6.656 \times 10^{-27}\,\text{kg} \times 15504\,\text{ms}^{-1}}$ [1]

$\quad = 6.4 \times 10^{-12}\,\text{m}$ [1] (*Allow ecf from* **Answer 14-1**)

14-3 The spacing between the nuclei in the gold foil is two orders of magnitude larger than the wavelength of the α-particles shot at the foil [1], so the conditions for diffraction are not met, as the wavelength of the incident particles should be of similar size to the gaps in the structure of the sample for diffraction to occur [1].

14-4 $\lambda = \dfrac{h}{mv} \rightarrow \dfrac{\text{Js}}{\text{kgms}^{-1}} = \dfrac{\text{kgms}^{-2}\text{ms}}{\text{kgms}^{-1}}$ [1]

$\quad = \dfrac{\text{kgms}^{-2}\text{ms}}{\text{kgms}^{-1}} = \text{kgm}^2\text{s}^{-2}\text{kg}^{-1}\text{m}^{-1}\text{s}^2 = \text{m}$ [1]

Topic review: particles and radiation (p.8)

1–1 Annihilation [1], the photons are γ-photons [1]

1–2 A pair is produced so the total momentum is conserved [1].

1–3 The photon will have energy $\dfrac{1.6 \times 10^{-13}\,\text{J}}{2 \times 1.60 \times 10^{-19}\,\text{J}}$ [1]

= 0.5 MeV, which is much bigger than 24.6 eV [1], so the helium atom will be ionised [1].

1–4 $\Delta E = hf = -1.6\,\text{eV} - (-24.6\,\text{eV}) = 23.0\,\text{eV}$ [1]

= $23.0 \times 1.6 \times 10^{-19}\,\text{J} = 3.7 \times 10^{-18}\,\text{J}$ [1]

$\lambda = \dfrac{hc}{\Delta E}$ [1] = $\dfrac{6.63 \times 10^{-34}\,\text{Js} \times 3.0 \times 10^{8}\,\text{ms}^{-1}}{3.7 \times 10^{-18}\,\text{J}}$

= 54 nm [1]

1–5 $hf = 4.8\,\text{eV} - 1.6\,\text{eV}$ [1] $= 3.2\,\text{eV}$

= $3.2 \times 1.6 \times 10^{-19}\,\text{J} = 5.1 \times 10^{-19}\,\text{J}$ [1] →

$\lambda = \dfrac{hc}{\Delta E} = \dfrac{6.63 \times 10^{-34}\,\text{Js} \times 3.0 \times 10^{8}\,\text{ms}^{-1}}{5.1 \times 10^{-19}\,\text{J}} = 388\,\text{nm}$ [1],

so the colour of the emitted photon is violet [1].

1–6 A proton is composed of two up (u) quarks and one down (d) quark (uud) [1] and the sum of their charges

is $+\dfrac{2}{3} + \dfrac{2}{3} - \dfrac{1}{3} = +1$ [1]; a neutron is composed of two

d quarks and one u quark (udd) [1] and the sum of their

charges is $+\dfrac{2}{3} - \dfrac{1}{3} - \dfrac{1}{3} = 0$ [1].

1–7 The two protons will repel each other with an electrostatic force [1], but the dimensions of the nucleus are very small and below distances of 3 fm [1] the strong nuclear force will attract all nucleons to each other with a force much stronger than the electrostatic force [1].

2–1 $E = h\dfrac{c}{\lambda} = \dfrac{6.63 \times 10^{-34}\,\text{m}^2\,\text{kg}\,\text{s}^{-1} \times 3 \times 10^{8}\,\text{ms}^{-1}}{1.849 \times 10^{-7}\,\text{m}}$

= $1.08 \times 10^{-18}\,\text{J}$ [1]

2–2 UV [1]

2–3 Photoelectrons emitted from the surface of the zinc plate will have maximum kinetic energy [1], but photoelectrons emitted from deeper in the zinc plate will have less kinetic energy [1], because some of their energy is transferred in moving to the surface [1].

2–4 $\phi = 4.33\,\text{eV} \times 1.6 \times 10^{-19}\,\text{J eV}^{-1} = 6.9 \times 10^{-19}\,\text{J}$ [1];

$E_{k\,max} = hf - \varnothing$ [1] $= 1.08 \times 10^{-18}\,\text{J} - 6.9 \times 10^{-19}\,\text{J}$

= $3.9 \times 10^{-19}\,\text{J}$ [1]

2–5 There would be no photoelectrons emitted by the zinc plate [1], because photons in the visible spectrum are below the threshold frequency of zinc [1].

Progressive and stationary waves (p.10)

Quick questions

1 $\phi = \dfrac{2\pi x}{\lambda} = \dfrac{2\pi \times 0.085\,\text{m}}{0.34\,\text{m}} = 2\pi \times 0.25 = \dfrac{2\pi}{4} = \dfrac{\pi}{2}$, i.e. $\dfrac{1}{4}$ of

a cycle phase difference.

2 An unpolarised wave could be drawn as a series of double arrows with the same centre in at least three different angles, and a polarised wave could be drawn as a single double arrow. Unpolarised waves have oscillations along a range of different planes, while polarised waves have oscillations only in one plane.

3 When two progressive waves travel along the string with the same frequency and amplitude, but in opposite directions, they will constructively interfere at the antinodes and destructively interfere at the nodes. For a standing wave to form in a string fixed at both ends, the length of the string needs to be a multiple of half wavelength of the progressive waves.

4 520 mm = 0.520 m

5 Because the oscillations in a longitudinal wave are always along the direction of energy transfer.

6 $c = f\lambda \rightarrow f = \dfrac{c}{\lambda} = \dfrac{340\,\text{ms}^{-1}}{0.5\,\text{m}} = 680\,\text{Hz}$

Exam-style questions

7–1 Stationary waves form when two progressive waves of the same amplitude and frequency moving in opposite directions superimpose [1]. At the nodes of a stationary wave, the particles do not oscillate because the amplitudes of the two progressive waves always cancel out [1], while at the antinodes the amplitudes add together and the thin cord is displaced between the peak and the trough during a whole cycle [1].

7–2 Rearrange $c = f\lambda$ to find the wavelength of the standing wave, $\lambda = c/f$ [1] = $17.5\,\text{ms}^{-1}/50\,\text{s}^{-1} = 0.35\,\text{m}$ [1] the distance between nodes is half wavelength, so 0.35/2 = 0.175 m [1].

7–3 Rearrange $f = \dfrac{1}{2l}\sqrt{\dfrac{T}{\mu}} \Rightarrow f^2 = \dfrac{1}{4l^2}\dfrac{T}{\mu}$ [1] $\Rightarrow \mu = \dfrac{1}{4l^2}\dfrac{T}{f^2}$ [1]

The tension T = $9.81\,\text{N kg}^{-1} \times 0.5\,\text{kg} = 4.9\,\text{N}$ [1]

Substitute in the equation

$\mu = \dfrac{1}{4 \times 1.2^2\,\text{m}^2}\dfrac{4.9\,N}{61.43^2\,\text{s}^{-2}} = \dfrac{4.9\,\text{kg m s}^{-2}}{21736\,\text{m}^2\,\text{s}^{-2}}$

= $2.25 \times 10^{-4}\,\text{kg m}^{-1}$ [1]

μ is the mass per metre, so

$\mu \times l = 2.25 \times 10^{-4}\,\text{kg m}^{-1} \times 1.2\,\text{m}$

= $2.71 \times 10^{-4}\,\text{kg}$ [1]

7–4 $0.1/4.9 \times 100 = 2.04\%$, $0.001/1.2 \times 100 = 0.08\%$, 0.02% [1]
% uncertainty on mass/metre = 2.04% + 0.08% + 0.02% = 2.14% [1]

Absolute uncertainty = $\dfrac{2.25 \times 10^{-4}\,\text{kg m}^{-1} \times 2.14}{100}$

= $\pm 0.05 \times 10^{-4}\,\text{kg m}^{-1}$ [1]

7–5 Repeat readings of frequency of first harmonic for different tensions (or different lengths, as long as only one variable is changed) [1], calculate the mean μ from the data collected [1], calculate the uncertainty as $\pm\dfrac{\text{range}}{2}$ [1].

8–1 Stationary waves are set up between the transmitter and the metal sheet [1], there is a node at the metal sheet and the distance between a node and its subsequent antinode

is $\dfrac{1}{4}$ wavelength [1], so the wavelength of the microwaves

is 5.5 cm × 4 = 22 cm [1].

8–2 The number of wavelengths between T and the metal sheet is 55/22 = 2.5 [1].

8–3 Convert 22 cm to 0.22 m [1], rearrange $c = f\lambda \rightarrow f = c/\lambda$ [1]

= $\dfrac{3 \times 10^{8}\,\text{ms}^{-1}}{0.22\,\text{m}} \approx 1.36 \times 10^{9}\,\text{Hz}$ [1]

8–4 The microwaves that are transmitted by the oven grid will be polarised along one plane only [1], while the microwaves along all other planes will be absorbed by the grid [1].

8–5 Sensor P would not detect any microwaves beyond the two grids [1], because the microwaves transmitted by the first grid will be perpendicular to the plane of polarisation of the second grid [1] (hence they will be absorbed).

9–1 At nodes, the particles in the medium are not displaced [1], but at the antinodes, the particles of the medium experience maximum displacement between peaks and troughs [1], so the polystyrene balls will be forced away from the antinodes [1] and collect at the nodes [1].

9–2 At this frequency, there are two antinodes and one single node in the centre, i.e. half wavelength [1], the wavelength of these standing waves is $\lambda = c/f = 330\,\text{ms}^{-1}/119\,\text{s}^{-1}$ [1] = 2.77 m, the distance between discs is 2.77 m/2 = 1.39 m [1]

9–3 If four nodes are visible, there are two wavelengths fitting inside the tube [1], so the wavelength of the stationary waves is 1.2 m/2 = 0.6 m [1], and the frequency is $f = c/\lambda$ = 330 m s⁻¹/0.6 m = 550 Hz [1].

Refraction, diffraction and interference (p.12)

Quick questions

1   It is the vector sum of the displacement of each wave.
2   Two waves are coherent when they have a fixed phase difference and the same frequency.
3   Diagram C
4   1
5   The light ray will travel inside the denser medium at a lower speed. This will cause the light ray to bend towards the normal to the boundary, if the incident ray is not perpendicular to the boundary between media.
6   When the size of the gap and the wavelength of the wave are comparable in size.

Exam-style questions

7–1 *Any six from:*
   • suitable labelled diagram of Young's double slit experiment [1]
   • suitable length for D (distance between slits and screen) – at least 1 m [1]
   • suitable spacing of slits (less than 1 mm) [1]
   • random selection of laser pointers from the batch [1]
   • multiple readings of fringe separation to calculate a mean separation [1]
   • using same conditions for each pointer tested [1]
   • using the equation $\lambda = ws/D$ to determine the answer [1].

7–2 Add % uncertainties 2.5% + 1.2% + 0.1% = 3.8% [1], calculate absolute uncertainty on mean wavelength
$630 \times 3.8/100 = \dfrac{630\,\text{nm} \times 3.8}{100} = \pm24\,\text{nm}$ [1], the wavelength on the laser pointers' label is within the range allowed by the uncertainty in the measurements of the H&S organisation, i.e. (630 ± 24) nm [1].

8–1 A laser beam is a coherent source of monochromatic light [1], so the electromagnetic waves from the double slits will be in phase and have the same frequency [1].

8–2 The separation of the fringes increases [1], but the fringes appear dimmer [1].

8–3 Rearrange $w = \dfrac{\lambda D}{s} \Rightarrow s = \dfrac{\lambda D}{w}$ [1]
Convert lengths to metres [1]
Substitute $s = \dfrac{6.35 \times 10^{-7}\,\text{m} \times 1.54\,\text{m}}{0.0024\,\text{m}} = 4 \times 10^{-4}\,\text{m}$ [1]

8–4 *Any two from (or other suitable suggestions):*
   • Do not point at people's eyes [1].
   • Mount the laser pointer below eye level [1].
   • Use screens that are not highly reflective to avoid strong reflections [1].
   • Ensure the laser power output is within safety regulations (e.g. <1 mW) [1].

8–5 Bright white central fringe [1] and repeated white light spectrum at the order of constructive interference fringes [1].

9–1 Path from L₁ to D = 40 m, using Pythagoras
$L_2D = \sqrt{(L_1L_2)^2 + (L_1D)^2}$ [1] $= \sqrt{9^2 + 40^2} = 41\,\text{m}$ [1],
path difference = 1 m [1]

9–2 The first minimum will be detected for a path difference of $(1/2)\lambda$ [1]
$\lambda = \dfrac{c}{f} = \dfrac{340\,\text{m s}^{-1}}{265\,\text{s}^{-1}} = 1.28\,\text{m}$ [1],
$\sin\theta = \dfrac{\left(\frac{1}{2}\right)\lambda}{L_1L_2} = \dfrac{\frac{1}{2} \times 1.28\,\text{m}}{9\,\text{m}} = 0.071$ [1]

$\tan\theta \approx \sin\theta$ for small angles
Distance of min from central max =
$\tan\theta \times 40\,\text{m} = 0.071 \times 40\,\text{m} = 2.8\,\text{m}$ (2 s.f.) [1]

9–3 The speaker at D is 40 m away from L₁, and the sound from L₁ will arrive at D with a short delay [1] because the electrical signal from the mixer to both speakers is almost instantaneous [1], so the signal to the speaker at D needs to be delayed or there will be an echoing effect/sound distortion [1].

9–4 t = distance/speed [1] = 40 m/340 m s⁻¹ = 0.118 s [1]

10–1 If the second maximum is at P, the fringe spacing
$w = 14\,\text{cm}/2 = 7\,\text{cm}$ [1]
Rearrange $w = \dfrac{\lambda D}{s} \Rightarrow \lambda = \dfrac{ws}{D} = \dfrac{7\,\text{cm} \times 8\,\text{cm}}{30\,\text{cm}} = 1.87\,\text{cm}$ [1]
Convert 1.87 cm = 0.0187 m,
$f = \dfrac{c}{\lambda} = \dfrac{3 \times 10^8\,\text{m s}^{-1}}{0.0187\,\text{m}} = 1.6 \times 10^{10}\,\text{s}^{-1}$ [1]

10–2 $w = \dfrac{\lambda D}{s} = \dfrac{1.87\,\text{cm} \times 50\,\text{cm}}{8\,\text{cm}}$ [1] = 11.7 cm [1]
(*Allow ecf on value of λ calculated in* **Answer 10–1**.)

10–3 The fringe spacing would increase [1] by a factor of 2 [1].
['*It would double' gives both marks*.]

11–1 The central maximum would become narrower [1].

11–2 The central maximum would become wider [1].

11–3 The central maximum will appear white [1] and the fringes will show a full spectrum of colours [1].

11–4 The gap of the door is of similar magnitude to the wavelength of the sound waves generated by the doctor's voice [1], but the gap is huge compared to the wavelength of light [1], so the sound from the doctor's voice will diffract around the wall and the light reflected from the doctor's body will not [1].

12–1 Convert 650 nm to 6.5×10^{-7} m
Rearrange $d\sin\theta = n\lambda \Rightarrow d = \dfrac{n\lambda}{\sin\theta}$ [1]
Calculate slit spacing for each order of maxima [1]

Order of maxima	Angle, ϑ/rad	Slit spacing/m
1	0.216	1.72×10^{-4}
2	0.490	1.52×10^{-4}
3	0.752	1.49×10^{-4}
4	1.076	1.38×10^{-4}

Calculate mean slit spacing
$\dfrac{1.72 + 1.52 + 1.49 + 1.38}{4} \times 10^{-4}\,\text{m} = 1.53 \times 10^{-4}\,\text{m}$ [1]
Number of slits per mm = $\dfrac{0.001\,\text{m}}{1.53 \times 10^{-4}\,\text{m}} \approx 6.5$ [1]
(*Accept 7*)

12–2 Use $\dfrac{\text{range}}{2} = \dfrac{1.72 - 1.38}{2}$ [1] $= \pm0.17$ [1]

12–3 From $d\sin\theta = n\lambda$ it can be seen that for a green laser beam the angle ϑ would decrease [1], because its wavelength λ is less than the wavelength of red light [1].

13–1 Using a ruler, measure the distance between maxima of the same order and divide by 2 to find the distance, s, between the central maximum (0th order) and the maximum of nth order, respectively $s_1 = 6.0 \times 10^{-2}\,\text{m}$, $s_2 = 1.2 \times 10^{-2}\,\text{m}$ and $s_3 = 1.7 \times 10^{-2}\,\text{m}$ [1]
$\tan\theta \approx \sin\theta$ for small angles
Calculate d for each order using $d\sin\theta = n\lambda \rightarrow d = \dfrac{n\lambda}{\sin\theta}$
[1], with $\sin\theta = \dfrac{s_n}{3.5\,m}$ [1]
$d_1 = \dfrac{\lambda}{\sin\theta} = \dfrac{5.40 \times 10^{-7}\,\text{m}}{1.7 \times 10^{-3}} = 3.18 \times 10^{-4}\,\text{m}$

$$d_2 = \frac{2\lambda}{\sin\theta} = \frac{2 \times 5.40 \times 10^{-7}\,\text{m}}{0.0034} = 3.18 \times 10^{-4}\,\text{m}$$

$$d_3 = \frac{3\lambda}{\sin\theta} = \frac{2 \times 5.40 \times 10^{-7}\,\text{m}}{0.0049} = 3.71 \times 10^{-4}\,\text{m}$$

$$d_{mean} = \frac{3.18 + 3.18 + 3.31}{3} \times 10^{-4}\,\text{m} = 3.22 \times 10^{-4}\,\text{m} \ [1]$$

13–2 $0.001\,\text{m}/(3.21 \times 10^{-4}\,\text{m})$ [1] $= 3.12$ [1] (*Allow ecf from Answer 13–1*)

13–3 The distance between maxima will increase [1], reducing the uncertainty on the measurements of s [1], but the maxima will become dimmer [1], which could make it more difficult to accurately find the centre of each maximum [1].

14–1 $\sin\vartheta_c = n_2/n_1 = 1/1.5$ [1], $\vartheta_c = a\sin\vartheta_c = 42°$ [1]

14–2 The angle of incidence is 45° because the incident ray is parallel to AB, use $n_1\sin\vartheta_1 = n_2\sin\vartheta_2$ to find $\vartheta_2 \rightarrow$

$$\sin\vartheta_2 = \frac{n_1}{n_2}\sin 45° = \frac{1}{1.5}\sin 45° \ [1] = 0.471 \Rightarrow \theta_2 = a\sin\theta_2$$

$$= 28° \ [1]$$

14–3 Draw correct path inside the prism [1]; $180 - 45 - 28 - 90 = 17$, so angle $= 90 - 17 = 73°$ [1]

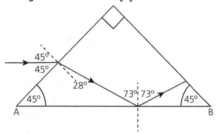

15–1 The cladding must have a lower refractive index than the core, because a critical angle must be exceeded [1] for total internal reflection (TIR) to occur [1].

15–2 Cladding protects the core of the optical fibre from scratches [1]; if there were no cladding and a scratch formed on the surface of the core, TIR might not occur at that point and some signal might be lost [1]. The refractive index of the cladding can be chosen to achieve a high critical angle [1]: this limits the path difference of different pulses due to the number of reflections inside the fibre [1].

15–3 Pulse broadening limits the maximum frequency of pulses that can be sent through the optical fibre [1], which limits the bandwidth available [1].

15–4 Material dispersion can cause pulse broadening because different wavelengths of light in the signal travel along the material of the optical fibre at different speeds [1], while modal dispersion happens when rays inside the optical fibre travel slightly different paths, leading to an increased duration of each pulse [1].

16–1 The refractive index of air is $n = 1$ [1] and $n = \frac{c}{c_{air}} = 1$ [1] $\rightarrow$ $c_{air} = c$ [1]

16–2 The limit for the speed of light is c (i.e. speed of light in a vacuum) [1], so the speed of light in a medium, c_s, will always be smaller than c, hence $n = \frac{c}{c_s} > 1$ [1]

16–3 $n = \frac{c}{c_{water}} = 1.33$ [1] $\rightarrow$ $c_{water} = \frac{c}{1.33} = 2.26 \times 10^8\,\text{ms}^{-1}$ [1]

16–4 Calculate $c_1 = \frac{3 \times 10^8\,\text{ms}^{-1}}{3} = 10^8\,\text{ms}^{-1}$ [1],

$$c_2 = \frac{c_1}{{}_1 n_2} \ [1] = \frac{10^8\,\text{ms}^{-1}}{0.4} = 2.5 \times 10^8\,\text{ms}^{-1}$$
$$[1]$$

Topic review: waves (p.16)

1–1 Longitudinal waves [1]

1–2 Transverse [1] and stationary waves [1]

1–3 Three antinodes [1] and two nodes [1] in the sketch

1–4 Convert 320 g to 0.320 kg

$T = mg = 0.320\,\text{kg} \times 9.82\,\text{N/kg} = 3.14\,\text{N}$ [1]

Rearrange

$$f = \frac{1}{2l}\sqrt{\frac{T}{\mu}} \rightarrow l = \frac{1}{2f}\sqrt{\frac{T}{\mu}} = \frac{1}{2 \times 147\text{s}^{-1}}\sqrt{\frac{3.14\,N}{2.3 \times 10^{-3}\,\text{kg m}^{-1}}}$$

$= 0.126\,\text{m}$ [1]

1–5 *Any four from:*
- students will hear loud sounds in certain spots and quiet (*or no*) sound in other spots along the line [1]
- mention of two coherent sources of sound (speakers) [1]
- mention of path difference [1]
- mention of two source interference [1]

Calculation using $w = \lambda D/s$ to show distance between loud spots $\rightarrow \lambda = c/f = 330\,\text{m s}^{-1}/147\,\text{s}^{-1} = 2.24\,\text{m}$ [1] $\rightarrow$

$$w = \frac{2.24\,\text{m} \times 8\,\text{m}}{1.2\,\text{m}} = 14.9\,\text{m} \ [1]$$

2–1 $n = \frac{c}{c_s} = \frac{3.0\,\text{ms}^{-1}}{2.3\,\text{ms}^{-1}} = 1.3$ [1],

$$\sin\theta_2 = \frac{\sin\theta_1}{n_2} = \frac{\sin 70°}{1.3} = 0.72 \rightarrow$$

$a\sin\theta_2 = \theta_2 = 46.1°$ [1]

2–2 Draw the beam inside the tank at an angle of 46.3° from the normal [1]

2–3 The laser beam will refract outside the tank at 70° from the normal [1], because it has not reached the critical angle [1] $\sin\theta_2 = \frac{n_2}{n_1} = \frac{1}{1.3} = 0.77° \rightarrow \theta_c = 50.4°$ [1].

2–4 The sugar will gradually dissolve from the bottom of the tank. This forms a density gradient in the water of the tank [1], with denser sugar solution at the bottom [1]. Therefore, there is a changing refractive index which causes the light beam to follow a bent path and reflect at the surface of the mirror [1].

3–1 Use $w = \frac{\lambda D}{s} \rightarrow \lambda = \frac{sw}{D}$ [1] $= \frac{4 \times 10^{-4}\,\text{m} \times 3.6 \times 10^{-3}\,\text{m}}{2.3\,\text{m}}$

$= 6.26 \times 10^{-7}\,\text{m}$ [1]

3–2 $\frac{0.001\,\text{m}}{330} = 3 \times 10^{-6}\,\text{m}$ [1]

3–3 The distance between maxima in a diffraction grating pattern is much wider [1], so giving more accurate results [1].

3–4 $\frac{\text{Range of } \lambda}{2} = \frac{635 - 626}{2}\,\text{nm} = \pm 4.5\,\text{nm}$ [1]

Force, energy and momentum (p.18)

Quick questions

1 *Any three suitable, examples are:* Vectors – forces, momentum, acceleration; scalars – energy, speed, mass.

2 An object is in equilibrium when the sum of all the forces and the sum of the turning moments on that object are zero.

3 The force on the side of the adult is twice the force on the child's side, so the adult should sit at half the distance from the pivot (centre of the seesaw) compared with the child's distance from the pivot.

4 The sum of all the clockwise moments must be the same as the sum of all the anticlockwise moments for a system to be in (rotational) equilibrium.

5 **C:** $s = ut + \frac{a}{2}t^2$

6 The speed of the object and its surface area.

Exam-style questions

7 Draw vertical force arrow from C pointing up and labelled 3.5 N (equal and opposite force to the pull from Newton's 3rd law), angle between AC and F = 30° [1], resolve

components along AC $F_{AC} = F \times \cos 30 = 3.03\,N$ and
BC $F_{BC} = F \times \sin 30 = 1.75\,N$ [1]

8–1 The mass of the balloon does not change, but its volume increases [1]. This increases the buoyancy force on the balloon and the resultant pushing force on the scale [1] which is given by the weight of the balloon and CO_2 – the upward buoyancy force on the balloon [1].

8–2 The balloon filled with CO_2 will reach the floor first, because CO_2 is denser than air and for the same volume the weight of the CO_2 balloon will be greater [1]. This means that the balloon will reach a higher terminal velocity and hit the floor first [1].

9–1 Weight = mg, normal = N and friction = D. [1]

9–2 The components of mg (W) along the ramp and perpendicular to the ramp must be equal and opposite to D and N, respectively, $W = mg = 5\,kg \times 9.81\,N\,kg^{-1} = 49\,N$ [1], $N = W_N = W \times \cos 20° = 49\,N \times \cos 20° = 46\,N$ [1], $D = W_p = W \times \sin 20° = 49\,N \times \sin 20° = 16.8\,N$ [1]

9–3 The resultant force on the box is $F = ma = 5\,kg \times 0.52\,m\,s^{-2} = 2.6\,N$, weight (and its components) and the normal will not change [1], but D will reduce to Wp – F, so $D = 16.8\,N - 2.6\,N = 14.2\,N$ [1].

9–4 Use $v^2 = u^2 + 2as$ to find v
$$v = \sqrt{u^2 + 2as}$$
$$= \sqrt{0 + 2 \times 0.52\,m\,s^{-2} \times 3.4\,s}$$
$$= 1.88\,m\,s^{-1} \ [1]$$
Use $F = \dfrac{\Delta(mv)}{\Delta t}$
$$= \frac{5\,kg \times 1.88\,m\,s^{-1}}{1.2\,s} = 7.8\,N \ [1]$$

9–5 We assume that the stopping force remains constant as the box decelerates [1].

10–1 Straight horizontal line until P [1], then parabolic trajectory down from P [1].

10–2 $v \times t = s = 0.32\,m\,s^{-1} \times 2.25\,s = 0.72\,m$ [1]

10–3 Convert 23 g to 0.023 kg, $a = \dfrac{F}{m} = \dfrac{0.13\,N}{0.023\,kg} = 5.65\,m\,s^{-2}$ [1], time to travel 165 cm 'vertically' → Rearrange
$$s = ut + \frac{at^2}{2} \rightarrow t^2 = 2\frac{s - ut}{a} = \frac{2s}{a}$$
$$= \frac{2 \times 1.65\,m}{5.65\,m\,s^{-2}} 58.4\,s^2 \rightarrow t = 0.76 \ [1]$$
(Allow ecf for **Answer 10–1**)
Distance travelled from $P = v \times t = 0.32\,m\,s^{-1} \times 0.76\,s = 0.244\,m$ [1],
Total distance from initial point = 0.244 m + 0.72 m = 0.964 m [1] (Allow ecf from **Answer 10–2**)

10–4 $v = u + at = at = 5.65\,m\,s^{-2} \times 0.76\,s = 4.29\,m\,s^{-1}$ [1]
Magnitude of velocity vector
$$v = \sqrt{0.32^2 + 4.29^2} \approx 4.3 \ [1] \rightarrow E_k$$
$$= \frac{1}{2}mv^2 = \frac{0.023\,kg \times 4.3^2\,m^2\,s^{-2}}{2} = 0.212\,J \ [1]$$

11–1 Convert 2.7 g to $2.7 \times 10^{-3}\,kg$ and 57.7 g to $5.77 \times 10^{-2}\,kg$
$F_t = ma = 5.77 \times 10^{-2}\,kg \times 7.28\,m\,s^{-2} = 0.42\,N$ [1]
$F_t = F_p$ due to Newton's 3rd law, $a_p = F_p/m_p$
$= 0.42\,N/2.7 \times 10^{-3}\,kg = 155.58\,m\,s^{-2}$ [1]

11–2 Both balls fall vertically down with the same acceleration (g) [1]; the ping pong ball will also accelerate towards the tennis ball with decreasing acceleration [1], while the tennis ball will barely move towards the ping pong ball (due to its higher inertia) [1].

11–3 $\Delta E_p = E_k = mg\Delta h = 5.77 \times 10^{-2}\,kg \times 9.81\,N\,kg^{-1} \times 1.84\,m$
$= 1.04\,J$ [1] $= \dfrac{1}{2}mv^2 \rightarrow$

$$v = \sqrt{\frac{2E_k}{m}} = \sqrt{\frac{2 \times 1.04\,J}{5.77 \times 10^{-2}\,kg}} = 6.01\,m\,s^{-1} \ [1]$$

11–4 $eff = \dfrac{\text{useful output energy}}{\text{input energy}} \rightarrow$ bounce height = 0.43×1.84
$= 0.79\,m$. [1]

12–1 Use area under v–t graph to find height of drop h_d and bounce $h_b \rightarrow$
$$h_d = \frac{1}{2}0.4\,s \times [0 - (-4)]\,m\,s^{-1} = 0.80\,m \ [1]$$
$$h_b = \frac{1}{2}(0.7 - 0.4)\,s \times [3 - 0]\,m\,s^{-1} = 0.45\,m \ [1]$$
$$\text{efficiency} = \frac{mgh_b}{mgh_d} = \frac{h_b}{h_d} = \frac{0.45\,m}{0.80\,m} = 0.56 \ (or \ 56\%) \ [1]$$

12–2 $m = w/g = 5.17\,N/9.81\,N\,kg^{-1} = 0.53\,kg$ [1], $v = 2.4\,m\,s^{-1}$
from graph, $E_k = \dfrac{1}{2}mv^2 = 0.5 \times 0.53\,kg \times 5.76\,m^2\,v^{-2}$
$= 1.53\,J$ [1]

12–3 d–t graph starting from 0.8 m downward parabola to zero metres at 0.4 s, then decreasing steepness to gradient zero for 0.7 s ... [1]; a–t graph straight horizontal line at $-9.81\,m\,s^{-2}$ [1]

13–1 $v = 92\,km/h = 25.6\,m\,s^{-1}$
Vertical velocity $v_v = v \times \sin 45° = 18.1\,m\,s^{-1}$ [1]
Use $v^2 = u^2 + 2as \rightarrow 0 = 16.7^2\,m^2\,s^{-2} + 2 \times (-9.82)\,m\,s^{-2} \times s$
$\rightarrow s = \dfrac{353.44\,m^2\,s^{-2}}{19.64\,m\,s^{-2}} \approx 16.7\,m$ [1]
Total height s + 1.30 m = 18.0 m [1]

13–2 Horizontal velocity $v_h = v_v = 18.1\,m\,s^{-1}$ [1]

13–3 Use $v = u + at \rightarrow 0 = 18.1\,m\,s^{-1} + (-9.81)\,m\,s^{-2} \times t \rightarrow t$
$+ (-9.81)\,m\,s^{-2} \times t \rightarrow t = \dfrac{18.1\,m\,s^{-1}}{9.81\,m\,s^{-2}} = 1.85\,s$ [1]

13–4 Find the velocity as the ball hits the ground
$v^2 = u^2 + 2as \rightarrow v = \sqrt{2gs} = \sqrt{2 \times 9.81\,m\,s^{-2} \times 18.0\,m}$
$= 18.8\,m\,s^{-1}$ [1] (allow ecf from **Answer 13–1**),
$v = u + at \rightarrow t = \dfrac{v}{a} = \dfrac{18.8\,m\,s^{-1}}{9.81\,m\,s^{-2}} = 1.92\,s$ [1]

13–5 Total time in the air = 1.9 s + 1.98 s = 3.88 s [1],
$s = vt = 18.8\,m\,s^{-1} \times 3.88\,s \approx 73\,m$ [1]

13–6 If the angle changed, the distance travelled by the ball horizontally would decrease [1].

14–1
- Crumple zones in the car frame reduce the impact force by increasing the time required by the car to stop [1].
- Airbags increase the time to stop the body of the driver and passengers, hence increasing the impulse [1].
- The seatbelts also reduce the force on the bodies by allowing them to stop in the same time it takes the crumple zones in the front of the car to buckle, hence reducing the force on the bodies [1].

14–2 $v = 83\,km\,h^{-1} = 23\,m\,s^{-1}$, $F = \dfrac{\Delta mv}{\Delta t} \rightarrow \Delta t = \dfrac{\Delta mv}{F}$ [1]
$$= \frac{4.5\,kg \times (23\,m\,s^{-1} - 0\,m\,s^{-1})}{130\,N} = 0.8\,s \ [1]$$

14–3 That the deceleration is uniform/constant [1]

15–1 A and C will topple over at the same time and before B [1]. This is because the centre of mass of A and C is in the geometrical centre of the carton [1] (i.e. the middle of the carton), while the centre of mass of B is lower, because most of its mass is concentrated below the middle of the carton [1].

15–2 $M_F = F \times$ perpendicular distance from pivot, perp distance = 0.270 m, $M_F = 0.83\,N \times 0.270\,m = 0.224\,N\,m$ [1]

15–3 $1.2\,\text{N m} - 0.224\,\text{N m} = 0.976\,\text{N m}$ (*this is the moment of the perpendicular component of the weight about the contact point of the carton, w_p*) (*Allow ecf from Answer 15–2*)
$w_p = 0.976\,\text{N m}/0.15\,\text{m} = 6.5\,\text{N}$ [1]
$w = w_p/\cos57° = 11.9\,\text{N}$ [1], so $m_A = 1.22\,\text{kg}$ [1]

15–4 A will return to its upright position, because the centre of mass has not passed the point of contact between the table and carton yet [1].

16–1 Couple [1]

16–2 Diameter = 2.60 m, $M = 56\,\text{N} \times 2.60\,\text{m} = 145.6\,\text{N m}$ [1]

16–3 Distance between forces is the radius now, $M = 28\,\text{N} \times 1.3\,\text{m} = 36.4\,\text{N m}$ [1]

16–4 The second force of the couple is applied at the centre of the roundabout (pivot) [1], it will be in the opposite direction to the girl's push and it will have the same magnitude of 28 N [1].

Materials (p.23)

Quick questions

1 The load above which a material is deformed permanently.

2 The density of the oil tanker is much lower than the solid steel sphere and the mass/volume ratio is lower than that of water for the oil tanker, but larger for the steel sphere.

3 By finding the area under the force–extension graph for that material.

4 energy stored $= \frac{1}{2}F\Delta L \rightarrow F = k\Delta L \rightarrow$
energy stored $= \frac{1}{2}k\Delta L \times \Delta L = \frac{1}{2}k\Delta L^2$

5 **C:** two springs in parallel

6 $E = \dfrac{stress}{strain} = \dfrac{F}{A} \times \dfrac{L}{\Delta L} \rightarrow \dfrac{N}{m^2} \times \dfrac{m}{m} = N\,m^{-2}$

Exam-style questions

7–1 The extension read by the students will 'only' be the projection of the actual extension on the plane of the table, so the actual extension $\Delta L_a = \Delta L / \cos2°$ [1]
Convert 2.3 mm to 2.3×10^{-3} m
$\Delta L_a = 2.3 \times 10^{-3}\,\text{m} / \cos2° = 0.0023014\,\text{m}$ [1]

7–2 Absolute uncertainty $= \dfrac{range}{2} = \dfrac{1.4 \times 10^{-6}\,\text{m}}{2}$
$= 0.7 \times 10^{-6}\,\text{m}$ [1],
% uncertainty $= \dfrac{abs\ unc}{value} \times 100\%$
$= \dfrac{0.7 \times 10^{-6}\,\text{m}}{2.3 \times 10^{-3}\,\text{m}} \times 100\% = 0.03\%$ [1]

7–3 This effect should be ignored, because the percentage uncertainty from the tilting of the wire is too small and can be neglected [1].

7–4 The force on the wire is $2.68\,\text{kg} \times 9.81\,\text{N kg}^{-1} = 26.29\,\text{N}$ [1]
Rearrange $F = k\Delta L \rightarrow \Delta L = \dfrac{F}{k} = \dfrac{26.26\,\text{N}}{6125\,\text{N m}^{-1}}$:
$= 4.3 \times 10^{-3}\,\text{m}$ [1]

8–1 Convert 3.2 cm = 0.032 m, 14.8 g = 0.0148 kg,
985.5 g = 0.9855 kg
Difference in mass = 0.9855 kg − 0.0148 kg = 0.907 kg
Force $= 0.907\,\text{kg} \times 9.81\,\text{N kg}^{-1} = 9.52\,\text{N}$ [1]
Rearrange $F = k\Delta L \rightarrow k = \dfrac{F}{\Delta L} = \dfrac{9.52\,\text{N}}{0.032\,\text{m}} = 298\,\text{N m}^{-1}$ [1]

8–2 Energy stored $= \frac{1}{2}k\Delta L^2 = 0.5 \times 298\,\text{N m}^{-1} \times 0.032^2\,\text{m}^2$
$= 0.153\,\text{J}$ [1]

8–3 Convert 62 cm = 0.62 m, $\Delta E_p = mgh$
$= 0.0148\,\text{kg} \times 9.81\,\text{N kg}^{-1} \times 0.62\,\text{m} = 0.090\,\text{J}$ [1],
Efficiency $= \dfrac{\Delta E_p}{energy\ stored} = \dfrac{0.090\,\text{J}}{0.153\,\text{J}} = 0.59$ [1] or 59%

8–4 The compression is greater, so there is more energy stored [1], hence the spring toy will jump higher [1].

9–1 $T = 34\,\text{N} / \cos80° = 196\,\text{N}$ [1]

9–2 Radius = 0.5 mm = 5.0×10^{-4} m, $A = \pi r^2 = 7.9 \times 10^{-7}\,\text{m}^2$ [1],
$\sigma = \dfrac{F}{A} = \dfrac{196\,\text{N}}{7.9 \times 10^{-7}\,\text{m}^2} \approx 2.5 \times 10^8\,\text{N m}^{-2}$ [1]

9–3 $210\,\text{GPa} = 2.1 \times 10^{11}\,\text{N m}^{-2}$,
Rearrange $E = \sigma\dfrac{L}{\Delta L} \rightarrow \Delta L = \dfrac{\sigma L}{E}$ [1]
$= \dfrac{2.5 \times 10^8\,\text{N m}^{-2} \times 0.24\,\text{m}}{2.1 \times 10^{11}\,\text{N m}^{-2}} = 2.9 \times 10^{-4}\,\text{m}$ [1]

Topic review: mechanics and materials (p.25)

1–1 Angle ACB = $180° - 65° - 25° = 90°$
Line BC = AB $\times \cos25° = 0.52\,\text{m} \times \cos25° = 0.47\,\text{m}$ [1]
Angle BCP $= \cos^{-1}\dfrac{CP}{BC} = \cos^{-1}\dfrac{0.36\,\text{m}}{0.47\,\text{m}} = 40°$
$\theta = 90° - 40° = 50°$ [1]

1–2 $T = W = mg = 0.320\,\text{kg} \times 9.81\,\text{N kg}^{-1} = 3.14\,\text{N}$ [1]
$T_{BC} = mg \times \cos\theta = 3.14\,\text{N} \times \cos50° = 2.02\,\text{N}$ [1]
$T_{AC} = mg \times \cos(90° - \theta) = 3.14\,\text{N} \times \cos40° = 1.55\,\text{N}$ [1]

1–3 $F = ma \rightarrow 1.55\,\text{N} = 0.320\,\text{kg} \times a \rightarrow$
$a = \dfrac{1.55\,\text{N}}{0.320\,\text{kg}} = 4.84\,\text{m s}^{-1}$ [1]

1–4 BP $= \sqrt{BC^2 - CP^2} = \sqrt{0.47^2 - 0.36^2}\,\text{m} = 0.30\,\text{m}$ (*Allow ecf from calculation of BC in Answer 1–1*)
$\Delta h = \text{BC} - \text{BP} = 0.47\,\text{m} - 0.30\,\text{m} = 0.17\,\text{m}$ [1]
$\Delta E_p = mg\Delta h = E_k$ (*kinetic energy at lowest point*)
$mg\Delta h = \frac{1}{2}mv^2 \rightarrow v = \sqrt{2g\Delta h}$
$= \sqrt{2 \times 9.82\,\text{N kg}^{-1} \times 0.17\,\text{m}} = 1.83\,\text{m s}^{-1}$ [1]

1–5 The ball will start falling in a parabolic motion [1].

1–6 Vertical motion $\rightarrow s = ut + \dfrac{at^2}{2} = \dfrac{at^2}{2} \rightarrow 0.85\,\text{m}$
$= \dfrac{9.82\,\text{m s}^{-2}t^2}{2} \rightarrow t = \sqrt{\dfrac{2 \times 0.85\,\text{m}}{9.82\,\text{m s}^{-2}}} = 0.42\,\text{s}$ [1];
horizontal motion is at constant velocity $\rightarrow s = vt$
$= 1.83\,\text{m s}^{-1} \times 0.42\,\text{s} = 0.77\,\text{m}$ [1]

2–1 Force on arrow $= 2 \times T\cos55° = 2 \times 400\,\text{N} \times \cos55°$
$= 459\,\text{N}$ [1]

2–2 The arrow is accelerated until the string loses contact with the arrow [1], at that point (0.016 s) the resultant force is zero and the arrow travels at constant velocity [1].

2–3 Large triangle used [1]; gradient
$= a = \dfrac{\Delta y}{\Delta x} = \dfrac{(74 - 30)\,\text{m s}^{-1}}{(0.014 - 0.009)\,\text{s}} = 6285\,\text{m s}^{-2}$ [1]

2–4 $F = ma = 0.060\,\text{kg} \times 6285\,\text{m s}^{-2} = 377\,\text{N}$
(*Accept ecf from Answer 2–1*) [1]

2–5 As the bow relaxes, the tension of the string decreases, so the force on the arrow is not constant [1].

2–6 Calculate the average force $= \dfrac{377\,\text{N} + 459\,\text{N}}{2} = 418\,\text{N}$ [1]
Impulse $= F \times t = 418\,\text{N} \times 0.016\,\text{s} = 6.7\,\text{kg m s}^{-1}$ [1]

2–7 Impulse from graph $= m\Delta v = 0.060\,\text{kg} \times (82.5 - 0)\,\text{m s}^{-1}$
$= 5.0\,\text{kg m s}^{-1}$ [1] $\rightarrow \dfrac{5.0\,\text{kg m s}^{-1}}{6.6\,\text{kg m s}^{-1}} = 0.75 \rightarrow$ there is a
discrepancy of 25% between the two estimated values of the impulse [1]

3–1 $T_v = T\cos20° = 50.7\,\text{N}$ [1];
$F_v = T_v - W = 50.7\,\text{N} - 0.250\,\text{kg} \times 9.81\,\text{N kg}^{-1} = 48.3\,\text{N}$ [1];
$T_h = F_h = T\sin20° = 18.5\,\text{N}$ [1];
$F = \sqrt{F_h^2 + F_v^2} = 51.7\,\text{N}$ [1]

3–2 Rearrange $s = ut + \dfrac{1}{2}at^2 \rightarrow t^2 = \dfrac{2s}{g}$ [1] $\rightarrow t = 3.0\,\text{s}$ [1]

3–3 $s = vt = 8\,\text{m s}^{-1} \times 3.0\,\text{s} = 24.2\,\text{m}$ [1] (*ecf on incorrect value of t*)

Current electricity (p.27)

Quick questions

1 If the resistance of an ammeter were significant, it would affect/reduce the amount of current flowing in the circuit, hence give an incorrect reading of the current.

2 p.d. is the work done per unit charge (coulomb) through the components in a circuit, while e.m.f. is the work done per coulomb as it flows through a source of power for the circuit (e.g. a cell).

3 Electrical resistance drops to zero below the critical temperature, so a current set up in a superconductor will flow indefinitely. They also exclude magnetic fields.

4 **C**: 8 : 1 : 4

5 The shape of the curve is the same, but on the left-hand side the current is less negative than in Figure 2, and on the right-hand side it is more negative.

6 $P = I^2R = \dfrac{V^2}{R}$

Exam-style questions

7–1 Incorrect parts: Statement 1: Each socket is connected in series with the cable [1]
Statement 2: the total resistance of the block of sockets increases [1]
Statement 3: The potential difference is shared across each appliance [1]

7–2 *Any four from:*
- Each socket connected in parallel with the mains [1].
- Each device plugged in draws a current $I = V/R$ [1].
- Current through each device adds up at the cable [1].
- For the same potential, a greater and greater current builds up as more appliances are connected, which increases the power $P = IV$ [1].
- If too much current is drawn, there could be a power surge [1].
- The safety mechanism opens the circuit if too much current is drawn, preventing cable overheating and causing electrical fire [1].

7–3 Use $I = V/R$ [1] to calculate current drawn; add currents from different appliances [1]; correct selection Heater + Drill + Radio + Grinder [1] $\rightarrow \left(\dfrac{230}{15} + \dfrac{230}{115} + \dfrac{230}{50} + \dfrac{230}{44} \right) \dfrac{V}{\Omega}$ [1]

7–4 230 V/27.2 A = 6247 W [1] (*or any correctly calculated power from incorrect selection of four appliances*)

7–5 Resistance of each appliance adds up 115 Ω + 44 Ω + 50 Ω = 209 Ω [1],
$I = V/R = 230\,V/209\,\Omega = 1.1\,A$ [1],
p.d. across each resistance = current × resistance [1],
$V(drill) = 1.1\,A \times 115\,\Omega = 126.5\,V,$
$V(grinder) = 1.1\,A \times 44\,\Omega = 48.4\,V,$
$V(radio) = 1.1\,A \times 50\,\Omega = 55\,V$ [1]

7–6 The current through each appliance is too low to operate the devices correctly (e.g. cause the motor of the grinder to spin) [1].

8–1 Correct potential divider circuit drawn [1] with voltmeter/ voltage output open circuit across the fixed resistor [1], correct circuit symbol for LDR [1], correct values for p.d. and fixed resistor, e.g. 5 V and 340 kΩ [1]

8–2 Resistance for fixed resistor in the range of 30 to 200 kΩ [1].
Any three from:
- The range of resistance of an LDR for 300 to 500 lux is 60 to 200 kΩ [1].
- If the resistance of a fixed resistor is too high, not much p.d. change will be measured for change of light intensity at that range [1].

- A resistance of 60 to 200 kΩ is comparable to the resistance of the LDR for those light intensities [1].
- So, p.d. is shared more evenly between the two components [1].
- Leading to clearer readings/intervals of p.d. for light intensity changes [1].

9–1 No current flows through R_5 [1], because the p.d. across it is zero [1]. In fact, R_2 is half R_1 and R_4 is half R_3 [1] and the p.d. drop across R_2 is the same as the p.d. drop across R_4 [1].

9–2 We can ignore R_5 because the p.d. across it is zero
Add $R_1 + R_2 = 225\,\Omega$ and $R_3 + R_4 = 300\,\Omega$ [1]
$\dfrac{1}{R} = \dfrac{1}{R_1 + R_2} + \dfrac{1}{R_3 + R_4} = \dfrac{1}{225\,\Omega} + \dfrac{1}{300\,\Omega} \approx 0.008\,\Omega$ [1],

$R = \dfrac{1}{0.008\,\Omega} = 129\,\Omega$ [1]

9–3 The current drawn by R_1 and R_2 is
$I = \dfrac{V}{R_1 + R_2} = \dfrac{9V}{225\,\Omega} = 0.04\,A$ [1],
$V_{R_1} = IR_1 = 0.04\,A \times 150\,\Omega = 6\,V$ [1]

10–1 The p.d. read will remain 9 V [1], because the resistance of the digital voltmeter is extremely high [1] and the variable resistor is connected in series with the voltmeter [1], so the p.d. across the voltmeter will always be almost 9 V and the p.d. across the resistor will be negligible [1].

10–2 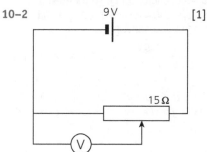 [1]

10–3 5 Ω [1], because 3 V is $\dfrac{1}{3}$ of 9 V (i.e. e.m.f.) [1], so a p.d. drop of $\dfrac{1}{3}$ will be noticed when the resistance to the left of the sliding contact is $\dfrac{1}{3}$ of the resistance to its right [1].

10–4 $I = $ e.m.f./$R_{tot} = 9\,V/15\,\Omega$ [1] = 0.6 A, $V_{Right} = $ e.m.f. $- V_{Left}$ = 9 V−7.3 V = 1.7 V [1], so $R_{Right} = V_{Right}/I = 1.7\,V/0.6\,A$ = 2.8 Ω [1]

11–1
- Use a thermometer, beaker and ice in water, Bunsen burner, clamps and stands, circuit set up as in diagram [1].
- Measure the temperature of the water with the thermometer [1].
- Place the thermistor in the water and ice [1].
- Warm the water up and record temperature changes [1].
- Record corresponding p.d. across R_2 for each temperature [1].
- Plot a graph of temperature vs p.d. across R_2 [1].

11–2 If $R_{T1} = R_2$ the initial p.d. across R_2 will be 4.5 V [1] (shared equally between the two components), and as the thermistor cools down, its resistance will increase [1], meaning that more and more p.d. will drop across the thermistor than R_2, so p.d. across R_2 will decrease [1].

11–3 Convert to $1.5 \times 10^5\,\Omega$ and $2.2 \times 10^5\,\Omega$
$R_{tot} = (1.5 + 2.2) \times 10^5\,\Omega = 3.7 \times 10^5\,\Omega$ [1],
$I = V/R_{tot} = 9\,V/(3.7 \times 10^5\,\Omega) = 2.4 \times 10^{-5}\,A$ [1]
$V_{R_2} = IR_2 = 2.4 \times 10^{-5}\,A \times 2.2 \times 10^5\,\Omega = 5.4\,V$ [1]

11–4 The analogue voltmeter now has a resistance that is comparable to the resistance of R_2, so it needs to be taken into account in the calculations. The analogue voltmeter and R_2 are two resistors in parallel, so their combined resistance

157

$$R_C = \left(\frac{1}{R_2} + \frac{1}{2.0 \times 10^5 \, \Omega}\right)^{-1} = \left(\frac{1}{2.2 \times 10^5 \, \Omega} + \frac{1}{2.0 \times 10^5 \, \Omega}\right)^{-1}$$

$$= 104\,762\,\Omega \; [1]$$

The current drawn by the circuit will be

$$I = \frac{9\,V}{(104\,762\,\Omega + 150\,000\,\Omega)} = 3.5 \times 10^{-5}\,A \; [1],$$

$$V = IR_C = 3.5 \times 10^{-5}\,A \times 104\,762\,\Omega = 3.7\,V \; [1]$$

11-5 The analogue voltmeter has a comparable resistance to R_2 [1], so it will affect the p.d. across R_2 [1]. In fact, the p.d. across R_2 with the digital voltmeter is higher than with the analogue voltmeter [1].

12-1 The superconductor's resistivity becomes zero [1] and it excludes magnetic fields inside it.

12-2 Production of strong magnetic fields [1] and reduction of energy loss in transmission of electric power [1].

12-3 $63\,mA = 0.063\,A$ [1], $I_1 = \dfrac{V}{R_s + R_1} = \dfrac{6\,V}{415\,\Omega + 45\,\Omega}$

$$= 0.013\,A \; [1],$$

$I_2 = I_{tot} - I_1 = 0.050\,A$ [1], $R_2 = V/I_2 = 6\,V/0.050\,A = 120\,\Omega$ [1]

12-4 The boiling point of liquid nitrogen is below the critical temperature of superconductor S [1], so its resistance will be zero [1], so $I_1 = 6\,V/45\,\Omega = 0.130\,A$ [1] and $I_{tot} = 0.130\,A + 0.050\,A = 0.180\,A$ [1].

13-1 $I = \dfrac{2 \times e.m.f.}{R + 2r}[1] = \dfrac{2 \times 1.5\,V}{55\,\Omega + 2 \times 1.2\,\Omega} = 0.052\,A$ (2 s.f.) [1]

13-2 $V = 2\varepsilon - 2rI$ [1] $= 3\,V - 2 \times 1.2\,\Omega \times 0.052\,A = 2.87\,V$ [1]
(*Alternative answer: p.d. = current × resistance. Using the rounded figure 0.052 A for the current, this will give 2.86 V.*)

13-3 $2.87\,V/2 = 1.44\,V$ [2]

14-1

Correct scale for V and I and appropriately labelled with units [1], all plots correctly drawn [2 – *deduct 1 mark for each incorrect plot*], straight line of best fit with negative gradient [1] extended to intersect V axis [1]

14-2 $V = -rI + \varepsilon$ [1]

14-3 Gradient $= -r = \Delta V/\Delta I$ [1], with large ΔI and ΔV chosen from line of best fit and NOT plots [1], indicative value of r = 0.8 found as $r = -$ gradient [1]

14-4 Extrapolate from V intercept [1], ε about 1.55 V [1]

15-1 0.29 mm [1]

15-2 Calculate mean diameter

$$= \frac{0.27 + 0.30 + 0.29 + 0.30 + 0.32 + 0.26}{6}\,mm$$

$$= 0.29\,mm \; [1]$$

Convert to $2.9 \times 10^{-4}\,m$ [1], $A = \pi r^2$

$$= \pi\left(\frac{2.9 \times 10^{-4}\,m}{2}\right)^2 = 6.6 \times 10^{-8}\,m^2 \; [1]$$

15-3 Mean diameter becomes $2.7 \times 10^{-4}\,m$ [1], $A = \pi r^2$

$$= \pi\left(\frac{2.7 \times 10^{-4}\,m}{2}\right)^2 = 5.7 \times 10^{-8}\,m^2 \; [1]$$

15-4 $\rho = \dfrac{RA}{L}$ [1] $= \dfrac{4.0\,\Omega \times 5.7 \times 10^{-8}\,m^2}{0.500\,m}$ [1]

$$= 4.6 \times 10^{-7}\,\Omega m \qquad [1]$$

15-5 % uncertainty on length $= 0.001\,m/0.500\,m = 0.2\%$ [1]
% uncertainty on resistance $= 0.1\,\Omega/4.0\,\Omega = 2.5\%$ [1]
Absolute uncertainty of mean diameter =

$$\frac{range}{2} = \frac{3.0 - 2.4}{2} \times 10^{-4}\,m = \pm 3 \times 10^{-5}\,m$$

% uncertainty on diameter =

$$\frac{abs\ uncertainty}{mean\ value} = \frac{3 \times 10^{-5}\,m}{2.7 \times 10^{-4}\,m} \times 100\% = 11.1\% \; [1]$$

Total % uncertainty $= 0.2\% + 2.5\% + 2 \times 11.1\% = 24.9\%$

Absolute uncertainty $= 4.6 \times 10^{-7}\,\Omega m \times \dfrac{24.9}{100}$

$$= \pm 1.1 \times 10^{-7}\,\Omega m \; (2 \text{ s.f.}) \; [1]$$

15-6 It is acceptable [1], because the accepted value of $4.9 \times 10^{-7}\,\Omega m$ is within the range of confidence of the value obtained using the students' data [1].

Topic review: electricity (p.34)

1-1 Take, for example, $6\,V \rightarrow eff(2.5\,W) = \dfrac{IV}{2.5\,W}$

$$\frac{IV}{2.5\,W} = \frac{0.38\,A \times 6\,V}{2.5\,W} = \frac{2.28}{2.5} = 0.912 \; [1]$$

$$eff(1.0\,W) = \frac{IV}{1.0\,W} = \frac{0.15\,A \times 6\,V}{1.0\,W} = 0.90 \; [1]$$

so the 2.5 W bulb is more efficient [1].

1-2 The resistance increases as the p.d. increases [1] by smaller and smaller increments [1].

1-3 The 1.0 W bulb has the higher resistance [1] as the current is lower at the same p.d. [1]

1-4 The resistance of the LDR for 1000 lux is $100\,\Omega$ [1]
So $I(100\,\Omega + R_L) = 16\,V$ [1]; but $R_L = V_L/I$ [1]
$\rightarrow I \times 100\,\Omega + V_L = 16\,V$ [1]
The values from the I–V graph of the light bulb
for $V_L = 1\,V$ gives $I = 0.15\,A \rightarrow$

$0.15\,A \times 100\,\Omega + 1\,V = 16\,V \rightarrow 15\,V + 1\,V = 16\,V$ [1]

2-1 Circuit diagram with play-dough shape, a variable resistor, ammeter and cell in series [1] and voltmeter in parallel with the shape of play-dough [1].
Any four indicative content from:
- Calculate mean cross-sectional area [1].
- Vary the p.d. across the play-dough by changing the resistance of the variable resistor [1].
- Record values of V and I [1].
- Plot an I–V graph and calculate gradient [1].
- $R = 1/\text{gradient}$ [1]

2-2 $h = L = s \times \sin 78° = 0.245\,m$ [1]
Convert 0.8 cm to 0.008 m and 0.6 cm to 0.006 m [1]

$$\text{mean cross-sectional area} = \pi\left(\frac{0.008\,m + 0.006\,m}{2}\right)^2$$

$$= 1.13 \times 10^{-4}\,m^2 \; [1]$$

Rearrange $\rho = \dfrac{RA}{L} \rightarrow R = \dfrac{\rho L}{A}$ [1]

$$R = \frac{0.24\,\Omega m \times 0.245\,m}{1.13 \times 10^{-4}\,m^2} = 520\,\Omega \; [1]$$

2-3 $\varepsilon = I(R + r) \rightarrow r = \dfrac{\varepsilon - IR}{I}$ [1];

$$= \frac{4.5\,V - 8.6 \times 10^{-3}\,A \times 520\,\Omega}{8.6 \times 10^{-3}\,A} = 3.3\,\Omega \qquad [1]$$

3–1 $V_{tot} = 10.5\,V - 6\,V = 4.5\,V$ [1]; $R_{tot} = V/I = 4.5\,V/0.220\,A$
     $= 20.45\,\Omega$ [1];

$$R_{tot} = 2.5 + 2 + \left(\frac{1}{25} + \frac{1}{50}\right)^{-1} + \left(\frac{1}{150 + 45} + \frac{1}{R}\right)^{-1} \text{[1]}$$

$$\rightarrow \quad 20.45 = 4.5 + 4.17 + \frac{195R}{R + 195} \text{ [1]}$$

$$\rightarrow \quad 11.78R + 2297.1 = 195R \rightarrow \quad 183.22R = 2297.1$$

$$\rightarrow \quad R = 12.54\,\Omega \text{ [1]}$$

3–2   The current would decrease [1], because R is in parallel with two other resistors, so by removing R, the resistance of that block would increase [1], increasing the total resistance of the circuit [1].

Periodic motion (p.36)

Quick questions

1   $2\pi\,\text{rad} = 360°$, so $1\,\text{rad} = 360°/2\pi$ and $2\,\text{rad}$

    $= 2 \times \dfrac{360°}{2\pi} = 114.6°$

2   A mass-spring system period is $T = 2\pi\sqrt{\dfrac{m}{k}}$ and a simple

    pendulum period is $T = 2\pi\sqrt{\dfrac{l}{g}}$ so the time period will

    increase for the pendulum, because it is dependent on $\dfrac{1}{g}$,

    but it will remain the same for the spring system, because it is not dependent on g.

3   The amplitude will decrease, but the frequency will remain unchanged.

4   **A:** The pendulum will have maximum E_k and zero ΔE_p.

5   The driving oscillations frequency must match the natural frequency of the oscillating system.

6   The statement is incorrect, because the rest position of the pendulum is the centre of oscillation, if it is allowed to stop, so the acceleration must be zero at that point. Whereas, the acceleration is maximum at the extremes of oscillation because that is where the pendulum is decelerated to a halt and changes direction.

Exam-style questions

7–1 $33\frac{1}{3} \times 2\pi/60 = 3.5\,\text{rad s}^{-1}$ [2]

7–2   Radius $= 0.30\,\text{m}/2 = 0.15\,\text{m}$, $v = \omega \times r = 3.5\,\text{rad s}^{-1} \times 0.15\,\text{m}$
     $= 0.53\,\text{m s}^{-1}$ [1]

7–3   The tangential velocity will be higher on the circumference because that spot will travel a longer distance in the same time [1]. The angular speed will be the same for both points because they both describe the same angle in the same time [1].

8–1   Arrows F, F_c and N added, all pointing from the point where the tyre is in contact with the road surface [1]: N equal and opposite to W, pointing upwards, F_c pointing along the ground to the left, F pointing diagonally upwards to the left, through the centre of mass of the motorbike [1].

8–2 $T = 3.5\,\text{s} \rightarrow \omega = \dfrac{2\pi}{3.5\,\text{s}} = 1.8\,\text{s}^{-1}$ [1]

    $\omega = \dfrac{v}{r} \rightarrow r = \dfrac{v}{\omega} = \dfrac{25\,\text{m s}^{-1}}{1.8\,\text{s}^{-1}} = 13.9\,\text{m}$ [1]

8–3 $a = \dfrac{v^2}{r} = \dfrac{25^2\,\text{m}^2\,\text{s}^{-2}}{13.9\,\text{m}}$ [1] $= 45\,\text{m s}^{-2}$

    (*Accept ecf from Answer 8–2*),
    $m = 2651\,\text{N}/9.81\,\text{N kg}^{-1} = 270\,\text{kg}$
    $F_c = ma = 270\,\text{kg} \times 45\,\text{m s}^{-2} = 12\,150\,\text{N}$ [1]

8–4   The friction between the tyres and the road surface [1].

8–5 $\tan\theta = \dfrac{N}{F_c} = \dfrac{W}{F_c} = \dfrac{2651\,\text{N}}{12\,150\,\text{N}} = 0.22$ [1]

    $\theta = \text{atan}\theta = 12.3°$ [1]

9–1   Angle between weight and radius $AF = 30°$
    Weight $W = 0.430\,\text{kg} \times 9.82\,\text{N kg}^{-1} = 4.2\,\text{N}$
    Perpendicular component of weight $W_p = W\cos30° = 3.7\,\text{N}$ [1]
    $F_c = N + W_p = 23.9\,\text{N} + 3.7\,\text{N} = 27.6\,\text{N}$ [1]
    $a = F_c/m = 27.6\,\text{N}/0.430\,\text{kg} = 64\,\text{m s}^{-2}$ [1]

9–2 $a = \dfrac{v^2}{r} \rightarrow v^2 = a \times r$, $v = \sqrt{a \times r}$ [1] $= \sqrt{64\,\text{m s}^{-2} \times 0.95\,\text{m}}$

    $= 7.8\,\text{m s}^{-1}$ [1] (*allow ecf on value of a from Answer 9–1*)

9–3 $\dfrac{v}{r} = 2\pi f \rightarrow f = \dfrac{v}{2\pi r} = 1.3\,\text{s}^{-1}$ [1]

    $T = 1/f = 1/1.3\,\text{s}^{-1} = 0.77\,\text{s}$ [1]

9–4   At C, draw W and N pointing vertically down and F_c as sum of $W + N$ [1]. At D, draw W pointing vertically down, N vertically up and longer than W, with F_c as vector sum of $N + W$, i.e. $N - W$ [1]. At E, draw W pointing vertically down and $N = F_c$ pointing horizontally to the centre of the circle [1].

10–1   The mass of an object is not relevant to the period of the oscillation of a pendulum [1], so the two swings will oscillate in phase, because they started at the same displacement [1].

10–2   For large angles from the rest position, the pendulum will stop oscillating in a simple harmonic motion, so the two swings will no longer oscillate in phase [1].

10–3 $T = 2\pi\sqrt{\dfrac{l}{g}} = 2\pi\sqrt{\dfrac{2.4\,\text{m}}{9.81\,\text{m s}^{-2}}}$

    $= 3.1\,\text{s}$ [1] $\rightarrow f = \dfrac{1}{T} = 0.32\,\text{s}^{-1}$ [1]

10–4   She would need to shorten the length of the chain, by

    $T^2 = 4\pi^2\dfrac{l}{g}$ [1] $\rightarrow l = \dfrac{gT^2}{4\pi^2} = \dfrac{9.81\,\text{m s}^{-2} \times 2.5^2\,\text{s}^2}{4\pi^2}$

    $= 1.55\,\text{m}$ [1]

11–1   Calculate the mean period

    $T_{mean} = \dfrac{0.32 + 0.29 + 0.31 + 0.32 + 0.33}{5}\,\text{s} = 0.31\,\text{s}$ [1],

    $a_{max} = \omega^2 A = \left(\dfrac{2\pi}{T_{mean}}\right)^2 A = \dfrac{4\pi^2}{0.31^2\,\text{s}^2} \times 0.025\,\text{m}$

    $= 10.3\,\text{m s}^{-2}$ [1]

11–2   The damping will force the amplitude of the oscillations to reduce at every cycle [1], so the vibration of the suspension will reduce more quickly than if it were left without damping [1]. The maximum acceleration depends on the maximum displacement A [1], so for damped oscillations the maximum displacement is reduced at every oscillation [1].

11–3   After three oscillations, the maximum displacement A will be $0.025\,\text{m} \times 0.915 = 0.023\,\text{m} \rightarrow 0.023\,\text{m} \times 0.915 = 0.021\,\text{m} \rightarrow 0.021\,\text{m} \times 0.915 = 0.019\,\text{m}$ [1],

    $v_{max} = \omega A = \dfrac{2\pi}{T_{mean}} A = 0.39\,\text{m s}^{-1}$ [1]

    (*Allow ecf on value of T_{mean} from Answer 11–1*)

11–4   Damping the oscillations reduces the amplitude of each oscillation and brings the front fork to a rest in a short time [1]; the damping also ensures that bumps in the road and other obstacles (like potholes) do not drive the oscillations of the spring at its natural frequency to avoid resonance [1], which could cause the vibrations of the fork to become too violent [1] and cause the rider to fall.

12–1 v–t graph: correct shape starting from origin and with negative gradient [1], trough at 0.1 s and peak at 0.3 s with $v = 0$ at 0.2 s and 0.4 s, amplitude of peak and trough

    at $\pm 2\pi fA = \pm\dfrac{2\pi}{0.4\,\text{s}} \times 0.1\,\text{m} = \pm 1.6\,\text{m s}^{-1}$ [1]

    a–t graph: correct shape starting from $-(2\pi f)^2A$, trough and peaks in the same position as in **Figure 4** [1], amplitude of peaks and trough at $\pm(2\pi f)^2A$

$$= \pm \frac{4\pi^2}{0.4^2\,\text{s}^2} \times 0.1\,\text{m} = \pm 24.7\,\text{ms}^{-2}\ [1]$$

12–2 Combined spring constant for springs in parallel = $2k$ (with k as the spring constant for one spring), force of spring $F = -2kx = ma$ [1], considering extension = maximum displacement $\rightarrow -2kA = ma_{max} = -m(2\pi f)^2$

$$A\ [1] \rightarrow k = \frac{2\pi^2}{0.4^2\,\text{s}^2} \times 0.850\,\text{kg}\ [2] = 105\,\text{kgs}^{-2}$$

(*1 mark for rearranging equation and 1 mark for converting to 0.850 kg.*)

13–1 The starting position is the centre of oscillation, because the graph starts with $E_k = \text{max}$ [1].

13–2 $T = 3.3\,\text{s} \rightarrow f = 1/T = 1/3.3\,\text{s} = 0.3\,\text{s}^{-1}$ [1]

13–3 $A = 4.8\,\text{cm} = 0.048\,\text{m}$, total energy $\approx 3.8\,\text{J}\ \frac{1}{2}kA^2$ [1] $\rightarrow$

$$k = \frac{2 \times total\ energy}{A^2} = \frac{2 \times 3.8\,\text{J}}{0.048^2\,\text{m}^2} = 3299\,\text{Jm}^{-2}\ [1]$$

13–4 At the centre of oscillation, $E_k = $ total energy, and $v_{max} = \omega A = 2\pi f A = 2\pi \times 0.3\,\text{s}^{-1} \times 0.048\,\text{m}$ [1]
$= 0.09\,\text{m s}^{-1}$ (*Allow ecf on value of f*)

$$\frac{1}{2}m(v_{max})^2 = 3.8\,\text{J} \rightarrow m = \frac{2 \times 3.8\,\text{J}}{0.09^2\,\text{m}^2\,\text{s}^{-2}}$$

$$= 938\frac{\text{kgm}^{-2}\,\text{m}}{\text{m}^2\,\text{s}^{-2}} = 938\,\text{kg}\ [1]$$

13–5 E_p intersects E_k when $E_p = E_k \rightarrow \frac{1}{2}mv^2 = \frac{1}{2}kx^2$ [1]

but $v = \pm\omega\sqrt{A^2 - x^2} \rightarrow m 4\pi^2 f^2 \left(A^2 - x^2\right)$ [1]

$$= kx^2\ [1] \rightarrow x^2 \left(k + 4m\pi^2 f^2\right) = 4m\pi^2 f^2 A^2 \rightarrow$$

$$x^2 = \frac{4m\pi^2 f^2 A^2}{k + 4m\pi^2 f^2}\ [1] \rightarrow x = \sqrt{\frac{4m\pi^2 f^2 A^2}{k + 4m\pi^2 f^2}}$$

$$= \sqrt{\frac{4\pi^2 \times 928\,\text{kg} \times 0.09\,\text{s}^{-2} \times 2.3 \times 10^{-3}\,\text{m}^2}{3299\,\text{Jm}^{-2} + 4\pi^2 \times 928\,\text{kg} \times 0.09\,\text{s}^{-2}}}$$

$$= 0.034\,m\ [1]$$

Thermal physics (p.39)

Quick questions

1 Energy transfer from a <u>hotter region</u> to a cooler region.

2 Internal energy is the energy that is stored in the vibration of molecules.

3 −273 °C

4 It is a gas that is:
   - monatomic (single atoms)
   - at low temperature
   - at low pressure.

5 You would see the smoke particles jiggling about <u>randomly</u>. This is due to the <u>changes of momentum</u> of the particles as air molecules <u>collide</u> with them. The air molecules are moving randomly. They bombard the particles randomly from different directions.

6 Avogadro's constant has the value 6.02×10^{23}.

7

Symbol	Meaning	Units
Q	Energy	J
m	Mass	kg
c	Specific heat capacity	J kg⁻¹ K⁻¹ or J kg⁻¹°C⁻¹ NOT J/kg/°C
$\Delta\theta$	Temperature <u>change</u>	K or °C

8

Symbol	Meaning	Units
p	Pressure	N m⁻² or Pa
V	Volume	m³
n	Number of moles	No units
R	Universal gas constant	J mol⁻¹ K⁻¹
T	Temperature	K (*NOT °C*)

9 $Q = mL$
$= 0.55\,\text{kg} \times 2.26 \times 10^6\,\text{J kg}^{-1}$
$= 1.2 \times 10^6\,\text{J}$ (2 s.f.)

10 Temperature change $= 45\,°\text{C} - 20\,°\text{C} = 25\,\text{K}$ (*°C acceptable*)
$\Delta Q = mc\Delta\theta = 0.25\,\text{kg} \times 1790\,\text{J kg}^{-1}\text{K}^{-1} \times 25\,\text{K}$
$\Delta Q = 11.2 \times 10^3\,\text{J} = 11\,\text{kJ}$ (*2 s.f. as the data is to 2 s.f.*)
Either J or kJ are acceptable, as long as the answer is consistent

11 Temperature change $= 100\,°\text{C} - 15\,°\text{C} = 85\,\text{K}$ (*or °C*)
Energy required to heat the water:
$\Delta Q = 0.25\,\text{kg} \times 4200\,\text{J kg}^{-1}\text{K}^{-1} \times 85\,\text{K} = 89\,250\,\text{J}$
Energy required to heat the cup:
$\Delta Q = 0.225\,\text{kg} \times 1000\,\text{J kg}^{-1}\text{K}^{-1} \times 85\,\text{K} = 19\,125\,\text{K}$
Total energy $= 89\,250\,\text{J} + 19\,125\,\text{J} = 108\,375\,\text{J}$

$$\text{Time taken} = \frac{energy}{power} = \frac{108\,375\,\text{J}}{800\,\text{W}} = 135\,\text{s}$$

(*= 2 minutes and 15 seconds*)
(*The answer should be 140 s to 2 s.f.*)

12 The same energy leaves the iron as enters the water.
$\Delta Q = 2.5\,\text{kg} \times 438\,\text{J kg}^{-1}\text{K}^{-1} \times -\Delta\theta_1$
$= 15\,\text{kg} \times 4200\,\text{J kg}^{-1}\text{K}^{-1} \times \Delta\theta_2$
Now the end temperature is T. It will be the same for the iron and the water.
For the iron:
$-\Delta\theta_1 = T - 600\,\text{K}$ (*it will be negative because it's a temperature drop*)
For the water:
$\Delta\theta_2 = T - 300\,\text{K}$ (*it will be positive because it's a temperature rise*)
Now substitute:
$2.5\,\text{kg} \times 438\,\text{J kg}^{-1}\text{K}^{-1} \times -(T - 600\,\text{K})$
$= 15\,\text{kg} \times 4200\,\text{J kg}^{-1}\text{K}^{-1} \times (T - 300\,\text{K})$
$1095\,\text{J K}^{-1} \times (-T + 600\,\text{K})$
$= 63\,000\,\text{J K}^{-1} \times (T - 300\,\text{K})$
$-1095\,T + 6.57 \times 10^5\,\text{J}$
$= 63\,000\,T - 1.89 \times 10^7\,\text{J}$
$-63\,000\,T - 1095\,T$
$= -1.89 \times 10^7\,\text{J} - 6.57 \times 10^5\,\text{J}$
$64\,095\,T$
$= 1.8243 \times 10^7\,\text{J}$ (*Minus signs cancel out*)
$T = 295\,\text{K}$ (3 s.f.)

13a) $\bar{c} = \dfrac{3.2 + 4.7 + 3.4 + 5.8 + 7.2 + 9.3 + 2.6}{7} = 5.17\ \text{m s}^{-1}$

b) $\bar{c}^2 = 26.3\ \text{m}^2\,\text{s}^{-2}$

c) $\overline{c^2} = \dfrac{3.2^2 + 4.7^2 + 3.4^2 + 5.8^2 + 7.2^2 + 9.3^2 + 2.6^2}{7}$
$= 31.80\ \text{m}^2\,\text{s}^{-2}$

d) $\sqrt{\left(\overline{c^2}\right)} = \sqrt{31.80} = 5.64\ \text{ms}^{-1}$

e) The difference is $0.47\ \text{ms}^{-1}$.

14 Formula: $p = \frac{1}{3}\rho \overline{c^2}$

Rearrangement: $\overline{c^2} = \frac{3p}{\rho}$

Substitution: $\overline{c^2} = \frac{3 \times 1.013 \times 10^5\,\text{Pa}}{1.23\,\text{kg m}^{-3}} = 2.47 \times 10^5\,\text{m}^2\,\text{s}^{-2}$

$\sqrt{\overline{c^2}} = 497\,\text{m s}^{-1}$

15 Pressure difference $= 1.0 \times 10^5\,\text{Pa} - 2.64 \times 10^4\,\text{Pa}$
$= 73\,600\,\text{Pa}$
Force = pressure × area $= 73\,600\,\text{Pa} \times 1.2\,\text{m}^2$
$= 88\,320\,\text{N} = 8.8 \times 10^4\,\text{N}$

16 Equation: $\frac{1}{2}m\overline{c^2} = \frac{3}{2}kT$

Rearrangement: $\overline{c^2} = \frac{3kT}{m}$

Substitute: $\overline{c^2} = \frac{3 \times 1.38 \times 10^{-23}\,\text{J K}^{-1} \times 500\,\text{K}}{4.8 \times 10^{-26}\,\text{kg}}$

$\overline{c^2} = 4.31 \times 10^5\,\text{m}^2\,\text{s}^{-2}$

$c_{\text{RMS}} = 660\,\text{m s}^{-1}$

Exam-style questions

17–1 Graph B, as $V \propto \dfrac{1}{p}$ [1]

17–2 Graph C as $p \propto T$ [1]

18 **B:** Xenon [1] (*It's a monatomic gas. All the others consist of two or more atoms.*)

19 Graph C (*Since $E \propto T$*) [1]

20 Graph B (*Since $T \propto c^2$*) [1]

21 **C:** It does not depend at all on the mass. It increases proportionally with Kelvin temperature. [1]

22–1 Latent heat of fusion is the energy required for a <u>unit mass</u> of a substance to <u>change state from solid to liquid</u> [1].
Latent heat of vaporisation is the energy required for a unit mass of a substance <u>to change state from liquid to gas</u> [1].

22–2 There is <u>zero temperature change</u> [1] in latent heat, while there is a <u>temperature change</u> [1] in specific heat capacity.

22–3 1. $\Delta Q = mL = 0.67\,\text{kg} \times 2.26 \times 10^6\,\text{J kg}^{-1} = 1.51 \times 10^6\,\text{J}$
$= 1.5 \times 10^6\,\text{J}$
Calculation must be shown for the mark. [1]

2. Time $= \dfrac{1.51 \times 10^6\,\text{J}}{2700\,\text{J s}^{-1}}$ [1]

Time $= 560\,\text{s}$ (*or 9 minutes and 20 seconds*) [1]
3. The temperature will rise, eventually the element will glow red hot/melt/catch fire [1].

23–1 $Q = mL_f = 0.50\,\text{kg} \times 1790\,\text{kJ kg}^{-1}$ [1]
$Q = 895 \times 10^3\,\text{J} = 8.95 \times 10^5\,\text{J}$ [1]

23–2 $\Delta Q = 3265\,°\text{C} - 1414\,°\text{C} = 1851\,°\text{C}$ [1]
$E = mc\Delta\theta = 0.50\,\text{kg} \times 712\,\text{J kg}^{-1}\,\text{k}^{-1} \times 1851\,\text{K}$
$E = 658\,956\,\text{J} = 6.59 \times 10^5\,\text{J}$ [1]

23–3 $Q = mL_v = 0.50\,\text{kg} \times 12\,800 \times 10^3\,\text{J kg}^{-1}$ [1] $= 6.4 \times 10^6\,\text{J}$ [1]

23–4 Total energy $= 8.95 \times 10^5\,\text{J} + 6.59 \times 10^5\,\text{J} + 6.4 \times 10^6\,\text{J}$
$= 7.99 \times 10^6\,\text{J} = 8.0 \times 10^6\,\text{J}$ [1] (2 s.f.)

24–1 Power (W) $= \dfrac{\text{energy transferred (J)}}{\text{time (s)}}$ [1]

$\Delta Q = mc\Delta\theta$ [1]
Therefore divide the equation by time:

$\dfrac{\Delta Q}{t} = \dfrac{mc\Delta\theta}{t}$ [1]

We can write this as:

$P = \dfrac{mc\Delta\theta}{t}$ [1]

Rearrange to make $\Delta\theta$ the subject:

$\Delta\theta = \dfrac{Pt}{mc}$ [1]

This mark is dependent on the previous step. A simple quotation of the formula without the correct argument will gain no marks.

24–2 5 minutes = 300 s

Use: $\Delta\theta = \dfrac{Pt}{mc}$

$\Delta\theta = \dfrac{(48\,\text{W} \times 300\,\text{s})}{(1.0\,\text{kg} \times 385\,\text{J kg}^{-1}\,\text{K}^{-1})}$ [1]

$\Delta\theta = 37.4\,\text{K}$ [1]
Final temperature $= 21\,°\text{C} + 37.4\,°\text{C} = 58.4\,°\text{C}$
Final temperature $= 58\,°\text{C}$ to 2 s.f. [1]

25–1 *Any five from:*
- Clamp the clamp stand to the bench to stop it rocking [1].
- Use a micrometer to measure the diameter of the plunger. Convert the reading into metres [1].
- Calculate the area of the plunger. ($A = \pi D^2/4$) [1]
- Replace the plunger and draw in about 4 to 5 cm³ of air [1].
- Place the rubber tubing over the nozzle. Tighten up the thumbscrew clip (pinch-clip) [1].
- Clamp the syringe. This should not be too tight, as it will distort the syringe and could make it stick. It should not be too loose either, because the weight of the slotted masses could pull it away from the clamp [1].
- The slotted masses should be attached to the plunger using a loop of copper wire which can be wrapped around the plunger [1].
- Move the plunger up and down to make sure it moves freely [1].
- Add 200 g at a time to a maximum of 1000 g and measure the new volume [1].
- Do at least one more set of repeat readings [1].

25–2 To get the pressure, we need to use: $p = \dfrac{F}{A}$ [1]
Any three from:
- Area will have been worked out as in **Answer 25–1** [1].
- Area in square metres [1].
- Force, $F = mg$ (use $g = 9.8\,\text{N kg}^{-1}$) [1].
- Pressure in Pa [1].
- Data recorded in a table, with at least one repeat [1].

25–3 *Graph looks like this:*

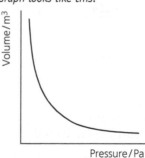

OR you could plot a graph of 1/V against P to get a straight line. Only a sketch graph needed. Graph drawn [1]. Axes to be labelled [1].

25–4 This is not a particularly effective method [1], as
Any four from (1 mark for each two correct points, max 2 marks):
- There is a lot of friction between the rubber plunger and bore of the syringe.
- The syringe can be squashed in the clamp, increasing the friction. If the clamp is not tight enough, the syringe will fall out.
- The graduations on the syringe are rather coarse, making it difficult to read with any precision.
- The rubber tube and pinch screw may leak.

- Air is not an ideal gas. (*Argon could be used, but not many schools or colleges have an argon supply.*)
- The data is likely to have considerable uncertainty as a result.

26–1 There are lots of randomly moving molecules. Each molecule will have an x, y and z component to its motion. Since there are so many molecules, we can assume that $\frac{1}{3}$ of them move in each direction. [1]

26–2 N is the number of the molecules, m is the total mass of each molecule, V is the volume [1].

Rearrange the equation to: $p = \frac{1}{3}\dfrac{Nm(c_{rms})^2}{V}$ [1]

Nm is the total mass of the gas [1].

Density = $\dfrac{mass}{volume}$ [1]

Hence: $p = \frac{1}{3}\rho(c_{rms})^2$

Steps must be shown to achieve the marks.

26–3 Rearrangement: $(c_{rms})^2 = \dfrac{3p}{\rho}$ [1]

$(c_{rms})^2 = \dfrac{(3 \times 1.013 \times 10^5\,Pa)}{0.166\,kg\,m^{-3}}$ [1]

$= 1.83 \times 10^6\,m^2\,s^{-2}$ [1] $c_{rms} = 1353\,m\,s^{-1}$ [1]

27–1 In 1 m³ the number of moles is $\dfrac{1\,m^3}{0.0244\,m^3\,mol^{-1}} = 41.0\,mol$ [1].

Number of particles = 41.0 mol × 6.02 × 10²³ mol⁻¹
$= 2.47 \times 10^{25}$ [1]

Mass of each particle = $\dfrac{1.2\,kg}{2.47 \times 10^{25}}$

$= 4.9 \times 10^{-26}\,kg$ [1]

27–2 $E_k = \frac{3}{2}kT$

$E_k = 3/2 \times 1.38 \times 10^{-23}\,J\,K^{-1} \times 293\,K$
$= 6.07 \times 10^{-21}\,J = 6.1 \times 10^{-21}\,J$ (2 s.f.) [1]

27–3 $(c_{rms})^2 = \dfrac{3p}{\rho}$

Substitution: $(c_{rms})^2 = \dfrac{3 \times 1.013 \times 10^5\,Pa}{1.2\,kg\,m^{-3}}$ [1]:

$= 253\,000\,m^2\,s^{-2}$ [1]
RMS speed = 503 m s⁻¹ = 500 m s⁻¹ (2 s.f.) [1]

27–4 $E_k = \frac{1}{2} \times 4.9 \times 10^{-26}\,kg \times$

$250\,000\,m^2\,s^{-2}$ [1] = $6.13 \times 10^{-21}\,J$ [1]

27–5 The answers are about the same [1].
The kinetic energy for all gas molecules is the same for a given temperature [1].

28–1 The volume of the box = (0.15 m)³ = 3.375 × 10⁻³ m³. [1]

Number of moles = $\dfrac{3.375 \times 10^{-3}\,m^3}{0.0224\,m^3}$

$= 0.151\,mol$ [1]
Number of particles = 0.151 mol × 6.02 × 10²³ mol⁻¹
$= 9.07 \times 10^{22}$ particles. [1]

28–2 Mass = density × volume = 0.937 kg m⁻³ × 3.375 × 10⁻³ m³
$= 3.16 \times 10^{-3}\,kg$ [1]

Mass per particle = $\dfrac{3.16 \times 10^{-3}\,kg}{9.07 \times 10^{22}}$

$= 3.49 \times 10^{-26}\,kg$ [1]

28–3 Rearrangement: $(c_{rms})^2 = \dfrac{3p}{\rho}$ [1]

Substitute: $(c_{rms})^2 = \dfrac{3 \times 1.013 \times 10^5\,Pa}{0.937\,kg\,m^{-3}}$

$= 3.24 \times 10^5\,m^2\,s^{-2}$ [1]

$\sqrt{(c_{rms})^2} = 570\,m\,s^{-1}$ [1]

28–4 Momentum change:
$\Delta p = 2m\sqrt{(c_{rms})^2}$ [1] $= 2 \times 3.49 \times 10^{-26}\,kg \times 570\,m\,s^{-1}$ [1]
$\Delta p = 3.98 \times 10^{-23}\,kg\,m\,s^{-1}$ [1]

29–1 1. Molar mass = 2 × 14.0 × 10⁻³ kg mol⁻¹
$= 28.0 \times 10^{-3}\,kg\,mol^{-1}$ [1]

2. Density = $\dfrac{mass}{volume}$ = $\dfrac{28 \times 10^{-3}\,kg\,mol^{-1}}{0.0244\,m^3\,mol^{-1}}$ [1]
Density = 1.15 kg m⁻³ [1]

29–2 1. Use: $pV = \frac{1}{3}Nm(c_{rms})^2$

Rearrangement and substitution:
$c^2 = \dfrac{3 \times 1.013 \times 10^5\,Pa}{1.15\,kg\,m^{-3}}$ [1] $= 264\,261\,m^2\,s^{-2}$ [1]

RMS speed = 514 m s⁻¹ [1]

2. The kinetic energy is proportional to the temperature [1]. If we double the speed, the kinetic energy goes up by 4 times, therefore the temperature goes up 4 times [1].
20 °C = 293 K [1]
New temperature = 1172 K = 1200 K (2 s.f.) [1]

29–3 *Unqualified answers like kinetic energy and speed will not gain marks.*
1. Similarities (*any 1 from*):
   - motion is random [1]
   - range of speeds [1].
2. Differences (*any 1 from*):
   - <u>mean</u> kinetic energy [1]
   - <u>root</u> mean square speeds [1]
   - <u>frequency</u> of collisions is greater in the warmer tyre [1].

30–1 1. *The graphs look like this:*

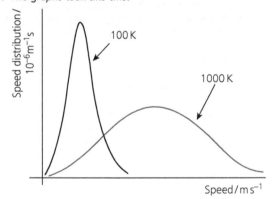

Any three from:
- axes labelled [1]
- 100 K graph [1]
- 1000 K graph [1].

2. *Any five from:*
   - These graphs are called <u>normal distributions</u> [1].
   - They show that the total kinetic energy [1] is shared <u>randomly</u> [1] throughout the molecules in the gas.
   - While the root mean square speed of the gas at 100 K is lower [1] than the gas at 1000 K, there is a <u>probability</u> [1] that the fastest molecules at 100 K can travel faster than the slowest molecules at 1000 K [1]. The slowest molecules at 1000 K have the same speed as the slowest molecules at 100 K [1].

30–2 1. Use: $(c_{rms})^2 = \dfrac{3p}{\rho}$ [1]

Substitution: $(c_{rms})^2 = \dfrac{3 \times 1.013 \times 10^5\,Pa}{1.66\,kg\,m^{-3}}$ [1]

$= 183072 \, m^2 \, s^{-2}$

RMS speed $= 428 \, m \, s^{-1} = 430 \, m \, s^{-1}$ [1]

2. 1 mole occupies $0.0244 \, m^3$.
    Therefore $1 \, m^3$ is occupied by $0.0244^{-1} = 41.0 \, mol$ [1].
    41 mol has a mass of 1.66 kg.
    Therefore 1 mol has a mass of 0.0405 kg [1].

    Mass of 1 argon atom $= \dfrac{0.0405 \, kg}{6.02 \times 10^{23}}$ [1]

    $= 6.73 \times 10^{-26} \, kg$ [1]

3. Use: $\dfrac{1}{2}m\left(c_{rms}\right)^2 = \dfrac{3}{2}kT$ [1]

    Rearrangement: $T = \dfrac{m\left(c_{rms}\right)^2}{3k}$ [1]

    RMS speed is now $860 \, m \, s^{-1}$.

    $T = \dfrac{\left(6.73 \times 10^{-26} \, kg \times \left(860 \, m \, s^{-1}\right)^2\right)}{\left(3 \times 1.38 \times 10^{-23} \, JK^{-1}\right)} = 1200K$ [1]

Topic review: further mechanics and thermal physics (p.45)

1–1

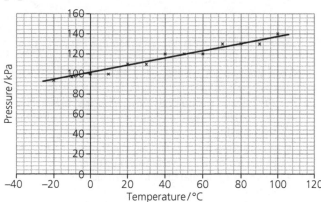

Any five from:

-   sensible scales [1]
-   axes labelled [1]
-   the data is plotted accurately, +/− 1 mm [1]
-   line drawn with a ruler [1]
-   line of best fit [1]

1–2

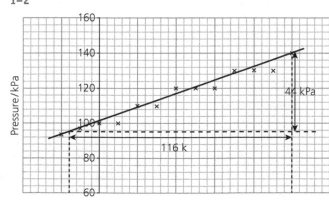

Any four from:

-   rise $= 44 \times 10^3 \, Pa$ [1]
-   run = 116 K [1]
-   calculation shown [1]
-   gradient $= \dfrac{44 \times 10^3 \, Pa}{116K} = 379$ [1] $Pa \, K^{-1}$ [1]

1–3 The graph fits the pattern of $y = mx + c$ [1]
Therefore $p = 379 \, \theta + 102 \times 10^3$
Value for gradient shown [1].
Allow error carried forward from **Answer 1–2**. y–axis intercept must have correct power of 10 [1].

1–4 Substitute: $0 = 379 \, \theta + 102 \times 10^3$ [1]
$379 \, \theta = -102 \times 10^3$ [1]
$\theta = -269 \, °C$ *(Allow +/− 12 °C for the final answer)* [1]

2–1 The energy transfer needed to raise <u>unit mass</u> of a material <u>by unit temperature change</u> [1].

2–2
-   A lab heater is connected to a circuit [1] that includes a <u>voltmeter and ammeter</u> [1]. (*These marks can be obtained for an appropriate circuit diagram. Both voltmeter and ammeter need to be mentioned for the second mark.*)
-   <u>Heater and thermometer</u> is inserted into an aluminium block [1].
-   Measure temperature at regular time intervals, e.g. 10s [1].
-   Basic measurements include voltage and current (to work out power) [1].
-   Time, initial temperature and temperature at time intervals should be recorded [1].

OR Allow the last 3 marks if use of a data logger is correctly described.

2–3 The graph of such an experiment is shown in **Answer 2–4**. To show direct proportionality, the graph needs to be a straight line of <u>positive gradient</u> [1] and <u>go through the origin</u> [1].

2–4

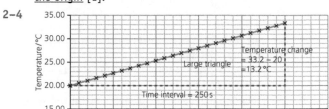

Large triangle (smallest dimension 8 cm) [1]

Gradient $= \dfrac{\text{rise}}{\text{run}} = \dfrac{\left(33.2 \, °C - 20 \, °C\right)}{\left(250 \, s - 0 \, s\right)} = 0.0528 \, Ks^{-1}$ [1]

Gradient $= \dfrac{P}{(mc)}$ [1]

$c = \dfrac{48 \, W}{\left(1.0 \, kg \times 0.0528 \, Ks^{-1}\right)} = 910 \, Jkg^{-1}K^{-1}$ [1]

Allow between 900 and 915 Jkg^{-1}K^{-1}.

2–5 Insulate the block with bubble wrap and put a lid on top to reduce heat loss from the block [1].
Two readings of the thermometer to reduce parallax error [1].
Use class results of data to get averages [1].
To get each mark, both procedure and explanation are needed. It is not enough to say 'take repeat readings' as the block will be hot, and it's not practicable to cool it back down to room temperature within the time for a school/ college physics lab session.

3–1 Specific latent heat of vaporisation is the amount of energy needed to change a <u>unit mass</u> of liquid to its vapour with zero change in temperature [1].

3–2 Energy is required to break the bonds between water molecules to allow single molecules to escape [1].
The energy used reduces the internal energy, leading to a lower temperature [1].

3–3 $P = \dfrac{mL}{t}$ *(mark allowed if this equation is implied in the calculation)* [1]

$P = \dfrac{\left(1.5 \times 10^{-4} \, kg \, s^{-1} \times 2.264 \times 10^6 \, Jkg^{-1}\right)}{1 \, s}$

$= \underline{339.6 \, W} = 340 \, W$ [1]
For second mark, answer of 339.6 W must be shown.

3–4 The energy source comes from the internal energy of the water inside [1].

3–5 $P = mc\dfrac{\Delta\theta}{t}$ where $\dfrac{\Delta\theta}{t}$ is the rate of change of temperature [1].

Rate of change of temperature

$$= \frac{339.6\,\text{W}}{\left(1.5\,\text{kg} \times 4200\,\text{Jkg}^{-1}\text{K}^{-1}\right)} = 0.054\,\text{K s}^{-1}\,[1]$$

Answer is given to 2 s.f. [1]
First mark can be implied in the numerical calculation. Full marks if 340 J s⁻¹ is used.

3–6 *Any three from:*
- At the start, the temperature will fall at an approximately constant rate [1].
- After a period of time, the temperature will level out/become steady [1] to a constant lower level [1].
- Eventually all the water is used up, so the temperature will rise to that of the surroundings [1].

4–1 SHM occurs when the acceleration of a body is directly <u>proportional</u> to the <u>displacement from the rest position</u> and is always directed <u>towards the rest position</u> [1].

4–2 $k = \dfrac{F}{\Delta l}$

Total weight $= 1500\,\text{kg} \times 9.81\,\text{N kg}^{-1} = 14\,715\,\text{N}$
Weight on each spring $= 14\,715\,\text{N} \div 4 = 3679\,\text{N}$ [1]
$k = 3679\,\text{N} \div 0.020\,\text{m}$ [1] $= 1.84 \times 10^5\,\text{N m}^{-1} = 1.8 \times 10^5\,\text{N m}^{-1}$

4–3 $f^2 = \dfrac{1}{4\pi^2}\left(\dfrac{k}{m}\right)$

There are 4 springs, so the mass on the spring is
$1500\,\text{kg} \div 4 = 375\,\text{kg}$ [1]
$f^2 = (4\pi^2)^{-1} \times (1.84 \times 10^5\,\text{N m}^{-1} \div 375\,\text{kg}) = 12.4\,\text{Hz}^2$
$f = 3.53\,\text{Hz} = 3.5\,\text{Hz}$ [1]

4–4 The <u>frequency applied by the bumps</u> is about the <u>same as the natural frequency</u> [1]. This gives the conditions for <u>resonance</u> [1]. The amplitude of the <u>oscillations will increase</u>, so that the car may lose contact with the road [1] (and become uncontrollable).

4–5 $E_k = \dfrac{1}{2}m(2\pi f)^2 A^2$

$A = 0.28\,\text{m} - 0.27\,\text{m} = 0.01\,\text{m}$

$E_k = \dfrac{1}{2} \times 1500\,\text{kg} \times (2 \times \pi \times 3.5\,\text{Hz})^2 \times (0.01\,\text{m})^2$

$= 36.3\,\text{J}$ [1]

4–6 $p = \dfrac{\Delta W}{\Delta V}$

Work done, $\Delta W = 36.9\,\text{J} \times 0.10 = 3.96\,\text{J}$
$\Delta V = 1.52 \times 10^{-3}\,\text{m}^2 \times 0.010\,\text{m} = 1.52 \times 10^{-5}\,\text{m}^3$
$p = 3.63\,\text{J} \div 1.52 \times 10^{-5}\,\text{m}^3 = 2.39 \times 10^5\,\text{Pa}$
$= 2.4 \times 10^5\,\text{Pa}$ [1]

4–7 Total volume when not compressed
$= 0.25\,\text{m} \times 1.52 \times 10^{-3}\,\text{m}^2 = 3.80 \times 10^{-4}\,\text{m}^3$
Volume when compressed
$= 3.80 \times 10^{-4}\,\text{m}^3 - 1.52 \times 10^{-5}\,\text{m}^3 = 3.65 \times 10^{-4}\,\text{m}^3$ [1]
At constant temperature: $p_1V_1 = p_2V_2$
$p_2 = (2.40 \times 10^5\,\text{Pa} \times 3.65 \times 10^{-4}\,\text{m}^3) \div 3.80 \times 10^{-4}\,\text{m}^3$
$p_2 = 2.31 \times 10^5\,\text{Pa}$ [1]

4–8 1. Use: $pV = nRT$
$n = (2.3 \times 10^5\,\text{Pa} \times 3.80 \times 10^{-4}\,\text{m}^3) \div$
$(8.31\,\text{J mol}^{-1}\text{K}^{-1} \times 293\,\text{K})$
$n = 0.0359\,\text{mol}$ [1]
2. $m = 0.0359\,\text{mol} \times 0.028\,\text{kg mol}^{-1} = 1.005 \times 10^{-3}\,\text{kg}$
$\rho = 1.005 \times 10^{-3}\,\text{kg} \div 3.80 \times 10^{-4}\,\text{m}^3 = 2.64\,\text{kg m}^{-3}$ [1]

3. Use: $<c^2> = \dfrac{3p}{\rho}$

$<c^2> = (3 \times 2.3 \times 10^5\,\text{Pa}) \div 2.64\,\text{kg m}^{-3}$
$= 2.61 \times 10^5\,\text{m}^2\,\text{s}^{-2}$
$<c> = (2.61 \times 10^5\,\text{m}^2\,\text{s}^{-2})^{0.5} = 511\,\text{m s}^{-1}$
$= 510\,\text{m s}^{-1}$ (2 s.f.) [1]

5–1 Air is not an ideal gas because:
- It is a mixture of gases.
- Nitrogen, oxygen and carbon dioxide molecules are made up of two or three atoms.
- Collisions happen between molecules.

- There is energy within the bonds of diatomic and triatomic molecules.

Any two for 1 mark each.

5–2 1. Rearrange $pV = nRT$ [1]

$V = \dfrac{nRT}{p}$

$V = \dfrac{\left(5.5\,\text{mol} \times 8.31\,\text{J mol}^{-1}\text{K}^{-1} \times 300\,\text{K}\right)}{2.00 \times 10^5\,\text{Pa}}$

$= 0.0686\,\text{m}^3$
$= 0.069\,\text{m}^3$ (2 s.f.) [1]

2. Equation:

$\dfrac{P_1V_1}{T_1} = \dfrac{P_2V_2}{T_2}$

Rearrange to make V_2 the subject:

$V_2 = \dfrac{P_1V_1T_2}{T_1P_2}$

List of quantities:
- $P_1 = 3.1 \times 10^6\,\text{Pa}$
- $V_1 = 0.150\,\text{m}^3$
- $T_1 = (-10\,°\text{C} + 273) = 263\,\text{K}$ (*all temperatures must be in K*)
- $P_2 = 1.01 \times 10^5\,\text{Pa}$
- $V_2 = ?$
- $T_2 = 296\,\text{K}$

$V_2 = (3.1 \times 10^6\,\text{Pa} \times 0.150\,\text{m}^3 \times 296\,\text{K}) \div (263\,\text{K} \times 1.01 \times 10^5\,\text{Pa})$ [1]
Therefore $V_2 = 5.18\,\text{m}^3$ [1] $= 5.2\,\text{m}^3$ (2 s.f.) [1]

5–3 *Any two from this list for 1 mark each:*
- All molecules are identical. They are assumed to be spherical.
- The volume of the molecules is very much smaller than the container.
- All collisions are perfectly elastic. There is no friction as the molecules move between collisions.
- There are no interactions between molecules between collisions.
- There is no attraction between molecules.
- The average distance between molecules is much greater than the molecular diameter.
- Newton's laws apply.
- The motion of the molecules is random.
- There is a random distribution of molecular velocities.
- The collisions are between the molecules and the walls of the container.
- The time taken between collisions is very much bigger than the time of the collisions.

5–4 $pV = nRT = \dfrac{1}{3}Nmc^2$

and $E_k = \dfrac{1}{2}mc^2$

Therefore [1] $nRT = \dfrac{2}{3}N\left(\dfrac{1}{2}mc^2\right)$

Rearrange to [1]: $\dfrac{1}{2}mc^2 = \dfrac{3}{2}\dfrac{nRT}{N}$

The term n/N = N_A, the Avogadro constant [1]
The term R/N_A = k, the Boltzmann constant [1]

Therefore: $\dfrac{1}{2}mc^2 = \dfrac{3}{2}kT$

6–1 Use $v = 2\pi f\sqrt{a^2 - x^2}$
$\to v_{\text{max}} = 2\pi fA = 2\pi \times 3.2\,\text{s}^{-1} \times 2.8 \times 10^{-2}\,\text{m} = 0.56\,\text{m s}^{-1}$ [1];

$E_k = \dfrac{1}{2}mv^2 \to 12\,\text{J} = 0.5 \times m \times 0.56^2\,\text{m}^2\,\text{s}^{-2}$

$\to m = \dfrac{24\,\text{J}}{0.31\,\text{m}^2\,\text{s}^{-2}} = 77\,\text{kg}$ [1];

mass of cyclist = m – mass of bike = 77 kg – 0.820 kg
= 76.2 kg [1]

For spring constant, use $f = \dfrac{1}{2\pi}\sqrt{\dfrac{k}{m}}$

$\to 4\pi^2 \times 3.2\,\text{s}^{-2} = \dfrac{k}{77\,\text{kg}}$ [1]

$\to k = 31\,128\,\text{N m}^{-1}$ [1]

6–2 We assumed that the whole mass of the bike is supported by the back suspension. [1]

Gravitational fields (p.49)

Quick questions

1 A field is a region in which a force is felt.

2 Every particle of matter in the Universe attracts every other particle with a gravitational force that is proportional to the products of the masses and inversely proportional to the square of the distance between them.

3

Term	Meaning	SI unit
F	Force	N
G	Universal constant of gravitation	$N\,m^2\,kg^{-2}$
m_1	Point mass 1	kg
m_2	Point mass 2	kg
r	radius	m

4 Gravity is always attractive. (*It is never repulsive.*)

5 $F = (-)\ (6.67 \times 10^{-11}\,N\,m^2\,kg^{-2} \times 75\,kg \times 75\,kg) \div (1.25\,m)^2$
$= (-)\ 2.4 \times 10^{-7}\,N$

6 Gravitational forces:
- occur between all objects that have mass
- are attractive (*See* **Question 4**.)
- are weak
- have infinite range
- are vectors.

7 $r^2 = (-)G\dfrac{m_1 m_2}{F}$

$r^2 = (-)\ (6.67 \times 10^{-11}\,N\,m^2\,kg^{-2} \times 1.5 \times 10^7\,kg \times 2.5 \times 10^7\,kg) \div (-)\ 2.7\,N$
$r^2 = 9264\,m^2$
$r = 96\,m$
(*Note that the minus signs cancel out. The force is negative as it's attractive.*)

8 Equation: $g = G\dfrac{m_1}{r^2}$

$g = (6.67 \times 10^{-11}\,N\,m^2\,kg^{-2} \times 1.99 \times 10^{30}\,kg) \div (6.96 \times 10^8\,m)^2$
$g = 274\,N\,kg^{-1}\ (or\ m\,s^{-2})$

9

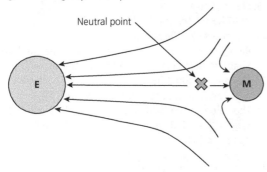
Neutral point
E   M

10 0 m: 0 J; 20 m: 400 J; 40 m: 800 J; 60 m: 1200 J; 80 m: 1600 J; 100 m: 2000 J
Example: Use $E_p = mg\Delta h = 5\,kg \times 4\,N\,kg^{-1} \times 120\,m = 2400\,J$

11

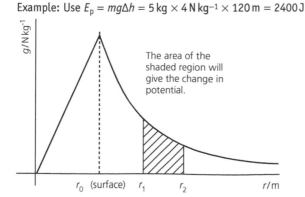

The area of the shaded region will give the change in potential.

$g/N\,kg^{-1}$ (vertical axis); r_0 (surface) r_1 r_2 r/m (horizontal axis)

$h = 4.90 \times 10^7\,m - 6.37 \times 10^6\,m = 4.26 \times 10^7\,m$
$(= 42\,600\,km)$

Exam-style questions

12 **B:** It is defined as force per unit mass. [1]

13 **A:** The centripetal force is provided by the gravity of the planet. [1]

14 **B:** The equation above is only true if the field is uniform. [1]

15 **D:** $W = 0$ [1]

16–1 Gravitational field strength is defined as the force due to gravity per unit mass. [1]

16–2 $g = \dfrac{F}{m}$ [1]

16–3 1. $F = mg = 250\,kg \times 18\,N\,kg^{-1} = 4500\,N.$ [1]
2. The force is the weight. [1]

16–4 Direction of arrows [1]; highest value [1]; lowest value [1]

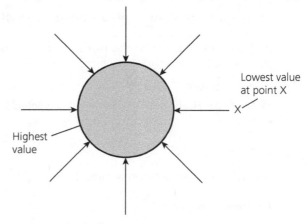
Lowest value at point X
X
Highest value

17–1 Diagram similar to answer for **Question 11**, with same axes and graph line: correct axes [1]; correct line shape [1]; r_0 marked correctly [1]
Note: this graph may be drawn upside down, if the negative sign is taken into account.

17–2 Equation: $g = G\dfrac{m_1}{r^2}$ [1]

$g = (6.67 \times 10^{-11}\,N\,m^2\,kg^{-2} \times 5.97 \times 10^{24}\,kg) \div (6.37 \times 10^6\,m)^2$
$g = 9.81\,N\,kg^{-1}\ (or\ m\,s^{-2})$ [1]

17–3 $g = \dfrac{F}{m}$

We know from Newton's second law that: $a = \dfrac{F}{m}$ [1]

So we can write: $a = \dfrac{F}{m} = g$

Therefore gravitational field strength is acceleration [1].

18–1 Gravitational potential is defined as gravitational potential energy per unit mass [1].

18–2 Unit: $J\,kg^{-1}$ [1]

18–3 Equation: $\Delta V = \dfrac{mg\Delta h}{m} = g\Delta h$ [1]

18–4 $E_p = 5.0\,kg \times 4.0\,N\,kg^{-1} \times 120\,m = 2400\,J$ [1]
$V_g = 2400\,J \div 5.0\,kg = 480\,J\,kg^{-1}$ [1]
Alternative answer: $\Delta V = 4.0\,N\,kg^{-1} \times 120\,m = 480\,J\,kg^{-1}$

18–5 The equation only applies to a uniform field. [1] Gravity fields are only uniform over short ranges. Over longer ranges, the field is radial [1].

19–1 Equation: $V_g = (-)G\dfrac{m_1}{r}$ [1]

19–2 1. At 250 m equipotential:
$V_g = 8.5\,N\,kg^{-1} \times 250\,m = 2125\,J\,kg^{-1}$ [1]
At the 110 m equipotential:
$V_g = 8.5\,N\,kg^{-1} \times 110\,m = 935\,J\,kg^{-1}$ [1]
$\Delta V_g = 2125\,J\,kg^{-1} - 935\,J\,kg^{-1}$
$= 1190\,J\,kg^{-1}$ [1]

2. Work is got out as the object is moving down (to the surface of the planet) and the direction of the field is downwards [1].

20–1 At 450 km, radius = 6.82×10^6 m
At 550 km, radius = 6.92×10^6 m [1]
At 450 km, $V_g = -(6.67 \times 10^{-11}\,\text{N}\,\text{m}^2\,\text{kg}^{-2} \times 5.97 \times 10^{24}\,\text{kg})$
$\div\ 6.82 \times 10^6$ m
$= -58.4 \times 10^6\,\text{J}\,\text{kg}^{-1}$ [1]
At 550 km, $V_g = -(6.67 \times 10^{-11}\,\text{N}\,\text{m}^2\,\text{kg}^{-2} \times 5.97 \times 10^{24}\,\text{kg})$
$\div\ 6.92 \times 10^6$ m
$= -57.5 \times 10^6\,\text{J}\,\text{kg}^{-1}$ [1]
$\Delta V_g = -57.5 \times 10^6\,\text{J}\,\text{kg}^{-1} - -58.4 \times 10^6\,\text{J}\,\text{kg}^{-1} = +9.00 \times 10^5\,\text{J}\,\text{kg}^{-1}$ [1]

20–2 The mass is not needed [1].

20–3 The gravitational potential energy of a point is defined as the work done on a mass in moving it to that point from infinity [1].

20–4 Energy = potential × mass = $+9.00 \times 10^5\,\text{J}\,\text{kg}^{-1} \times 4500\,\text{kg}$
$= 4.05 \times 10^9\,\text{J}$ [1]

20–5 <u>Minimum</u> velocity that is needed to move a rocket, mass m, from the Earth's <u>surface</u> to <u>infinity</u> [1].

20–6 Change in potential: $\Delta V = \dfrac{GM}{r}$
We know that kinetic energy is converted into potential energy:

$E_p = m\Delta V = E_k = \dfrac{1}{2}mv^2$ [1]

The mass terms cancel out, so we get: $\Delta V = \dfrac{1}{2}v^2$ [1]

Therefore: $v^2 = 2\Delta V = 2\dfrac{GM}{r}$

Square root the equation: $v = \sqrt{\left(\dfrac{2GM}{r}\right)}$ [1]

20–7 $v^2 = 2\dfrac{GM}{r}$ [1]
$v^2 = (2 \times 6.67 \times 10^{-11}\,\text{N}\,\text{m}^2\,\text{kg}^{-2} \times 1.90 \times 10^{27}\,\text{kg}) \div (6.99 \times 10^7\,\text{m})$
$= 3.63 \times 10^9\,\text{m}^2\,\text{s}^{-2}$
$v = (3.63 \times 10^9\,\text{m}^2\,\text{s}^{-2})^{0.5}$
$= 6.02 \times 10^4\,\text{m}\,\text{s}^{-1}$ [1]

21–1 Centripetal force = gravitational force
$(-)\dfrac{GMm}{r^2} = (-)\dfrac{mv^2}{r}$ [1]

The minus signs cancel out, as do the m terms. So we get:
$\dfrac{GM}{r} = v^2$
Square root the equation: $v = \sqrt{\left(\dfrac{GM}{r}\right)}$ [1]

21–2 Orbital radius = $6.99 \times 10^7\,\text{m} + 2.25 \times 10^7\,\text{m}$
$= 9.24 \times 10^7\,\text{m}$ [1]
$v^2 = (6.67 \times 10^{-11}\,\text{N}\,\text{m}^2\,\text{kg}^{-2} \times 1.90 \times 10^{27}\,\text{kg}) \div (9.24 \times 10^7\,\text{m})$
$= 1.37 \times 10^9\,\text{m}^2\,\text{s}^{-2}$ [1]
$v = (1.37 \times 10^9\,\text{m}^2\,\text{s}^{-2})^{0.5}$
$= 3.70 \times 10^4\,\text{m}\,\text{s}^{-1}$ [1]

22–1 $v = \omega r$ or $v = 2\pi f r$ [1]

22–2 $v = \dfrac{2\pi r}{T}$ [1]

22–3 $v^2 = \dfrac{4\pi^2 r^2}{T^2} = \dfrac{GM}{r}$ [1]

Rearranging: $\dfrac{r^3}{T^2} = \dfrac{GM}{4\pi^2}$
Make T^2 the subject:

$T^2 = \left(\dfrac{4\pi^2}{GM}\right)r^3$ [1]

22–4 Rearrange the equation: $r^3 = \left(\dfrac{GM}{4\pi^2}\right)T^2$ [1]

Height = $4.90 \times 10^7\,\text{m} - 6.37 \times 10^6\,\text{m} = 4.26 \times 10^7\,\text{m}$
$T = 30\,\text{h} \times 3600\,\text{s}\,\text{h}^{-1}$
$= 1.08 \times 10^5\,\text{s}$ [1]
$r^3 = (6.67 \times 10^{-11}\,\text{N}\,\text{m}^2\,\text{kg}^{-2}$
$\times\ 5.97 \times 10^{24}\,\text{kg} \times (1.08 \times 10^5\,\text{s})^2) \div (4 \times \pi^2)$
$r^3 = 1.176 \times 10^{23}\,\text{m}^3$

$r = \left(1.176 \times 10^{23}\,\text{m}^3\right)^{\frac{1}{3}}$
$= 4.90 \times 10^7\,\text{m}$ [1]

23–1 Use $V = \dfrac{4}{3}\pi r^3$
$V = 4/3 \times \pi \times (3.5 \times 10^6\,\text{m})^3$
$= 1.796 \times 10^{20}\,\text{m}^3$ [1]
$m = \rho V = 5500\,\text{kg}\,\text{m}^{-3} \times 1.796 \times 10^{20}\,\text{m}^3 = 9.878 \times 10^{23}\,\text{kg}$
$= 9.9 \times 10^{23}\,\text{kg}$ [1]

23–2 1. Definition of gravitational field strength is <u>force</u> per <u>unit mass</u> [1]

2. Use: $g = \dfrac{GM}{r^2}$
$g = (6.67 \times 10^{-11}\,\text{N}\,\text{m}^2\,\text{kg}^{-2} \times 9.9 \times 10^{23}\,\text{kg}) \div (3.5 \times 10^6\,\text{m})^2$
$g = 5.39\,\text{N}\,\text{kg}^{-1}$ [1]
$= 5.4\,\text{N}\,\text{kg}^{-1}$ (2 s.f.) [1]

23–3 1. Use: $V = -\dfrac{GM}{r}$
$V = -(6.67 \times 10^{-11}\,\text{N}\,\text{m}^2\,\text{kg}^{-2} \times 9.9 \times 10^{23}\,\text{kg}) \div (3.5 \times 10^6\,\text{m})$
$V = -1.9 \times 10^7$ [1] $\text{J}\,\text{kg}^{-1}$ [1]
(*Inclusion of the mass of the probe is a physics error.*)

2. Equation: $v^2 = 2\dfrac{GM}{r}$
$v^2 = (2 \times 6.67 \times 10^{-11}\,\text{N}\,\text{m}^2\,\text{kg}^{-2} \times 9.9 \times 10^{23}\,\text{kg}) \div (3.5 \times 10^6\,\text{m})$
$v^2 = 3.77 \times 10^7\,\text{m}^2\,\text{s}^{-2}$ [1]
$v = 6140\,\text{m}\,\text{s}^{-1}$ (= 6100 m s^{-1} to 2.s.f.) [1]

24–1 r_0 shows the surface of the planet [1].

24–2 Draw a tangent to the graph at the distance r required [1].
Find the gradient of the tangent: $G = \Delta Vg \div \Delta r$ [1]

24–3 Radius = $6.37 \times 10^6\,\text{m} + 4.08 \times 10^5\,\text{m} = 6.778 \times 10^6\,\text{m}$ [1]
$g = (6.67 \times 10^{-11}\,\text{N}\,\text{m}^2\,\text{kg}^{-2} \times 5.97 \times 10^{24}\,\text{kg}) \div (6.778 \times 10^6\,\text{m})^2$
$g = 8.67\,\text{N}\,\text{kg}^{-1}$ (or m s^{-2}) [1]

25–1 1. See answer for **Question 16**
<u>Radial</u> fields shown for both the Earth and the Moon [1]
Shape of field lines around the neutral point [1]

2. The gravitational field strength is zero because:
Gravitational field strength is a <u>vector</u> [1].
The value of the field strength from the Moon is the <u>same</u> [1] as that from the Earth, but the directions are <u>opposing</u> [1].

25–2 1. $-\dfrac{GM_E}{r_E^2} = 0 = -\dfrac{GM_M}{r_M^2}$ [1]

The minus signs cancel.

The G terms cancel to give $\dfrac{M_E}{r_E^2} = \dfrac{M_M}{r_M^2}$ [1].

Rearrange to give: $\dfrac{M_E}{M_M} = \dfrac{r_E^2}{r_M^2}$ [1]

2. Substitution: $\dfrac{5.97 \times 10^{24}\,\text{kg}}{7.35 \times 10^{22}\,\text{kg}} = \dfrac{r_E^2}{r_M^2}$ [1]

Evaluation: $\dfrac{r_E^2}{r_M^2} = 81.22$

Square root: $\dfrac{r_E}{r_M} = \sqrt{81.22}$
$= 9.01$ [1]

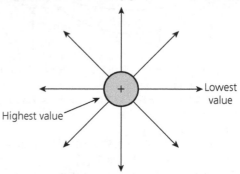

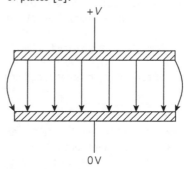

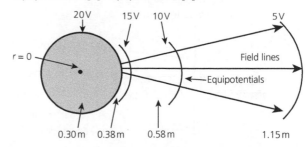

25–3 1. $r_E = 9.01\,r_M$

$r_E + r_M = 3.84 \times 10^8\,\text{m}$ [1]

$9.01\,r_M + r_M = 10.01\,r_M$

$= 3.84 \times 10^8\,\text{m}$

$r_M = 3.84 \times 10^7\,\text{m}$ [1]

2. $r_E = 9.01 \times 3.84 \times 10^7\,\text{m}$

$= 3.46 \times 10^8\,\text{m}$ [1]

3. $d = 3.46 \times 10^8\,\text{m} - 6.37 \times 10^6\,\text{m} = 3.39 \times 10^8\,\text{m}$ [1]

25–4 There would be no effect [1].

At the neutral point the <u>force per unit mass is zero</u> hence the <u>force will be zero</u> [1].

(Alternatively: $\dfrac{Gm_E m_s}{r_E^2}$

$= \dfrac{Gm_M m_s}{r_M^2}$ [2])

Electric fields (p.53)

Quick questions

1 a) A negatively charged body has an excess of electrons.

b) A positively charged body has a deficiency of electrons.

c) In a neutral body, the number of electrons is the same as the number of protons.

2 An electric field can be attractive (opposite charges) or repulsive (like charges). A gravity field is always attractive.

3 Coulomb's law states that the magnitude of attractive or repulsive force between two electrical charges is directly proportional to the magnitude of the charges and inversely proportional to the square of the distance between the charges.

4

Term	Meaning	SI unit
F	**Force**	**N**
ε_0	**Permittivity of free space**	(8.85×10^{-12}) $\text{C}^2\,\text{N}^{-1}\,\text{m}^{-2}$ (or $\text{F}\,\text{m}^{-1}$)
Q_1	**Charge 1**	**C**
Q_2	**Charge 2**	**C**
R	**Radius**	**m**

5 $k = (4 \times \pi \times 8.85 \times 10^{-12}\,\text{F}\,\text{m}^{-1})^{-1} = 8.99 \times 10^9\,\text{m}\,\text{F}^{-1}$ (or $\text{N}\,\text{m}^2\,\text{C}^{-2}$)

6 $F = (8.99 \times 10^9\,\text{m}\,\text{F}^{-1} \times +75 \times 10^{-9}\,\text{C} \times +75 \times 10^{-9}\,\text{C})$ $\div (1.25\,\text{m})^2 = (+)\,3.2 \times 10^{-5}\,\text{N}$

The positive sign tells us that the force is repulsive.

7 Electrostatic forces:
- occur between all objects that have charge
- can be attractive or repulsive
- are strong
- have infinite range
- are vectors.

8 $r^2 = k\dfrac{Q_1 Q_2}{F}$

$r^2 = (8.99 \times 10^9\,\text{m}\,\text{F}^{-1} \times 1.5 \times 10^{-7}\,\text{C} \times -2.5 \times 10^{-6}\,\text{C}) \div -0.27\,\text{N}$

$r^2 = 0.0125\,\text{m}^2$

$r = 0.11\,\text{m}$ (2 s.f.)

9

Property	Gravitational field	Electric field
Acts on	All objects with mass	Charged objects only
Exchange particle	**Graviton**	**Virtual photon**
Direct contact?	**No**	No
Range	Infinite	**Infinite**
Attractive/repulsive	Attractive only	**Attractive or repulsive**
Can be shielded?	**No**	Yes

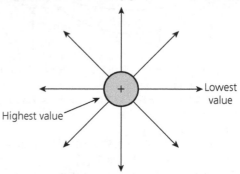

Exam-style questions

10 **A:** The field is radial, and can be attractive or repulsive. [1]

11 **B:** It is defined as force per unit charge. [1]

12 Diagram **B** [1]

13 Path **A** [1]

14 **A:** attractive force of $\dfrac{F}{4}$ [1]

15–1 Electric field strength is defined as the (Coulomb) force per unit charge [1].

15–2 $E = \dfrac{F}{Q}$ [1]

15–3 Highest and lowest value [1]; arrows with correct direction [1].

Highest value / Lowest value

15–4 Straight arrows between plates [1]; curved arrows at edge of plates [1].

$+V$

$0\,V$

15–5 The field is uniform [1].

16–1 $F_P = (8.99 \times 10^9\,\text{m}\,\text{F}^{-1} \times 1.5 \times 10^{-9}\,\text{C} \times +1.0 \times 10^{-9}\,\text{C}) \div (0.02\,\text{m})^2 = 3.37 \times 10^{-5}\,\text{N}$ [1]

$F_R = (8.99 \times 10^9\,\text{m}\,\text{F}^{-1} \times +2.0 \times 10^{-9}\,\text{C} \times +1.0 \times 10^{-9}\,\text{C}) \div (0.005\,\text{m})^2 = 7.19 \times 10^{-4}\,\text{N}$ [1]

$F_T = 3.37 \times 10^{-5}\,\text{N} + 7.19 \times 10^{-4}\,\text{N} = 7.5 \times 10^{-4}\,\text{N}$

The direction is to the left/towards P [1].

16–2 $E = 7.52 \times 10^{-4}\,\text{N} \div 1.0 \times 10^{-9} = 7.5 \times 10^5\,\text{N}\,\text{C}^{-1}$ [1]

16–3 The number of squares counted is 104. (Allow 90 to 120.) [1]

Each square represents $0.040\,\text{m} \times 0.40\,\text{N}\,\text{C}^{-1}$ $= 0.016\,\text{J}\,\text{C}^{-1}$ [1]

Total area $= 104 \times 0.016\,\text{J}\,\text{C}^{-1} = 1.664\,\text{J}\,\text{C}^{-1}$ $= 1.7\,\text{J}\,\text{C}^{-1}$ (2 s.f.) [1]

17–1 Tangent to curve drawn 0.60 m [1]

Gradient of tangent calculated

Electric field strength = 17.2 V m^{-1} (or N C^{-1}) [1]

17–2 The gradient of the graph increases closer to the surface of the sphere. This shows that potential gradient and the electric field strength increase.

17–3 Field lines [1]; calculation of correct distances for equipotentials [1]; equipotentials [1].

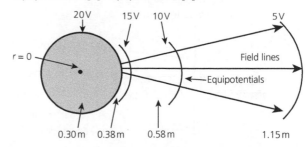

18–1 0.10 m: 150 V; 0.20 m: 300 V; 0.30 m: 450 V [*2 marks for all 3 correct; 1 mark for 2 correct*]

18–2 1. Change in potential = 450 V − 150 V = 300 V [1]
2. Work done = $Q\Delta V = 1.5 \times 10^{-9}$ C $\times 300$ V $= 4.5 \times 10^{-7}$ J [1]

19–1 Curved arrows going from proton to electron [1]; straight arrows pointing to electron and away from proton [1].

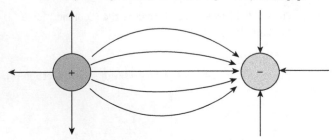

19–2 1. $V = (8.99 \times 10^9 \,\text{m F}^{-1} \times 1.60 \times 10^{-19}\,\text{C}) \div (5.0 \times 10^{-11}\,\text{m})$
   $= 28.8$ V [1]
2. $E_p = (8.99 \times 10^9 \,\text{m F}^{-1} \times 1.60 \times 10^{-19}\,\text{C} \times 1.60 \times 10^{-19}\,\text{C})$
   $\div (5.0 \times 10^{-11}\,\text{m})$
   $= 4.60 \times 10^{-18}$ J [1]
3. $E = (8.99 \times 10^9 \,\text{m F}^{-1} \times 1.60 \times 10^{-19}\,\text{C}) \div (5.0 \times 10^{-11}\,\text{m})^2 = 5.75 \times 10^{11}\,\text{V m}^{-1}$ [1]

20–1 $E = 2000$ V $\div 5.0 \times 10^{-2}$ m $= 40\,000$ V m^{-1}

20–2 1. $F = EQ = 40\,000$ V m$^{-1} \times +1.5 \times 10^{-9}$ C $= 6.0 \times 10^{-5}$ N [1]
2. $a = F/m = 6.0 \times 10^{-5}$ N $\div 3.0 \times 10^{-6}$ kg $= 20$ m s^{-2} [1]
   The direction is downwards (towards the bottom plate) [1].

20–3 Parallel horizontal lines [1] equally spaced (every 1.25 cm) [1].

21–1 Add the potentials of each of the charges together:
$$V_T = \frac{1}{4\pi\varepsilon_0}\left(\frac{Q+Q+Q+2Q}{d}\right) = \frac{5Q}{4\pi\varepsilon_0 d} \quad [1]$$

21–2 Consider the vector nature of the electric field strength: The fields from A and D both cancel out [1].
The fields from C and B add up as vectors:
$$E_T = 2\frac{kQ}{d^2} + -\frac{kQ}{d^2} = k\frac{Q}{d^2} \quad [1]$$

21–3 The direction of the resultant field is from C to B.

22–1 Potential at R = $(8.99 \times 10^9 \,\text{m F}^{-1} \times 6.3 \times 10^{-9}\,\text{C}) \div 0.100$ m $= 566.4$ J C^{-1} [1]
Potential at S = $(8.99 \times 10^9 \,\text{m F}^{-1} \times 6.3 \times 10^{-9}\,\text{C}) \div 0.115$ m $= 492.5$ J C^{-1} [1]
$\Delta V = 566.4$ J C$^{-1} − 492.5$ J C$^{-1} = 73.9$ J C^{-1} [1]

22–2 $W = \Delta V \times Q = 73.9$ J C$^{-1} \times 1.25 \times 10^{-9}$ C $= 9.24 \times 10^{-8}$ J [1]

Capacitance (p.59)

Quick questions

1 The battery stores energy as chemical energy. The capacitor stores it in an electric field.

2 **A:** Non-electrolytic
B: Electrolytic

3 *Relevant points to include:*
Advantage:
• Can hold much more charge.
Disadvantages:
• Needs to be connected to the correct polarity.
• Working voltage is lower.
• Charge stored leaks away.

4

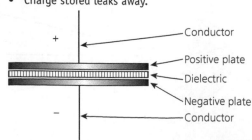

5

Prefix	Capacitance/F
μF	1×10^{-6}
nF	1×10^{-9}
pF	1×10^{-12}

6

Term	Meaning	SI unit
Q	Charge	C
C	Capacitance	F
V	Potential difference	V

7

Term	Name	Unit	Constant/variable
C	Capacitance	F	Variable
ε_0	Permittivity of free space	8.85×10^{-12} F m^{-1}	Constant
ε_r	Dielectric constant (*or relative permittivity*)	No units	Constant for the material
A	Area	m^2	Variable
d	Separation of plates	m	Variable

8 Energy from charge and voltage: $E = \frac{1}{2}QV$
Charge from capacitance and voltage: $Q = CV$
Combine the two: $E = \frac{1}{2}CV \times V = \frac{1}{2}CV^2$

9 Equation: $E = \frac{1}{2}CV^2$
Rearrangement: $V^2 = \frac{2E}{C}$
Substitution: $V^2 = (2 \times 3.0\,\text{J}) \div 2.2 \times 10^{-3}$ F $= 2727$ V^2
$V = (2727\,\text{V}^2)^{0.5} = 52$ V. (2 s.f.)

10 Equation: $V = V_0 e^{-\frac{t}{RC}}$
Take natural logarithms: $\ln V = \ln V_0 - \frac{t}{RC}$
Rearrange: $\ln V - \ln V_0 = -\frac{t}{RC}$
Rearrangement: $t = -RC(\ln V - \ln V_0)$
Time constant = 680×10^{-6} F $\times 5600\,\Omega = 3.81$ s
Substitution: $t = -3.81$ s $(\ln 4.5 - \ln 12.0) = -3.81$ s $\times (1.504 - 2.485) = -3.81$ s $\times -0.981$
Answer: $t = 3.74$ s

Exam-style questions

11 Graph **D** [1]

12 **B:** $\frac{E}{2}$ [1]

13 **B:** 72 ms [1]

14 **C:** 2.0 V [1]

15 **D:** $C = C_0 e^{-\frac{t}{RC}}$ [1]

16–1 1 farad is the capacitance of a conductor that has a potential difference of 1 volt when it carries a charge of 1 coulomb [1].

16–2 *Any three from:*
• Electrons move from the negative terminal of the cell to the plate which is connected to the negative side of the cell [1].
• These repel electrons from the other plate [1].
• The deficiency of electrons makes the plate positive [1].
• The electrons go to the positive terminal of the cell [1].

16–3 *Any three from:*
- There is an excess of electrons on the negative plate [1].
- They flow through the resistor [1].
- They move to the positive plate that has a deficiency of electrons [1].
- Therefore a current flows [1].

16–4 $d = \dfrac{V}{E}$

$d = 200\,V \div 12.0 \times 10^6\,Vm^{-1} = 16.7 \times 10^{-6}\,m = 16.7\,\mu m$ [1]

17–1 $Q = CV = 4.7 \times 10^{-6}\,F \times 2.5\,V = 1.175 \times 10^{-5}\,C$
$= 1.2 \times 10^{-5}\,C\ (2\ s.f.)$ [1]

17–2 Rearrange: $C = \dfrac{Q}{V}$ [1]

$C = 135 \times 10^{-9}\,C \div 17.6\,V = 7.67 \times 10^{-9}\,F = 7.67\,nF$ [1]

17–3 Rearrange: $V = \dfrac{Q}{C}$ [1]

$V = 3.5 \times 10^{-6}\,C \div 2.2 \times 10^{-6}\,F = 1.6\,V$ [1]

18–1 The second charge reading (339.5 μC) for 7.0 V. [1]

18–2 It will be discarded. [1]

18–3

Voltage/V	Average Q/μC
0.0	0
1.0	39.0
2.0	77.6
3.0	117.4
4.0	155.4
5.0	193.0
6.0	232.8
7.0	272.6
8.0	316.0
9.0	356.8
10	394.5

(2 marks all correct, 1 mark six or more values correct)

18–4 *Graph with:*
- axes labelled with sensible scales [1]
- plot (points ± 1 mm) [1]
- line of best fit that goes through the origin [1].

18–5 Gradient of line of best fit found, e.g. rise = $395 \times 10^{-6}\,C$, run = 10 V [1]
Gradient = $395 \times 10^{-6}\,C \div 10\,V = 39.5\,\mu F$ (or CV^{-1})
Value of the capacitor is 39.5 μF [1].

19–1 *Any four from:*
- The insulating gap between the plates of a capacitor is called the dielectric [1].
- Some insulating materials do not alter the capacitance at all, but others can increase the capacitance [1].
- The amount by which the capacitance is increased is called the relative permittivity or dielectric constant [1].
- The molecules in the dielectric are polarised [1].
- The positive charge and the negative charge form an electric field in the molecule [1].
- When these kinds of molecules are used in a capacitor, they align themselves with the electric field formed [1].

19–2 *Any four from:*
- As the current changes direction, the polarity changes [1].
- This makes the field between the plates change [1].
- The polarised molecules change orientation [1].
- As they do so, there is a greater chance of a collision between surrounding molecules [1].
- This increases the internal kinetic energy of the dielectric, making the capacitor warmer [1].

19–3 Use: $C = \dfrac{\varepsilon_0 \varepsilon_r A}{d}$ [1]

$C = (8.85 \times 10^{-12}\,Fm^{-1} \times 2.25 \times 9.0 \times 10^{-2}\,m^2) \div 40 \times 10^{-6}\,m = 4.5 \times 10^{-8}\,F\ (= 45\,nF)$ [1]

20–1 Find the area under the graph between two points on the voltage axis [1]. Area of triangle = $\dfrac{1}{2}QV$ [1].

20–2 Energy from charge and voltage: $E = \dfrac{1}{2}QV$ [1]

Voltage from charge and capacitance: $V = \dfrac{Q}{C}$ [1]

Combine the two: $E = \dfrac{1}{2}Q \times \dfrac{Q}{C} = \dfrac{1}{2}\dfrac{Q^2}{C}$ [1]

20–3 $E = \dfrac{1}{2} \times 5.0 \times 10^{-4}\,C \times 25\,V$
$= 6.25 \times 10^{-3}\,J$ [1]

21–1 $E = \dfrac{1}{2} \times 45 \times 10^{-9}\,F \times (4000\,V)^2$
$= 0.36\,J$ [1]

21–2 1. Equation: $E = \dfrac{1}{2}\dfrac{Q^2}{C}$

Rearrange: $C = \dfrac{1}{2}\dfrac{Q^2}{E}$ [1]

$C = (\dfrac{1}{2} \times (5.0 \times 10^{-3})^2) \div 3.0\,J$
$= 4.2 \times 10^{-6}\,F\ (= 4.2\,\mu F)$ [1]

2. $V = \dfrac{Q}{C}$

$V = 5.0 \times 10^{-3}\,C \div 4.2 \times 10^{-6}\,F = 1190\,V = 1200\,V$ [1]

21–3 1. Find the gradient at 0.40 s.
Draw a tangent to the curve at 0.40 s. [1]
Typical rise = $-6.0 \times 10^{-4}\,C - 2.14 \times 10^{-3}\,C$
$= -1.54 \times 10^{-3}\,C$ and run 2.0 s [1]
$\Delta Q/\Delta t = -1.54 \times 10^{-3}\,C \div 2.0\,s = -7.7 \times 10^{-4}\,A$ [1]
(Allow other reasonable readings of the graph, and therefore answers from −5 to −9 × 10^{-4}.)

2. The current is falling (as shown by the minus sign). This will be shown by the value of the gradient being less/gradient flattening out [1]. The reason for this is that the overall charge is being reduced as the charge is flowing around the circuit [1].

21–4 Sketch graph with voltage on vertical axis and time on horizontal axis [1]
Curve drawn should be an exponential decay curve, i.e. graph must change by the same proportion for each time interval [1]
V_0 marked [1]

21–5 This graph is called an exponential decay [1].

22–1 $I = -\dfrac{dQ}{dt}$ [1]

22–2 $\dfrac{dQ}{dt} = -\dfrac{V}{R}$ [1]

22–3 $Q = CV$ [1]

22–4 Substitute $V = Q/C$ into: $\dfrac{dQ}{dt} = -\dfrac{V}{R}$

To give: $\dfrac{dQ}{dt} = -\dfrac{Q}{CR}$ [1]

22–5 The term k corresponds with $1/CR$ [1].

22–6 Solution: $Q = Q_0 e^{-\frac{t}{RC}}$ [1]

23–1 Basic formulae for R and C: $R = V/I$ and $C = Q/V$
Therefore: $\tau = RC = V/I \times Q/V$
$= Q/I$ [1]
Units for $\tau = C\,A^{-1} = A\,s\,A^{-1} = s$
Therefore the unit of the time constant is second [1].

23–2 $\tau = RC = 470 \times 10^{-6}\,F \times 1500\,\Omega$
$= 0.705\,s = 0.71\,s\ (2\ s.f.)$ [1]

23–3 Equation: $V = V_0 e^{-\frac{t}{RC}}$

Time constant = $680 \times 10^{-6}\,F \times 5600\,\Omega = 3.81\,s$ [1]

Substitute: $V = 12.0\,V \times e^{-\frac{10\,s}{3.81\,s}}$
$V = 12.0\,V \times e^{-2.63} = 0.86\,V$ [1]

24–1 When $t = RC$: $Q = Q_0 e^{-\frac{RC}{RC}}$ [1]

Therefore: $Q = Q_0 \times e^{-1}$
$= Q_0 \times 0.3679$
$0.3679 = 0.37 = 37\%$ [1]

24–2 Equation: $V = V_0 \left(1 - e^{-\frac{t}{RC}}\right)$ [1]

$V = 12.0\,V \left(1 - e^{-\frac{2.0s}{2.71s}}\right)$

$V = 12.0\,V \times (1 - 0.739) = 12.0\,V \times 0.261 = 3.13\,V$
$= 3.1\,V$ (2 s.f.) [1]

24–3 Half-life of a capacitor is the time for a capacitor to discharge from its original charge to one half of its original charge.

24–4 At half voltage: $\frac{Q_0}{2} = Q_0 e^{-\frac{t}{RC}}$

$e^{-\frac{t}{RC}} = \frac{1}{2}$ [1]

Take natural logarithms: $-\frac{t}{RC} = \ln\frac{1}{2} = -0.693$

Rearrange: $t = 0.693\,RC$
Therefore the half-life time is about 69% of the time constant. [1]

25–1 Sketch graph with t on the x-axis and $\ln V$ on the y-axis [1]. Straight line going from top left to bottom right [1].

25–2 The gradient is $-1/RC$ (minus sign is important) [1].

25–3 The y-axis intercept is $\ln V_0$ [1].

25–4 The x-axis intercept shows the time taken for the voltage to fall to $1.0\,V$ (as $\ln 1 = 0$) [1].

26 *The temptation is to use the electric field equation $E = V/d$ to give the new electric field strength as $E/2$. This is not correct.*
$E = V/d$ and $V = Q/C$

Therefore: $E = \dfrac{Q}{Cd}$ [1]

Capacitance C is given by: $C = \varepsilon_0 \varepsilon_r \dfrac{A}{d}$ [1]

The relative permittivity of air is 1.0.

Therefore: $E = \dfrac{Qd}{\varepsilon_0 A d} = \dfrac{Q}{\varepsilon_0 A}$ [1]

Since no charge leaves the plates, the electric field remains the same. From this, the electric field is independent of the plate separation.

27–1 1. They consist of <u>polar molecules</u> which line up in the opposite direction to the capacitor's <u>electric field</u> [1]. This reduces the capacitor's electric field [1], so more electrons can crowd onto the negative plate and repel electrons from the positive plate [1].
(Mention of protons crowding is a physics error.)

2. Many insulating materials do not have polar molecules [1]. Therefore the relative permittivity is 1.0 [1], which is the same as air.

27–2 1. Equation: $C = \varepsilon_0 \varepsilon_r \dfrac{A}{d}$

$C = (8.85 \times 10^{-12}\,Fm^{-1} \times 32 \times 0.15\,m^2) \div 1.2 \times 10^{-4}\,m$ [1]
$C = 3.54 \times 10^{-7}\,F$ [1] $= 350\,nF$

2. $Q = CV$
$Q = 3.54 \times 10^{-7}\,F \times 6.0\,V$ [1] $= 2.124 \times 10^{-6}\,C$ [1]
$= 2.1 \times 10^{-6}\,C$ (2 s.f.) [1]

27–3 1. Sketch graph with Q on the y-axis and p.d. on the x-axis [1].
Straight line passing through the origin [1].

2. $E = \dfrac{1}{2}CV^2$

$E = \dfrac{1}{2} \times 3.54 \times 10^{-7}\,F \times (8.0\,V)^2$ [1] $= 1.13 \times 10^{-5}\,J$

$= 1.1 \times 10^{-5}\,J$ (2 s.f.) [1]
3. Work out the <u>gradient</u> [1].

28–1 $I_0 = 15\,V \div 4700\,\Omega = 3.2 \times 10^{-3}\,A$ [1]

28–2 $\tau = RC = 4700\,\Omega \times 680 \times 10^{-6}\,F = 3.196\,s = 3.2\,s$ [1]

28–3 Value for voltage read off graph:
At 3.2 s, $V = 9.4\,V$ (*Allow readings from 9.2 to 9.5*)
Percentage of the final voltage $= (9.4\,V \div 15\,V)$ [1]
$= 63\%$ (*Allow from 61% to 63% according to the graph reading*) [1]

28–4 Sketch graph: current on the y-axis and time, t, on the x-axis. Exponential decay curve sketched [1].
(*If an exponential rise is shown, zero marks.*)

28–5 The current is <u>greatest when the capacitor is uncharged</u> [1]. As the capacitor charges up, the electric field between the plates increases, which makes it harder for electrons to crowd on to the <u>negative</u> plate, leading to <u>reduction in current</u> [1].
(*For the second mark, the negative plate must be mentioned. Simple mention of the plates is not enough.*)

Magnetic fields (p.66)

Quick questions

1

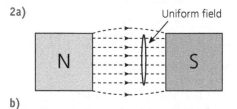

Strong magnetic field

Weak magnetic field

2a)

Uniform field

b)

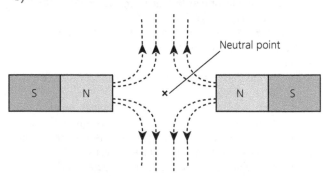

Neutral point

Neutral point is where the sum of magnetic forces is zero.

3 Current goes into the page

Current comes out of the page

4

Field lines should be consistent with orientation of the poles. Direction of force must be consistent with direction of the current.

5 Thumb, first finger and second finger held at right angles in three planes. First finger shows direction of field from N to S, second finger current, thumb shows direction of motion.

6a)

Symbol	Quantity	Unit	Vector or scalar
F	Force	Newton (N)	Vector
B	Flux density	Tesla (T)	Vector
I	Current	Ampere (A)	Scalar
l	Length of wire in the magnetic field	Metre (m)	Scalar

b) 1 Tesla (T) is the magnetic flux density that results in a force of 1 N when a conductor of unit length is at 90° across the field and carries a current of 1 A.

c) The equation becomes: $F = BIl \sin \theta$

7

Term	Symbol	What it means	Units	Equation
Magnetic flux density	B	The amount of magnetic flux through a unit area taken perpendicular to the direction of the magnetic flux.	Tesla (T) or weber per square metre (Wb m⁻²)	$B = \dfrac{\Phi}{A}$
Magnetic flux	Φ	**A measure of the quantity of magnetism, being the total number of magnetic lines of force passing through a specified area in a magnetic field.**	**Weber (Wb) or tesla square metre (T m²)**	$\Phi = BA\cos\theta$

Magnetic flux linkage	$N\Phi$ or λ	**The product of the magnetic flux and the number of turns in a given coil.**	**Weber turns**	$\lambda = N\Phi$ $N\Phi = BAN\cos\theta$

8 θ on x-axis; flux linkage on y-axis. Curve drawn in the shape of $\cos\theta$, i.e. starting at a maximum at 0, going to a minimum at π and back to a maximum at 2π.

9a) Work out f first: $f = 1200\,\text{min}^{-1} \div 60\,\text{s min}^{-1} = 20\,\text{Hz}$
Work out the angular velocity: $\omega = 2 \times \pi \times f = 2 \times \pi \times 20.0\,\text{Hz} = 126\,\text{rad s}^{-1}$
Now work out e.m.f.: $\varepsilon_0 = BAN\omega$
$E_0 = 0.050\,\text{T} \times 0.15\,\text{m}^2 \times 150\,\text{turns} \times 126\,\text{rad s}^{-1} = 142\,\text{V}$

b) Use: $\varepsilon = BAN\omega \sin\omega t$
In this case $BAN\omega = 141.75\,\text{V}$
$E = 141.75\,\text{V} \times \sin(126\,\text{rad s}^{-1} \times 0.065\,\text{s}) = 133.8\,\text{V} = 134\,\text{V}$
Remember: your calculator should be set to radians.

10

Control	What it does
On/off switch	Turns the instrument on
Brightness	**Makes the display brighter or dimmer**
Focus	**Makes the line sharp or fuzzy**
Time base	**Increases or decreases the period of a displayed waveform** (NOT increases or decreases the wavelength)
y-gain or voltage gain	**Increases or decreases the scaling of the signal; increasing y-gain causes each vertical waveform to represent a greater p.d.**

Exam-style questions

11 **D: V s⁻¹ [1]**

12 **C: $2r$ [1]**

13 Graph **B** [1]

14 **B: The core reduces eddy currents but does not eliminate them entirely. [1]**

15 **D: Z first, Y second and X third [1]**

16–1 Experiment method:
- Measure the length of the wire that is in the magnetic field [1].
- Find the current that produces a measurable change in the reading of the balance [1].
- Change the current for at least 5 different readings [1].
- Measure the change in the readings [1].

Safety:
You will be using large currents. You must turn the current off between readings, otherwise the rheostat will get hot enough to cause burns, or even burn out. To aid this, a push switch should be in the circuit. [1]

16–2 *Any four from:*
- The balance will read in grams (g) [1].
- The data taken will be current (A) and <u>change</u> in reading (g) [1].
- At least two readings taken for each current [1].
- Average is taken [1].
- The average change in reading will need to be converted to newtons (N) by multiplying by 0.0098 [1].
- A single table should show the raw data and the processed data [1].

16–3 Sketch graph: axes labelled [1], straight line through the origin [1].

16–4 Gradient = Bl; B = gradient ÷ length of the wire <u>in the magnetic field</u> [1]

16–5 Rearrange: $B = \dfrac{F}{Il} = (0.063\,\text{N}) \div (4.5\,\text{A} \times 0.050\,\text{m}) = 0.28\,\text{T}$ [1]

17–1 We know: $F = BIl$; $l = vt$; and $I = q/t$ [1].

Substitute into first equation: $F = B \times \dfrac{q}{t} \times vt$

The t terms cancel out to give: $F = Bqv$ [1].

17–2 The path is circular because the force acting on the particle is always at 90° [1].

17–3 Start off with: $F = \dfrac{mv^2}{r}$

Therefore: $F = \dfrac{mv^2}{r} = Bqv$ [1]

Therefore: $Bq = \dfrac{mv}{r}$

Rearrange to: $r = \dfrac{mv}{Bq}$ [1]

17–4 If the charge is changed, the direction of the circular path is changed [1].

17–5 Rearrange to: $B = \dfrac{mv}{rq}$ [1]

$B = (9.11 \times 10^{-31}\,\text{kg} \times 1.6 \times 10^6\,\text{m s}^{-1}) \div (0.055\,\text{m} \times 1.6 \times 10^{-19}\,\text{C}) = 1.7 \times 10^{-4}\,\text{T}$ [1]

17–6 Fleming's LH rule concerns conventional currents. The flow of electrons is in the opposite direction to a conventional current [1].

18–1 Rearrange for mass: $m = \dfrac{Bqr}{v}$ [1]

$m = (0.920\,\text{T} \times 1.60 \times 10^{-19}\,\text{C} \times 0.500\,\text{m}) \div 3.00 \times 10^7\,\text{m s}^{-1}$
$= 2.45 \times 10^{-27}\,\text{kg}$ [1]

18–2 $m_p / m_e = 2.45 \times 10^{-27}\,\text{kg} \div 9.11 \times 10^{-31}\,\text{kg} = 2700$ [1]

18–3 Equation: $f = \dfrac{Bq}{2\pi m}$

$f = (2.30\,\text{T} \times 1.60 \times 10^{-19}\,\text{C}) \div (2 \times \pi \times 1.67 \times 10^{-27}\,\text{kg})$
$= 3.51 \times 10^7\,\text{Hz}$ [1]

18–4 Rearrange equation for mass: $m = \dfrac{Bq}{2\pi f}$ [1]

$m = (1.50\,\text{T} \times 2 \times 1.60 \times 10^{-19}\,\text{C})$
$\div (2 \times \pi \times 2.00 \times 10^6\,\text{Hz}) = 3.82 \times 10^{-26}\,\text{kg}$ [1]

18–5 Mass of particle ÷ mass of proton = $3.82 \times 10^{-26}\,\text{kg}$
$\div 1.67 \times 10^{-27}\,\text{kg} = 23$ times [1]

19–1 Area of coil = $\pi r^2 = \pi \times (0.06\,\text{m})^2 = 0.0113\,\text{m}^2$ [1]
$\Phi = BA = 0.15\,\text{T} \times 0.0113\,\text{m}^2 = 1.7 \times 10^{-3}\,\text{Wb}$ [1]

19–2 $\Phi = BA \cos \theta = (0.15\,\text{T} \times 0.0113\,\text{m}^2) \times \cos 50°$
$= 1.1 \times 10^{-3}\,\text{Wb}$ [1]

19–3 $A = 4.0 \times 10^{-2}\,\text{m} \times 5.0 \times 10^{-2} = 2.0 \times 10^{-3}\,\text{m}^2$ [1]
$N\Phi = BAN \cos \theta = (0.125\,\text{T} \times 2.0 \times 10^{-3}\,\text{m}^2 \times 120\,\text{turns})$
$\times \cos 30°$
$N\Phi = 0.026\,\text{Wb turns. (2 s.f.)}$ [1]

20–1 The induced e.m.f. across a conductor is equal to the rate at which flux is cut [1].

20–2 The direction of any induced current is such as to oppose the flux change that caused it [1].

20–3 $\varepsilon = -N\dfrac{\Delta\Phi}{\Delta t}$ [1]

20–4 Area of the coil: $A = \pi \times (0.075\,\text{m})^2 = 0.0177\,\text{m}^2$ [1]
Work out the flux: $\Phi = BA = 0.15\,\text{T} \times 0.0177\,\text{m}^2$
$= 2.65 \times 10^{-3}\,\text{Wb}$ [1]
Rate of change of flux:
$\Delta\Phi/\Delta t = -2.65 \times 10^{-3}\,\text{Wb} \div 10\,\text{s} = -2.65 \times 10^{-4}\,\text{Wb s}^{-1}$
(*Minus sign because the magnetic field strength is decreasing to zero.*) [1]
e.m.f. $= -(-2.65 \times 10^{-4}\,\text{Wb s}^{-1} \times 500\,\text{turns}) = 0.133\,\text{V}$
$= 0.13\,\text{V (2 s.f.)}$ [1]

20–5 equations: $\varepsilon = -N\dfrac{\Delta\Phi}{\Delta t}$ and $\Delta\Phi = B\Delta A$

Therefore $\varepsilon = -N\dfrac{B\Delta A}{\Delta t}$ [1]

Rearrange for B: $B = -\dfrac{\varepsilon \Delta t}{N\Delta A}$

Change in area $\Delta A = 0 - 1.5 \times 10^{-4}\,\text{m}^2 = -1.5 \times 10^{-4}\,\text{m}^2$
(*Area in the B-field is decreasing.*)
$B = -(0.75\,\text{V} \times 0.3\,\text{s}) \div (2500\,\text{turns} \times -1.5 \times 10^{-4}\,\text{m}^2)$
$= 0.60\,\text{T}$ [1]

20–6 In the case of the copper tube, the magnetic field from the magnet induces <u>eddy currents</u>. The eddy currents form a magnetic field that opposes the movement of the magnet due to <u>Lenz's law</u> [1].
In the case of the PVC tube, the material does not conduct, so no eddy currents are formed. Therefore there is no opposition to the movement of the magnet [1].

21–1 $B_{\text{vert}} = 6.5 \times 10^{-5}\,\text{T} \times \cos 35 = 5.32 \times 10^{-5}\,\text{T} = 53\,\mu\text{T.}$ [1]

21–2 $E = 5.32 \times 10^{-5}\,\text{T} \times 150\,\text{m s}^{-1} \times 35\,\text{m} = 0.28\,\text{V}$ [1]

22–1 Direct current from a battery moves in one direction only, from positive to negative [1]. In alternating current, the direction is changing all the time [1].

22–2 1. $I_{\text{rms}} = \dfrac{5.0\,\text{A}}{\sqrt{2}} = 3.54\,\text{A}$ [1]

2. $V_{\text{rms}} = 3.54\,\text{A} \times 10\,\Omega = 35.4\,\text{V}$ [1]

3. $P_0 = (5.0\,\text{A})^2 \times 10\,\Omega = 250\,\text{W}$ [1]

4. $P = 250\,\text{W} \div 2 = 125\,\text{W}$ [1]

22–3 $V_{\text{pk}} = 12.0\,\text{V} \times \sqrt{2} = 17.0\,\text{V}$ [1]

23–1 6.5 divisions = $V_{\text{pk to pk}} = 6.5\,\text{cm} \times 0.50\,\text{V cm}^{-1} = 3.25\,\text{V}$ [1]
(*Allow a reading of 6.4 divisions and answer of 3.2 V.*)

23–2 $V_0 = 3.25\,\text{V} \div 2 = 1.625\,\text{V}$ [1]

23–3 $V_{\text{rms}} = \dfrac{1.625\,\text{V}}{\sqrt{2}} = 1.15\,\text{V}$ [1]

23–4 8.4 div for 3 whole waves.
Therefore period for 3 waves = $8.4 \times 2.0 \times 10^{-3}\,\text{s}$
$= 16.8 \times 10^{-3}\,\text{s}$ [1]
$T = 16.8 \times 10^{-3}\,\text{s} \div 3 = 5.6 \times 10^{-3}\,\text{s}$ [1]

23–5 $f = 1/T = (5.6 \times 10^{-3}\,\text{s})^{-1} = 179\,\text{Hz}$ [1]

24–1 Primary and secondary coils [1] wrapped round a laminated soft iron core, but not touching [1].

Laminated soft iron core

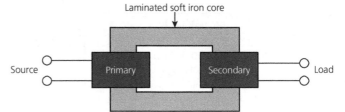

Source
Primary   Secondary
Load

24–2 $\dfrac{N_p}{N_s} = \dfrac{V_p}{V_s}$ [1]

24–3 Power in = power out so $V_P I_P = V_S I_S$ [1]

Therefore: $\dfrac{V_p}{V_s} = \dfrac{I_s}{I_p}$

Therefore: $\dfrac{N_p}{N_s} = \dfrac{I_s}{I_p}$ [1]

24–4 1. N_p/N_s = turns ratio
Turns ratio = $3600 \div 150 = 24{:}1$ [1]

2. Output voltage = $230\,\text{V} \div 24 = 9.6\,\text{V}$; output current
$= 1.5\,\text{A} \times 24 = 36\,\text{A}$ [1]

25–1 The energy losses arise from (*any three from*):
- <u>Resistance</u> in the coils. If the currents are large, the power loss can become large, as the heating effect is governed by $P = I^2 R$ [1].
- <u>Eddy currents</u> in the laminations of the core. These are reduced by lamination, but not stopped altogether [1].
- Work needs to be done to build up the magnetic field in the core. Not so much work is output when the domains randomise again. This is called <u>hysteresis</u> [1].
- At a certain current level, all domains are lined up. The magnet is <u>saturated</u>, and cannot become more magnetised. Therefore, a greater current will not pass across to the secondary coil [1].

25-2 $I = 200 \times 10^6\,\text{W} \div 15\,000\,\text{V} = 13\,333\,\text{A} = 13\,000\,\text{A}$ [1]

25-3 1. Power loss = $I^2R = (13\,333\,\text{A})^2 \times 0.75 = 133 \times 10^6\,\text{W}$ [1]

2. Power available to the factory
$= 200 \times 10^6\,\text{W} - 133 \times 10^6\,\text{W} = 67 \times 10^6\,\text{W}$ [1]

25-4 A step-up transformer would increase the voltage ten times. At 150 000 V, the current would be about 1300 A. The power loss would be 100 times less ($1.33 \times 10^6\,\text{W}$) so that there would be 199 MW available to the factory [1]. There would need to be a second transformer (a step-down transformer) to reduce the voltage to a level at which it can be used [1].

Topic review: fields and their consequences (p.73)

1-1 Orbits are <u>elliptical</u> [1].

1-2 Missing values are: r^3/m^3: 1.99×10^{26} [1]; T^2/s^2: 1.28×10^{11} [1]

1-3 *Graph with:*
- axes labelled [1]
- accurate plotting (± 1 mm) [1]
- line of best fit [1].

1-4 $T^2 \propto r^3$
$r^3 = (4.77 \times 10^8\,\text{m})^3 = 1.1 \times 10^{26}\,\text{m}^3$ [1]
Use of graph
$T^2 = 7.6 \times 10^{11}\,\text{s}^2$
$T = 8.7 \times 10^5\,\text{s}$ [1]
$T = 8.7 \times 10^5\,\text{s} \div 86\,400\,\text{s dy}^{-1} = 10\,\text{dy}$ [1]

1-5 Work out gradient:
Rise = $1.23 \times 10^{12}\,\text{s}^2$; Run = $1.80 \times 10^{26}\,\text{m}^3$
Gradient = $1.23 \times 10^{12}\,\text{s}^2 \div 1.80 \times 10^{26}\,\text{m}^3 = 6.83 \times 10^{-15}$ [1] $\text{s}^2\,\text{m}^{-3}$ [1]

1-6 Equation: $T^2 = \left(\dfrac{4\pi^2}{GM}\right) r^3$

Equation of a straight line is: $y = mx + c$

Therefore: Gradient $= \left(\dfrac{4\pi^2}{GM}\right)$

Rearrange equation to [1]: $M = \left(\dfrac{4\pi^2}{G \times \text{Gradient}}\right)$
Substitute:
$$M = \left(\dfrac{4\pi^2}{6.67 \times 10^{-11}\,\text{Nm}^2\,\text{kg}^{-2}\,\text{kg} \times 6.83 \times 10^{-15}\,\text{s}^2\,\text{m}^{-3}}\right)$$

$M = 8.7 \times 10^{25}\,\text{kg}$ [1] *(Allow ± 10% in student answers)*
If the value $7 \times 10^{-15}\,\text{s}^2\,\text{m}^{-3}$ is used, M = $8.5 \times 10^{25}\,\text{kg}$
(The quoted mass is $8.681 \times 10^{25}\,\text{kg}$.)

2-1 Gravitational force per unit point mass [1].

2-2

Height/m	Radius/m	g/N kg^{-1}	Log$_{10}(r)$	Log$_{10}(g)$
0.00	6.37×10^6	9.81	6.804	0.992
1.00×10^6	**7.37×10^6**	7.34	**6.867**	**0.866**
2.00×10^6	**8.37×10^6**	5.69	**6.923**	**0.755**
3.00×10^6	**9.37×10^6**	4.54	**6.972**	**0.657**
4.00×10^6	**1.04×10^7**	3.71	**7.016**	**0.569**
5.00×10^6	**1.14×10^7**	3.09	**7.056**	**0.490**
6.00×10^6	**1.24×10^7**	2.61	**7.092**	**0.417**
7.00×10^6	**1.34×10^7**	2.23	**7.126**	**0.348**
8.00×10^6	**1.44×10^7**	1.93	**7.157**	**0.286**
9.00×10^6	**1.54×10^7**	1.69	**7.187**	**0.228**
1.00×10^7	**1.64×10^7**	1.49	**7.214**	**0.173**

[3] *(1 mark for each correct column)*

2-3 *Graph with:*
- axes labelled [1]
- plotting [1]
- line of best fit [1].

2-4 Answer marked on the graph. Vertical line from lg (r) = 6.804 to intercept the line at lg (g) = 0.992. *(Do not award mark if 6.37 or 0.98 are used.)* [1]

2-5 1. Rise calculated = −0.8 [1] *(Must have a negative sign for the first of these marks)*
Run calculated = 0.4
Gradient = −2 [1]

2. It shows that: $g \propto \dfrac{1}{r^2}$ [1]

Explanation: $g = GM \times r^{-2}$
Taking logs: $\log_{10} g = \log_{10}(GM) + -2\log_{10} r$
Or: $\log_{10} g = -2\log_{10} r + \log_{10}(GM)$
Matches up with: $y = mx + c$
Therefore the gradient is −2 [1].

3-1 Force per unit charge [1]

3-2 $E = V/d = 1500\,\text{V} \div 5.0 \times 10^{-2}\,\text{m} = 30\,000\,\text{V m}^{-1}$ [1]

3-3 $F = EQ = 30\,000\,\text{V m}^{-1} \times 3.20 \times 10^{-19}\,\text{C} = 9.60 \times 10^{-15}\,\text{N}$ [1]

3-4 1. $a = F/m = 9.60 \times 10^{-15}\,\text{N} \div 6.64 \times 10^{-27}\,\text{kg}$
$= 1.45 \times 10^{12}\,\text{m s}^{-2}$ [1]

2. The direction is <u>vertically downwards</u> and the path is <u>parabolic</u> [1].

4-1 Equation: $V = \dfrac{1}{4\pi\varepsilon_0} \dfrac{Q}{r}$

Rearrange: $Q = 4\pi\varepsilon_0 Vr$
$Q = 4 \times \pi \times 8.85 \times 10^{-12}\,\text{F m}^{-1} \times 5000\,\text{V} \times 0.15\,\text{m}$
$= 83.4 \times 10^{-9}\,\text{C}$ [1]

4-2 Missing numbers are:

Radius, r/m	Potential, V/V	Log$_{10}$ (r/m)	Log$_{10}$ (V/V)
1.95	390	**0.290**	**2.59**
2.25	320	**0.352**	**2.51**
2.55	290	**0.407**	**2.46**
2.85	260	**0.455**	**2.41**
3.15	240	**0.498**	**2.38**

[2] *(1 mark for each correct column)*

4-3 *Graph with:*
- axes labelled [1]
- plot (± 1 mm) [1]
- line of best fit [1].

4-4 1. Working out of gradient shown on graph.
Rise determined, e.g. $2.30 − 3.70 = −1.4$
Run determined, e.g. $0.57 − -0.83 = 1.4$
Gradient = $-1.4/1.4$ [1] = −1 [1].

2. This gradient shows that potential and radius are related:
$V \propto \dfrac{1}{r}$ or $V = r^{-1}$ [1]

5-1

Time/s	Voltage/V	ln (V/V)
7.0	0.48	**−0.734**
8.0	0.31	**−1.171**
9.0	0.20	**−1.609**
10.0	0.13	**−2.040**

(1 mark for the correct column)

5-2 *Graph with:*
- axes labelled [1]
- plot [1]
- straight line drawn through points [1].

5-3 1. • Gradient: Large triangle shown [1]
- Gradient = $-3.50 \div 8.0\,\text{s} = -0.4375\,\text{s}^{-1}$ [1] *(Correct unit for the mark)*
- Time constant = −gradient = $(0.4375\,\text{s}^{-1})^{-1}$ = 2.29 s = 2.3 s [1]

2. $R = \tau/C = 2.29\,\text{s} \div 590 \times 10^{-6}\,\text{F} = 3874\,\Omega = 3900\,\Omega$ [1] *(to 2 s.f.)* [1].

5–4 There will be no change [1] to the time constant, because the resistance and the capacitance are both constant quantities [1], unaffected by the start voltage.

6–1

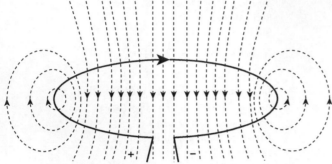

Magnetic field shown in the correct direction, with uniform magnetic field shown in the middle [1].

6–2 1.

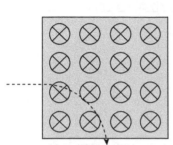

Magnetic field into page

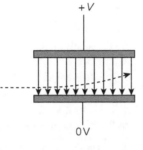

Uniform electric field

- Circular path for magnetic field and direction is downwards [1].
- Parabolic path for electric field and direction is upwards [1].

2. The magnetic field provides a force that always acts at 90° to the path of the electron, so the path is circular. The direction is downwards because the electrons are moving in the opposite direction to a conventional current [1]. (*Both points for the mark.*)
The path in the electric field is parabolic, because the electron accelerates along the electric field lines. The direction is upwards, because the electron is attracted to the positive plate [1]. (*Both points for the mark.*)

6–3 Equation to start with: $F = \dfrac{mv^2}{r} = Bqv$

Divide both sides by v: $Bq = \dfrac{mv}{r}$ [1]

Rearrange: $Bqr = mv$ [1]

Hence: $r = \dfrac{mv}{Bq}$

6–4 $r = (9.11 \times 10^{-31}\,\text{kg} \times 1.5 \times 10^7\,\text{m s}^{-1}) \div (1.75 \times 10^{-3}\,\text{T} \times 1.6 \times 10^{-19}\,\text{C})$
$r = 4.88 \times 10^{-2}\,\text{m}$ [1] $= 4.9\,\text{cm}$ (to 2 s.f.) [1]

6–5 The path would be spiral [1]. Collisions with air molecules would make the electrons lose energy, so the speed will be reduced hence the radius would be reduced [1].

7–1 1. [1]   2. [1]

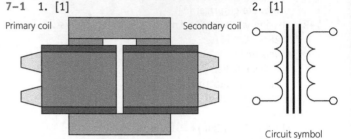

Primary coil   Secondary coil

Laminated soft iron core

Circuit symbol

7–2 1. There have been large eddy currents induced by the alternating magnetic field from the primary, which are much reduced with a laminated block [1].
2. The tutor has used a solid block of soft iron instead of a laminated block and the soft iron block should be replaced with a laminated one [1].

7–3 The coils will have a certain value of resistance. If the currents are large, the energy loss is large, since $P = I^2R$ [1]. (*Both points for the mark. Allow equation for second point. Allow alternative answer based on sound energy loss.*)
With the laminated soft iron core, there are still eddy currents, even though they are much reduced. These will heat up the core. [1]

7–4 1. Power of furnace $= 1000\,\text{V} \times 40\,000\,\text{A} = 4.0 \times 10^7\,\text{W}$ [1]
2. Power required by the transformer $= 4.0 \times 10^7\,\text{W} \div 0.95 = 4.21 \times 10^7\,\text{W}$
Current $= 4.21 \times 10^7\,\text{W} \div 1.32 \times 10^5\,\text{V} = 319\,\text{A} = 320\,\text{A}$ (2 s.f.) [1]
3. Power lost $= 4.21 \times 10^7\,\text{W} - 4.0 \times 10^7\,\text{W} = 0.21 \times 10^7\,\text{W} = 2.1 \times 10^6\,\text{W}$ [1]
4. The oil is cooled in heat exchangers that transfer the heat to the air [1].

Radiation and the nucleus (p.78)

Quick questions

1

Radiation	Description and charge	Penetration	Ionisation	Effect of E or B field
Alpha (α)	**Helium nucleus** **2p + 2n** $Q = +2e$	**Few cm air, thin paper**	**Intense, about 10^4 ion pairs per mm.**	**Slight deflection as a positive charge**
Beta (β)	**High speed electron** $Q = -1e$	**Few mm of aluminium**	**Less intense than** α, **about 10^2 ion pairs per mm.**	**Strong deflection in opposite direction to** α.
Gamma (γ)	**Very short wavelength EM radiation**	**Several cm lead, 2 m of concrete**	**Weak interaction about 1 ion pair per mm.**	**No effect**

2a) An isotope has the same number of protons, but different numbers of neutrons.

b) $^{226}_{88}\text{Ra}$

c) $^{226}_{88}\text{Ra} \rightarrow\, ^{222}_{86}\text{Rn} +\, ^4_2\text{He}$

d) Half-life is the time taken for half the original nuclei of a radioactive isotope to decay (or the time taken for the activity of the sample to halve).

3a) $^{24}_{11}\text{Na} \rightarrow\, ^{24}_{12}\text{Mg} +\, ^{\ 0}_{-1}e^- + \bar{v}_e$

b) $^{22}_{11}\text{Na} \rightarrow\, ^{22}_{10}\text{Ne} +\, ^0_1 e^+ + v_e$

4 *Any two from:*
- Most of the atom was empty space.
- The positive charge was concentrated in a very small space.
- The radius of the nucleus was in the order of $3 \times 10^{-14}\,\text{m}$.
- The alpha particles that were deflected through an angle of 180° (a very small proportion) had to be travelling in a line with the nucleus.

5 The cobalt-60 specimen is held in a metal container that has a window made of aluminium of about 3 mm thickness. This stops both alpha and beta radiation, allowing just gamma radiation to pass.

6a) $^{222}_{88}Rn \rightarrow ^{218}_{84}X + ^{4}_{2}He(+E)$ X is polonium (Po).

b) While the alpha decay from the radium is stopped by the skin, radon can be breathed in and this will expose the delicate tissues of the lungs to the alpha radiation. Alpha radiation is intensely ionising, leading to damage to DNA directly, or indirectly by making free radicals, which can damage DNA and other biological molecules.
The risk can be minimised by having ventilation systems that prevent build-up of radon gas.

7 Electrons have wave properties as well as particle properties. As the speed approaches the speed of light, the mass increases because of relativistic effects.

8 Equation: $r = r_0 A^{\frac{1}{3}}$

$r = 1.2 \times 10^{-15} \, m \times (59)^{\frac{1}{3}} = 4.67$ (2 s.f.)
$\times 10^{-15} \, m = 4.7 \times 10^{-15} \, m$

Exam-style questions

9 **C:** A beta-minus particle is emitted, followed by a gamma photon. [1]

10 **A:** $9.0 \times 10^{-14} \, m$ [1]

11 **C:** 11.4 fm [1]

12 **C:** $\frac{A}{16}$ [1]

13–1 Alpha particle A passes straight through the atom. Alpha particle B is reflected backwards. Alpha particle C is deflected slightly as it passes though the atom. (2 marks all 3 correct, 1 mark 2 correct, 0 marks 1 or 0 correct)

13–2 The nucleus is much smaller than the atom/most of the atom is empty space [1].
Most of the mass of the atom is contained in the nucleus [1].

13–3 1. Charge = proton number × electronic charge
$Q_\alpha = 2 \times 1.6 \times 10^{-19} C = 3.2 \times 10^{-19} C$
$Q_{Au} = 79 \times 1.6 \times 10^{-19} C = 1.264 \times 10^{-17} C$ [1] (Both calculations correct for mark.)
2. $E = 5.0 \times 10^6 \, eV \times 1.6 \times 10^{-19} \, J \, eV^{-1} = 8.0 \times 10^{-13} J$ [1]
3. Rearrange: $r = \frac{1}{4\pi\varepsilon_0} \frac{Q_\alpha Q_{Au}}{E}$
$r = (4\pi\varepsilon_0)^{-1} \times ((3.2 \times 10^{-19} C \times 1.264 \times 10^{-17} C) \div 8.0 \times 10^{-13} J)$
$r = 8.99 \times 10^9 \, m \, F^{-1} \times 5.056 \times 10^{-24} \, C^2 \, J^{-1}$
$r = 4.545 \times 10^{-14} \, m = 4.5 \times 10^{-14} \, m$ (2 s.f.) [1]

13–4 The minimum approach distance is not the same as the radius. The positive electromagnetic force from the nucleus repels the positive alpha particles at a point that is about 10 times the true radius of the nucleus [1]. The nucleus does not have a distinct boundary; its edge is fuzzy [1].

14–1 Any two from:
- people (who contain carbon-14, a beta-minus emitter) [1]
- wooden desks (see above) [1]
- rocks in the ground [1]
- cosmic rays [1]
- radioisotopes in the bricks and concrete in the school building [1]
- medical and industrial sources [1].

14–2 • Make sure that all sources are away in their containers [1].
• Leave the detector in the open lab connected to the counter. Reset the counter to zero [1].
• Time for a long period of time (e.g. 10 minutes) as the background decay events are random [1].
• Divide the count by time period (in seconds) to obtain a background count rate [1].

14–3 Count = 867 ÷ 900 s = 0.963 Bq [1]

14–4 1. Activity of the sample = 1800 min⁻¹ − 85 min⁻¹ = 1715 min⁻¹.
1715 ÷ 60 = 28.6 Bq = 29 Bq (2 s.f.) [1]
2. 20 h = 72 000 s; number of disintegrations = 72 000 s × 28.6 Bq = 2.06 × 10⁶ disintegrations [1]

14–5 Corrected count = 1000 cpm − 90 cpm = 910 cpm [1]
Equation: $\frac{I_1}{I_2} = \left(\frac{x_2}{x_1}\right)^2$ so $\frac{910}{I_2} = \left(\frac{60 \, cm}{15 \, cm}\right)^2$
New count = 910 cpm ÷ 16 = 57 cpm [1]
Expected count = 57 cpm + 90 cpm = 147 cpm [1]

15–1 The absorbed dose is defined as energy absorbed per unit mass. [1] Unit: $J \, kg^{-1}$ [1].

15–2 $1 \, Gy = 1 \, J \, kg^{-1}$ [1]

15–3 1. The nature of the radiation, e.g. gamma rays. This is given the term W_R (weighting factor – a dimensionless unit) [1]
2. The unit is Sievert (Sv). $1 \, Sv = 1 \, Gy = 1 \, J \, kg^{-1}$ [1]

15–4 Dose = energy ÷ mass = 12 J ÷ 0.150 kg = 80 Gy [1]

15–5 Equivalent dose = 80 Gy × 3 = 240 Sv [1]

16–1 $E = 268 \times 10^6 \, eV \times 1.60 \times 10^{-19} \, J \, eV^{-1} = 4.3 \times 10^{-11} J$ [1]

16–2 $p = \frac{E}{c}$
$p = 4.29 \times 10^{-11} J \div 3.00 \times 10^8 \, m \, s^{-1} = 1.43 \times 10^{-19}$ [1] $kg \, m \, s^{-1}$ [1]

16–3 $\lambda = \frac{h}{p}$
$\lambda = 6.63 \times 10^{-34} \, J \, s \div 1.43 \times 10^{-19} \, kg \, m \, s^{-1} = 4.64 \times 10^{-15} m$ [1]

16–4 $d = (1.22 \times 4.64 \times 10^{-15} m) \div \sin 55 = 6.91 \times 10^{-15} m$ [1]
$r = 6.91 \times 10^{-15} m \div 2 = 3.45 \times 10^{-15} m$ [1]

17–1 The gamma source is shielded from the outside by 3 mm aluminium to absorb beta-minus particles [1].
The anode in the Geiger–Müller tube is set back by about 1 cm from the mica window [1].

17–2 There is background radiation in the room [1].

17–3

Distance x/m	Count 1/s⁻¹	Count 2/s⁻¹	Corrected count 1/s⁻¹	Corrected count 2/s⁻¹	Average/s⁻¹	1/average⁰·⁵/s⁰·⁵
0.05	8.82	8.93	8.43	8.54	**8.49**	**0.343**
0.1	3.32	3.1	2.93	2.71	**2.82**	**0.595**
0.15	1.75	1.76	1.36	1.37	**1.37**	**0.856**
0.2	1.26	1.23	0.87	0.84	**0.86**	**1.081**
0.25	0.95	1.1	0.56	0.71	**0.64**	**1.255**
0.3	0.87	0.82	0.48	0.43	**0.46**	**1.482**
0.35	0.72	0.71	0.33	0.32	**0.33**	**1.754**
0.4	0.64	0.62	0.25	0.23	**0.24**	**2.041**

[2] (1 mark for each correct column)

17–4 *Graph with:*
- axes labelled [1]
- sensible scales [1]
- plot (± 1 mm) [1]
- line of best fit [1].

17–5
- Evidence of extrapolation of the graph below the count rate axis [1].
- Estimation of the distance to give c (*using graph from* **Answer 17–4**) [1].

18–1 Equation: $\sin\theta = 1.22\dfrac{\lambda}{d}$

1. The angle of the first minimum would increase, because the de Broglie wavelength increases, therefore $\sin\theta$ will increase [1].
2. The angle of the first minimum would increase, because the diameter will decrease, making $\sin\theta$ increase [1].

18–2 1. Work out energy: $E = 550 \times 10^6\,\text{eV} \times 1.60 \times 10^{-19}\,\text{J eV}^{-1}$
$= 8.80 \times 10^{-11}\,\text{J}$ [1]

Work out momentum: $p = \dfrac{E}{c}$

$p = 8.80 \times 10^{-11}\,\text{J} \div 3.00 \times 10^8\,\text{m s}^{-1}$

$= 2.93 \times 10^{-19}\,\text{kg m s}^{-1}$ [1]

Now work out λ: $\lambda = \dfrac{h}{p}$

$\lambda = 6.63 \times 10^{-34}\,\text{J s} \div 2.93 \times 10^{-19}\,\text{kg m s}^{-1}$
$= 2.26 \times 10^{-15}\,\text{m} = 2.3 \times 10^{-15}\,\text{m}$ (2 s.f.) [1]

2. $d = \dfrac{1.22\lambda}{\sin\theta}$

$d = (1.22 \times 2.3 \times 10^{-15}\,\text{m}) \div (\sin(10.6°))$
$= 1.50 \times 10^{-14}\,\text{m}$ [1]
(*Use of 2.3 × 10⁻¹⁵ m gives 1.53 × 10⁻¹⁴ m*)

19–1 1. $m = 58.93\,\text{u} \times 1.661 \times 10^{-27}\,\text{kg u}^{-1} = 9.79 \times 10^{-26}\,\text{kg}$ [1]

2. Assuming a sphere: $V = \dfrac{4}{3}\pi r^3$

$V = 4 \times \pi \times (4.67 \times 10^{-15}\,\text{m})^3 \div 3 = 4.27 \times 10^{-43}\,\text{m}^3$ [1]
$\rho = 9.79 \times 10^{-26}\,\text{kg} \div 4.27 \times 10^{-43}\,\text{m}^3 = 2.29 \times 10^{17}\,\text{kg m}^{-3} = 2.3 \times 10^{17}\,\text{kg m}^{-3}$ (2 s.f.) [1]

19–2 1.

Element	Nucleon	$A^{\frac{1}{3}}$	r/fm
K	39	3.39	4.07
Ca	40	3.42	4.10

[2] (*1 mark for each correct row*)

2.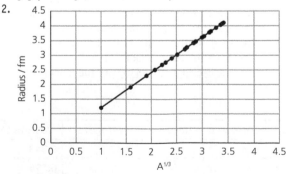

Graph with:
- suitable axes and scales [1]
- points plotted [1]
- line of best fit drawn [1].

3. Graph with A on the horizontal axis and ρ on the vertical axis; no values needed [1].
Straight horizontal line drawn labelled 2.3 × 10¹⁷ kg m⁻³. Note that the line should have its lower limit at $A = 1$ [1].

4. Nuclear density is a constant at $2.3 \times 10^{17}\,\text{kg m}^{-3}$ [1]

20–1 $\dfrac{R_1}{\left(A_1\right)^{\frac{1}{3}}} = r_0 = \dfrac{R_2}{\left(A_2\right)^{\frac{1}{3}}}$

Therefore: $\dfrac{R_1}{\left(A_1\right)^{\frac{1}{3}}} = \dfrac{R_2}{\left(A_2\right)^{\frac{1}{3}}}$ [1]

Therefore: $\dfrac{R_1}{R_2} = \dfrac{\left(A_1\right)^{\frac{1}{3}}}{\left(A_2\right)^{\frac{1}{3}}}$ [1]

20–2 Substitute: $\dfrac{R_1}{3.7 \times 10^{-15}\,\text{m}} = \left(\dfrac{99}{28}\right)^{\frac{1}{3}}$

$R_1 = 3.7 \times 10^{-15} \times 1.523 = 5.6 \times 10^{-15}\,\text{m}$ (2 s.f.) [1]

Radioactive decay (p.83)

Quick questions

1a) Number of moles = $1.0\,\text{kg} \div 0.014\,\text{kg mol}^{-1} = 71.4\,\text{mol}$
Number of particles = $71.4\,\text{mol} \times 6.02 \times 10^{23}$
$= 4.284 \times 10^{25}\,\text{kg}^{-1}$
Number of C-14 particles = $4.284 \times 10^{25}\,\text{kg}^{-1} \times 1.4 \times 10^{-12}$
$= 6.0 \times 10^{13}\,\text{kg}^{-1}$

b) $\Delta N/\Delta t = -260\,\text{Bq kg}^{-1}$
$-260\,\text{Bq kg}^{-1} = -\lambda \times 6.0 \times 10^{13}\,\text{kg}^{-1}$
$-\lambda = (260\,\text{Bq kg}^{-1} \div 6.0 \times 10^{13}\,\text{kg}^{-1}) = 4.3 \times 10^{-12}\,\text{s}^{-1}$

2a) Activity is defined as the number of emissions per second (or the number of nuclei that decay every second).

b) Number of particles = 3.96×10^{21}
$A = 4.29 \times 10^{-12}\,\text{s}^{-1} \times 3.96 \times 10^{21} = 1.7 \times 10^{10}\,\text{Bq}$

3a) Work out time in s:
$t = 250\,\text{y} \times 365\,\text{d y}^{-1} \times 86\,400\,\text{s d}^{-1} = 7.884 \times 10^9\,\text{s}$
$\lambda = (0.693 \div 7.884 \times 10^9\,\text{s}) = 8.79 \times 10^{-11}\,\text{s}^{-1}$

b) Work out time in s:
$t = 60\,\text{y} \times 365\,\text{d y}^{-1} \times 86\,400\,\text{s d}^{-1} = 1.89 \times 10^9\,\text{s}$
Use: $A = A_0 e^{-\lambda t}$
$\ln A - \ln A_0 = -\lambda t$
$\ln A - \ln(5000) = -1.89 \times 10^9\,\text{s} \times 8.79 \times 10^{-11}\,\text{s}^{-1} = -0.116$
$\ln A = -0.116 + 8.517 = 8.401$
$A = \ln^{-1}(8.401) = 4452\,\text{Bq} = 4500\,\text{Bq}$

4a) 22 minutes = 1320 s
$\lambda = \dfrac{\ln 2}{t_{\frac{1}{2}}} = \ln(2) \div 1320\,\text{s} = 5.25 \times 10^{-4}\,\text{s}^{-1}$

b) 120 minutes = 7200 s
$\lambda t = 5.25 \times 10^{-4}\,\text{s}^{-1} \times 7200 = 3.78$
$A = A_0 e^{-\lambda t}$
$A = 2000\,\text{Bq} \times e^{-3.78} = 45.6\,\text{Bq}$

Exam-style questions

5 **D:** Radioactivity is an entirely random process. [1]

6 **B:** $1.5 \times 10^8\,\text{s}$ [1]

7 **D:** 5.2 Sv [1]

8–1 The decay constant is the probability of a nucleus decaying per unit time [1].

8–2 $\dfrac{1}{6}$ [1]

8–3 Work out the number of particles:
$\dfrac{0.250}{0.226} \times 6.0 \times 10^{23} = 6.64 \times 10^{23}\,\text{atoms}$ [1]

Use: $\dfrac{\Delta N}{\Delta t} = -\lambda t$

$-9.0 \times 10^{12}\,\text{s}^{-1} = -\lambda \times 6.64 \times 10^{23}$ [1]
$\lambda = 1.4 \times 10^{-11}\,\text{s}^{-1}$ [1]

9–1 Draw a tangent at $t = 0\,\text{s}$. $A = -\dfrac{\Delta N}{\Delta t} = \text{Gradient}$

Gradient $= -2.70 \times 10^{12} \div 650\,\text{s}$ [1] $= -4.2 \times 10^9\,\text{Bq}$
Initial activity $= 4.2 \times 10^9\,\text{Bq}$ [1]. (*Allow between 4.0 and 4.5 × 10⁹ Bq.*)

9–2 Half-life = 460 s from the graph [1].
Answers in range 450 s–480 s are acceptable.

9–3 Use half-life: $\lambda = \dfrac{\ln 2}{t_{\frac{1}{2}}}$

$\lambda = 0.693 \div 460\,\text{s} = 1.5 \times 10^{-3}\,\text{s}^{-1}$ [1]
(*Allow ecf from* **Answer 9–2**.)
(*Alternative method using $A = -\lambda N$ gives $1.6 \times 10^{-3}\,s^{-1}$ gains the mark.*)

9–4 The unstable nuclide needs to be at the top right, above the end of the stable nuclides line [1].
Alpha decay occurs when $Z > 82$ [1]; it should be close to the limit, as its decay constant is high (therefore making it a very unstable nuclide) [1].
(*No marks for saying the nuclide is unstable, as this is in the stem of the question.*)

10–1 $^{87}_{37}\text{Rb} \rightarrow\ ^{87}_{38}\text{Sr} +\ ^{0}_{-1}e^- + \bar{v}_e$ [1]

10–2 $t_{\frac{1}{2}} = \dfrac{\ln 2}{\lambda} = \ln(2) \div 1.42 \times 10^{-11}\,\text{a}^{-1} = 4.88 \times 10^{10}\,\text{a}$ [1]

10–3 1. Mass and number are proportional, so we can write:
$m = m_0 e^{-\lambda t}$

Rearrange to: $m_0 = \dfrac{m}{\left(e^{-\lambda t}\right)} = m\left(e^{-\lambda t}\right)^{-1} = m\,e^{+\lambda t}$ [1]

$\lambda t = 1.42 \times 10^{-11}\,\text{a}^{-1} \times 4.5 \times 10^9\,\text{a} = 0.0639$ [1]
$m_0 = 2.50 \times 10^{-6}\,\text{kg} \times e^{0.0639} = 2.66 \times 10^{-6}\,\text{kg} = 2.66 \times 10^{-3}\,\text{g}.$ [1]

2. Convert the decay constant to s^{-1}.
$\lambda = (1.42 \times 10^{-11}\,\text{a}^{-1}) \div (365\,\text{d a}^{-1} \times 86\,400\,\text{s d}^{-1}) = 4.50 \times 10^{-19}\,\text{s}^{-1}$ [1]
$A = -\lambda N$
$N = \text{mass} \div 87\,\text{u} = 2.50 \times 10^{-6}\,\text{kg} \div (87 \times 1.661 \times 10^{-27}\,\text{kg}) = 1.73 \times 10^{19}$ nuclei [1]
$A = 4.50 \times 10^{-19}\,\text{s}^{-1} \times 1.73 \times 10^{19} = 7.8\,\text{Bq}$ (2 s.f. – as the time is given to 2 s.f.) [1]

11–1 $^{A}_{Z}\text{X} \rightarrow\ ^{A-4}_{Z-2}\text{Y} +\ ^{4}_{2}\text{He}(+Q)$ [1]

11–2 Work out the area of the source:
$A = 4\pi r^2 = 4 \times \pi \times (6.74 \times 10^{-3}\,\text{m})^2 = 5.71 \times 10^{-4}\,\text{m}^2$ [1]
Intensity = counts ÷ area
$I_1 = 1.7 \times 10^9\,\text{Bq} \div 5.71 \times 10^{-4}\,\text{m}^2 = 2.98 \times 10^{12}\,\text{Bq m}^{-2}$ [1]

11–3 Use: $\dfrac{I_1}{I_2} = \left(\dfrac{x_2}{x_1}\right)^2$

$\dfrac{2.98 \times 10^{12}}{I_2} = \left(\dfrac{5.0\,\text{m}}{6.74 \times 10^{-3}\,\text{m}}\right)^2$ [1]

$I_2 = 5.41 \times 10^6\,\text{Bq m}^{-2}$ [1]

11–4 Area of the detector = $\pi r^2 = \pi \times (5.0 \times 10^{-3}\,\text{m})^2 = 7.85 \times 10^{-5}\,\text{m}^2$
Area at 5.0 m = $4 \times \pi \times (5.0\,\text{m})^2 = 314\,\text{m}^2$
Fraction of this area occupied by the detector = $7.85 \times 10^{-5}\,\text{m}^2 \div 314\,\text{m}^2 = 2.47 \times 10^{-6}$ [1]
Count rate = $2.47 \times 10^{-6} \times 5.41 \times 10^6\,\text{Bq} = 13.4\,\text{Bq}$
Corrected count = $1.5\,\text{Bq} + 13.4\,\text{Bq} = 14.9\,\text{Bq}$ [1]
1 minute count = $14.9\,\text{Bq} \times 60\,\text{s min}^{-1} = 892\,\text{min}^{-1} = 890\,\text{min}^{-1}$ (2 s.f.) [1]

Nuclear instability (p.85)

Quick questions

1 The daughter nucleus is left in an excited state, which means it has more energy than in the ground state. This excess energy is emitted as a gamma photon.

2 A metastable nucleus remains in an excited state for a much longer time than expected. In some cases the half-life can be hours (e.g. Tc-99m).

3 Most excited nuclei lose energy almost instantly. A metastable nucleus remains in an excited state for a longer period than would be expected. It loses its excited state entirely randomly. Therefore a large sample of metastable nuclei will decay exponentially like a radioactive isotope with a half-life.

4

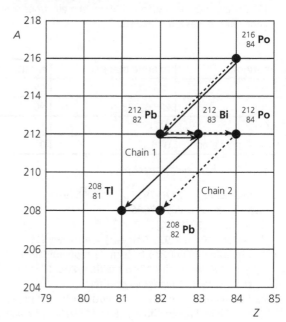

5a) $^{228}_{90}\text{Th} \rightarrow\ ^{224}_{88}\text{Ra} + \alpha$

b) $227.97932\,\text{u} \rightarrow 223.971888\,\text{u} + 4.00150\,\text{u}$
$\Delta m = 227.97932\,\text{u} - 227.973388\,\text{u}$
$\Delta m = 5.932 \times 10^{-3}\,\text{u}$
$E = 5.932 \times 10^{-3}\,\text{u} \times 931.6\,\text{MeV u}^{-1} = 5.53\,\text{MeV} = 5.5\,\text{MeV}$

Exam-style questions

6 Graph **C** (*Note that* **B** *is wrong because it starts at the origin.*) [1]

7 **D:** Beta-plus [1]

8–1 Z drawn on horizontal axis, going from 87 to 91 [1]; A on the vertical scale going from 223 to 229 [1]; arrow drawn from (90, 228) to (88, 224) [1].

8–2 1. to 4.

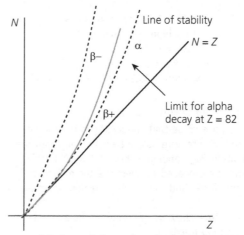

[4] (*1 mark for each correct part*)

8–3 1. The nucleus captures an electron from the bottom (K) shell. One proton is turned into a neutron. [1]
2. $^{57}_{24}\text{Cr} +\ ^{0}_{-1}e^- \rightarrow\ ^{57}_{23}\text{V} + v_e$ [1]

9–1 $^{64}_{28}\text{Ni}$ drawn at (28, 64) [1]; arrow drawn from (29, 64) to (28, 64) and labelled 'K-capture and β-plus' [1]

9–2 Positron emission: $^{64}_{29}\text{Cu} \rightarrow\ ^{64}_{28}\text{Ni} +\ ^{0}_{1}e^+ + v_e$ [1]

Electron capture: $^{64}_{29}\text{Cu} +\ ^{0}_{-1}e^- \rightarrow\ ^{64}_{28}\text{Ni} + v_e$ [1]

9–3 The copper-64 nucleus has too few neutrons (*or too many protons*) for it to be stable [1].

9–4 Electron (or K-capture) involves the capture of an electron from the closest electron shell into the nucleus [1].

10–1 The atomic masses have the mass of the electrons included [1]. While a single electron has a small mass (0.000549 u), for a large atom, there would be a reasonable mass in the electrons. The mass defect would be larger than expected. [1] For example, in a uranium atom, the electrons would contribute a mass of 0.0505 u.

11–1

Component	Atomic mass / u	Number of electrons	Mass of electrons / u	Nuclear mass / u
Actinium	228.03310	89	0.048861	227.984239
Thorium	228.02873	90	0.04941	227.97932
Beta-minus	0.000549	1	0.000549	0.000549
Electron antineutrino	0	0	0	0

[3] (*1 mark for each correct row*)

11–2 $227.984239\,u \rightarrow 227.97932\,u + 0.000549\,u$
$227.984239\,u \rightarrow 227.979869\,u$
$\Delta m = 4.37 \times 10^{-3}\,u$ [1]
$E = 4.37 \times 10^{-3}\,u \times 931.5\,MeV\,u^{-1} = 4.07\,MeV$ [1]

11–3 1. Alpha: polonium-216 to lead-212; polonium-212 to lead-208; bismuth-212 to thallium-208. (*2 marks all three correct, 1 mark two correct*)
2. Beta-minus: lead-212 to bismuth-212; bismuth-212 to polonium-212; thallium-208 to lead-208. (*2 marks all three correct, 1 mark two correct*)

Mass and energy (p.88)

Quick questions

1a) The atomic mass unit is exactly $\frac{1}{12}$ the mass of a carbon-12 atom.

b) $m = 1.661 \times 10^{-27}\,kg$.
$E = 9.315 \times 10^8\,eV = 931.5\,MeV = 1.495 \times 10^{-10}\,J$

2a) Mass of nucleons:
$m = (20 \times 1.00728\,u) + (20 \times 1.00867\,u) = 40.319\,u$
Mass defect = mass of nucleons − actual mass
$\Delta m = 40.319\,u − 39.9626\,u = 0.3564\,u$.

b) Binding energy = $0.3564\,u \times 931.5\,MeV\,u^{-1} = 332.0\,MeV$

3a) Mass of nucleons:
$m = (2 \times 1.00728\,u) + (2 \times 1.00867\,u) = 4.0319\,u$
Mass defect = mass of nucleons − actual mass
$\Delta m = 4.0319\,u − 4.00151 = 0.03039\,u$.

b) Binding energy = $0.03039\,u \times 931.5\,MeV\,u^{-1} = 28.31\,MeV$
Binding energy per nucleon = $28.31\,MeV \div 4 = 7.08\,MeV$ per nucleon

4a) $^2_1H + ^3_1H \rightarrow ^4_2He + ^1_0n + energy$

b) The conditions have to be that the hydrogen is very hot ($1 \times 10^8\,K$) and the pressure has to be very high to ensure that the density is high enough (about $1.5 \times 10^5\,kg\,m^{-3}$). These conditions are needed to overcome the repulsive electromagnetic force, and the repulsive region of strong nuclear force.
If nuclei can do this, they are captured by the attractive region of the strong force.

5 $m = 5.0 \times 10^{13}\,J \div (3.0 \times 10^8\,m\,s^{-1})^2 = 5.6 \times 10^{-4}\,kg$

6a) Equation: $E = \dfrac{hc}{\lambda}$
Rearrange: $\lambda = \dfrac{hc}{E}$
$\lambda = (6.63 \times 10^{-34}\,J\,s \times 3.00 \times 10^8\,m\,s^{-1}) \div (1.0\,eV \times 1.60 \times 10^{-19}\,J\,eV^{-1}) = 1.2 \times 10^{-6}\,m$

b) Equation: $E_k = \dfrac{1}{2}mv^2$

10–2 $228.02873\,u \rightarrow 224.02020\,u + 4.00260\,u$
$\Delta m = 228.02873\,u − (224.02020\,u + 4.00150\,u)$
$\Delta m = 7.03 \times 10^{-3}\,u$. [1]
$E = 7.03 \times 10^{-3}\,u \times 931.5\,MeV\,u^{-1} = 6.54\,MeV$ [1]

10–3 The alpha particle is a helium <u>nucleus</u>. The second data item was for a helium <u>atom</u> [1].

Rearrange: $v^2 = \dfrac{2E_k}{m}$
$v^2 = (2 \times 1.60 \times 10^{-19}\,J) \div 1.675 \times 10^{-27}\,kg = 1.91 \times 10^8\,m^2\,s^{-2}$
$v = (1.91 \times 10^8\,m^2\,s^{-2})^{0.5} = 14\,000\,m\,s^{-1}$ (2 s.f.)

7a) For X, $A = 216$, $Z = 84$; for α, $A = 4$, $Z = 2$

b) X is unstable, as its proton number is greater than 82 (Pb). Helium is stable as it has a particularly high binding energy per nucleon.

c) $^{220}_{86}Rn \rightarrow ^{216}_{84}Po + ^4_2He + energy$
$219.916886\,u \rightarrow 213.902874\,u + 4.00150\,u + \Delta m \rightarrow 217.904374\,u$
$\Delta m = 219.916886\,u − 217.904374\,u = 2.012512\,u$
$\Delta E = 2.012512\,u \times 931.5\,MeV\,u^{-1} = 1875\,MeV$
Convert to J: $\Delta E = 1875 \times 10^6\,eV \times 1.60 \times 10^{-19}\,J\,eV^{-1} = 3.0 \times 10^{-10}\,J$ (2 s.f.)

8a) ^3_2He

b) $^1_1p + ^2_1H \rightarrow ^4_2He + energy$
$1.00728\,u + 2.01355\,u \rightarrow 3.01493\,u + \Delta E$
$\Delta E = (1.00728\,u + 2.01355\,u) − 3.01493\,u = 5.9 \times 10^{-3}\,u$
$\Delta E = 5.9 \times 10^{-3}\,u \times 931.5\,MeV\,u^{-1} = 5.496\,MeV = 5.5\,MeV$ (2 s.f.)

9 $^7_4Be + ^0_{-1}e^- \rightarrow ^7_3Li + v_e$
$7.014727\,u + 0.000549\,u \rightarrow 7.014356\,u$
$\Delta m = 9.20 \times 10^{-4}\,u$
$\Delta E = 9.20 \times 10^{-4}\,u \times 931.5\,MeV\,u^{-1} = 0.857\,MeV$

Exam-style questions

10–1 1. Nucleon number = 9; neutron number = 9 − 4 = 5. [1]
2. Mass of nucleons: $m = (4 \times 1.00728\,u) + (5 \times 1.00867\,u) = 9.07247\,u$ [1]
Mass defect = mass of nucleons − actual mass
$\Delta m = 9.07247\,u − 9.01218\,u = 0.06029\,u$ [1]

10–2 Beryllium-8 is very unstable with a short half-life [1]. (*Allow explanations of beta-plus decay as there are too few neutrons.*)

10–3 1. Mass of nucleons: $m = (4 \times 1.00728\,u) + (4 \times 1.00867\,u) = 8.0638\,u$ [1]
Mass defect = mass of nucleons − actual mass
$\Delta m = 8.0638\,u − 8.00531\,u = 0.05849\,u$ [1]
2. Binding energy = $0.05849\,u \times 931.5\,MeV\,u^{-1} = 54.48\,MeV$ [1]
3. Binding energy per nucleon = $54.48\,MeV \div 8 = 6.81\,MeV$ [1]

10–4 1. Energy of the two alpha particles = $2 \times 28.31\,MeV = 56.62\,MeV$ [1]
2. Energy change = $56.62\,MeV − 54.48\,MeV = 2.34\,MeV$ [1]

11–1 1. Put in the masses in u into the equation:
$2.01355\,u + 2.01355\,u \rightarrow 3.01493\,u + 1.00867\,u$
Therefore: $4.02710\,u \rightarrow 4.02360\,u$
$\Delta m = 4.02710\,u − 4.02360\,u = 3.50 \times 10^{-3}\,u$ [1]
2. $\Delta E = 3.50 \times 10^{-3}\,u \times 931.5\,MeV\,u^{-1} = 3.26\,MeV$ [1]

11-2　1. Put in the masses in u into the equation:
$2.01355\,u + 2.01355\,u \rightarrow 3.01493\,u + 1.00782\,u$
Therefore: $4.02710\,u \rightarrow 4.02275\,u$
$\Delta m = 4.02710\,u - 4.02275\,u = 4.35 \times 10^{-3}\,u$ [1]

2. $\Delta E = 4.35 \times 10^{-3}\,u \times 931.5\,MeV\,u^{-1} = 4.05\,MeV$ [1]

11-3　1. Put in the masses in u into the equation:
$2.01355\,u + 3.01550\,u \rightarrow 4.00151\,u + 1.00867\,u$
Therefore: $5.02905\,u \rightarrow 5.01018\,u$
$\Delta m = 5.02905\,u - 5.01018\,u = 1.887 \times 10^{-2}\,u$ [1]

2. $\Delta E = 1.887 \times 10^{-2}\,u \times 931.5\,MeV\,u^{-1} = 17.58\,MeV$ [1]

11-4　1. Put the masses in u into the equation:
$2.01355\,u + 3.01493\,u \rightarrow 4.00151\,u + 1.00782\,u$
Therefore: $5.02848\,u \rightarrow 5.00933\,u$
$\Delta m = 5.02848\,u - 5.00933\,u = 1.915 \times 10^{-2}\,u$ [1]

2. $\Delta E = 1.915 \times 10^{-2}\,u \times 931.5\,MeV\,u^{-1} = 17.84\,MeV$ [1]

11-5　1. $3.26\,MeV + 4.05\,MeV + 17.58\,MeV + 17.84\,MeV$
$= 42.7\,MeV$ [1]

2. $E = 42.7 \times 10^{6}\,eV \times 1.6 \times 10^{-19}\,J\,eV^{-1} \times 6.0 \times 10^{23}$
$= 4.102 \times 10^{12}\,J = 4.1 \times 10^{12}\,J$ (2 s.f.) [1]

3. Equation rearranged: $m = \dfrac{E}{c^2}$

$m = 4.102 \times 10^{12}\,J \div 9.0 \times 10^{16}\,m^2\,s^{-2} = 4.6 \times 10^{-5}\,kg$ [1]

12-1　The statement is incorrect. Nuclear fission has nothing to do with radioactive decay [1]. Fission only happens when there are fissile nuclei [1]. The most common of these are specific isotopes of uranium and plutonium. They interact with slow (thermal) neutrons [1].

12-2　A uranium-235 nucleus captures a slow moving neutron [1]. The nucleus becomes U-236 which is very unstable [1]. The nucleus usually splits into two fission fragments and two or three neutrons, releasing energy [1].

12-3　When a fissile nucleus splits, it releases, on average, 3 neutrons. The neutrons are fast and will pass through adjacent nuclei [1]. The neutrons have to be slowed down to be absorbed into further nuclei. For a chain reaction to occur, at least one further nucleus has to absorb one of the neutrons [1]. This requires a critical mass of about 11 – 15 kg of the material [1]. In an uncontrolled chain reaction, 1 fission event causes 3 fission events. These in turn cause 9 fission events, and so on [1].

13-1　1. $E = mc^2 = 3.1 \times 10^{-28}\,kg \times (3.0 \times 10^{8}\,m\,s^{-1})^2$
$= 2.79 \times 10^{-11}\,J$ [1]

2. Number of nuclei $= (2.79 \times 10^{-11}\,J)^{-1} = 3.58 \times 10^{10}$
$= 3.6 \times 10^{10}$ (2 s.f.) [1]

13-2　1. $\Delta m = Zm_p + (A - Z)m_n - M$ [1]

2. $E = \dfrac{\Delta m c^2}{A}$ [1]

Topic review: nuclear physics (p.91)

1-1　$^{60}_{27}Co \rightarrow {}^{60}_{28}Ni + {}^{0}_{-1}e^- + {}^{0}_{0}\bar{\nu}_e$ [1]
(Allow symbol β for the electron. Accept $\bar{\nu}_e$. Must be electron antineutrino.)

1-2　1. The source is surrounded by 3 mm aluminium [1].

2. The nucleus is excited [1] and loses energy by emitting a gamma photon [1].
('Loses energy' is not sufficient for the second mark.)

1-3　$P = 2.505\,MeV - 1.332\,MeV = 1.173\,MeV$ [1]
$Q = (1.332\,MeV + 1.480\,MeV) - 2.505\,MeV = 0.307\,MeV$ [1]

1-4　1. $A = 4 \times \pi \times (0.05\,m)^2 = 0.0314\,m^2$
$I_1 = 1.0 \times 10^9\,Bq \div 0.0314\,m^2 = 3.18 \times 10^{10}\,Bq\,m^{-2}$ [1]

2. $I_2 = \left(\dfrac{x_1}{x_2}\right)^2 I_1$

$I_2 = (0.050\,m \div 0.60\,m)^2 \times 3.18 \times 10^{10}\,Bq\,m^{-2}$
$= 2.21 \times 10^8\,Bq\,m^{-2}$ [1]

3. $A = 2.21 \times 10^8\,Bq\,m^{-2} \times (0.05\,m)^2 = 5.52 \times 10^5\,Bq$ [1]

1-5　1. $E = 1.25 \times 10^6\,eV \times 1.6 \times 10^{-19}\,J\,eV^{-1} = 2.0 \times 10^{-13}\,J$
$P = 2.0 \times 10^{-13}\,J \times 5.52 \times 10^5\,Bq = 1.104 \times 10^{-7}\,W$ [1]

2. $D = E/m = 1.104 \times 10^{-7}\,W \div 1.5 \times 10^{-4}\,kg$
$= 7.36 \times 10^{-4}\,J\,kg^{-1}(s^{-1})$
Dose per minute $= 7.36 \times 10^{-4}\,J\,kg^{-1}\,s^{-1} \times 3600\,s\,h^{-1}$
$= 2.65\,Gy\,h^{-1}$ [1]

3. $H = W_R D = 2.65\,Gy \times 1.0 = 2.65$ [1] Sv [1]

2-1　1. $^{59}_{27}Co + {}^{1}_{0}n \rightarrow {}^{60}_{27}Co$ [1]

2. This happens in a nuclear reactor [1]. The neutron has to have the right energy to be absorbed by the nucleus [1].

2-2　1. N labelled 31, 32 and 33; Z labelled 24–30 [1].

2. Vertical arrow to cobalt 60 isotope [1].

3. Arrow going from (27, 59) to (28, 33) [1].

2-3　Half-life $= 5.27\,y \times 365\,d\,y^{-1} \times 86\,400\,s\,d^{-1} = 1.66 \times 10^8\,s$
$\lambda = \ln (2) \div 1.66 \times 10^8\,s = 4.17 \times 10^{-9}\,s^{-1}$ [1]
Probability $= 4.17 \times 10^{-9}\,s^{-1} \times 3600\,s = 1.50 \times 10^{-5}$ [1]
(1 mark only if answer is given as $0.13\,y^{-1}$)

2-4　1. Molar mass $= 60 \times 10^{-3}\,kg\,mol^{-1}$
Number of moles $= 60 \times 10^{-9}\,kg \div 60 \times 10^{-3}\,kg\,mol^{-1}$
$= 1.0 \times 10^{-6}\,mol$
Number of atoms $= 1.0 \times 10^{-3}\,mol \times 6.02 \times 10^{23}\,mol^{-1}$
$= 6.0 \times 10^{17}\,atoms$ [1]

2. $A = (-)\lambda N = 4.17 \times 10^{-9}\,s^{-1} \times 6.0 \times 10^{17}$
$= 2.5 \times 10^9\,Bq$ [1]

3. $15\,y = 15\,y \times 365\,d\,y^{-1} \times 86\,400\,s\,d^{-1} = 4.73 \times 10^8\,s$
$\lambda t = 4.17 \times 10^{-9}\,s^{-1} \times 4.73 \times 10^8\,s = 1.972$
$A = A_0\,e^{-\lambda t} = 2.5 \times 10^9\,Bq \times e^{(-1.972)} = 3.48 \times 10^6\,Bq$
$= 3.5 \times 10^6\,Bq$ [1].

3-1　1. Critical mass is the minimum mass in which a chain reaction can be sustained [1].

2. A U-235 nucleus absorbs a thermal (slow-moving) neutron to form U-236 [1]. The U-236 becomes like a 'wobbly drop' (or very unstable) [1]. It splits into two (on average) fission fragments releasing three (on average) fast neutrons [1]. (Any mention of alpha decay is a physics error and gains no marks.)

3-2　Moderator slows the fast neutrons so that they have a low kinetic energy with the same energy as an infrared photon [1]. Control rods slow (or stop) the nuclear reactions by absorbing neutrons [1]. Coolant transfers thermal energy to the heat exchanger where energy is transferred to water [1].

3-3　Mass of protons and neutrons $= (92 \times 1.00728\,u) + (143 \times 1.00867\,u) = 236.90957\,u$
$\Delta m = 236.90957\,u - 235.044\,u = 1.8656\,u$ [1]
$E = 1.8656\,u \times 931.5\,MeV\,u^{-1} = 1738\,MeV$
Binding energy per nucleon $= 1738\,MeV \div 235 = 7.39\,MeV$ [1]

3-4　1. U-235 marked towards the right-hand end of the graph line; fission fragments at positions on the line which are about halfway along the A axis. [1]
(For the mark, both the U-235 and the fission fragments should be shown.)

2. The stability in the fission fragments is greater than the U-235 [1]. Therefore, the binding energy per nucleon is greater [1]. Therefore, the energy difference is released as heat [1].

3-5　Advantages: Much more energy than coal or oil; Zero carbon dioxide emissions [1].
Disadvantages: Nuclear waste is dangerous; High-profile accidents leading to serious contamination (e.g. Fukushima Daiichi) [1].
(Two advantages and disadvantages for each mark. Other valid points may contribute.)

4-1　1. Binding energy per nucleon for U-235 = 7.8 MeV [1].

2. Binding energy per nucleon for Sr-90 = 8.5 MeV [1].

3. Binding energy per nucleon for Xe-144 = 8.3 MeV [1].

4-2　$^{235}_{92}U \rightarrow {}^{144}_{56}Xe + {}^{90}_{38}Sr + 2{}^{1}_{0}n$
$235 \times 7.8\,MeV \rightarrow (144 \times 8.3\,MeV) + (90 \times 8.5\,MeV)$ [1]
$1833\,MeV \rightarrow 1195\,MeV + 765\,MeV$
$1833\,MeV \rightarrow 1960\,MeV$
Difference in the binding energy = 127 MeV [1]

4–3 1. The moderator is usually made from graphite. The moderator reduces the speed of the neutrons so that they can be captured by nuclei [1].
The reduction in speed is achieved by repeated collisions with the molecules of the moderator. Energy is lost by excitation of nuclei and heating due to the increase in internal energy of the moderator [1].

2. The control rods absorb excess neutrons to control the chain reaction [1]. The materials in the control rods can absorb neutrons without fission. Boron or cadmium are used commonly [1]. The rods are moved up and down to control the reaction. They can be dropped into the reactor fully to stop the reaction completely [1].

3. The coolant in the primary circuit removes thermal energy from the reactor core [1]. The coolant transfers the thermal energy to a heat exchanger where water is turned into steam [1].

5–1 *Any three from:*
- The steel pressure vessel is surrounded by 5 m concrete [1].
- This absorbs neutrons and gamma rays [1].
- The reactor building is designed to contain any leak of radiation [1].
- Control rods are held by electromagnets so that they drop immediately into the reactor in the event of an emergency [1].
- The building is carefully monitored throughout at all times [1].

5–2 *Any two from:*
- Back-up pumps are used to keep the coolant circulating. These can be driven by diesel generators. If these fail, there are back-up batteries [1].
- Even when the reactor is off, cooling pumps must be used to ensure heat flows away from the reactor [1].
- The reactor is very hot, even though it has been turned off. There is a risk that control rods and fuel rods could melt. It is even possible, in pressurised water reactors, that the water gets so hot it decomposes to form hydrogen and oxygen gases (highly dangerous) [1].

5–3 1. Low level, intermediate level and high level [1].

2. • Low level waste forms the bulk of the waste. It consists of disposable overalls, paper and rags. It is compacted and encased in cement [1].
- It is stored in licensed sites until any radioactivity decreases to an acceptable level. Then it can be dealt with as normal waste [1].
- Intermediate waste is mostly produced by nuclear waste processing or the decommissioning of old plant [1].
- It is encased in concrete and stored in steel drums in underground repositories. Short half-life material may be stored in deep trenches covered by several metres of soil [1].
- High level waste mostly consists of spent fuel rods. Since they are hot, they have to be cooled in deep tanks of water [1].
- For long-term storage, the waste is mixed with molten glass (vitrification), before being poured into stainless steel drums and allowed to cool. They are stored in deep repositories [1].

5–4 The unused fuel rods consist of uranium isotopes. These have long half-lives. Therefore, the probability of a decay is low [1]. Additionally, they are alpha emitters. Alpha radiation is stopped by the outside cladding. The fuel rods can be handled safely [1].

Spent fuel rods are full of many unstable isotopes which are results of fission. The half-lives are short, so the activity is high [1]. The fission fragments are beta emitters. The excited nuclei from the beta emissions emit gamma radiation. Some fragments may emit neutrons. The cladding itself may have absorbed neutrons and electrons to form radioactive isotopes [1].

6–1 $Z = 92 - 56 - 0 = 36$
$A = 236 - 144 - 3 = 89$ [1] (*Both correct for the mark*)

6–2 1. $235.995061 \, u \rightarrow 143.892211 \, u + 88.898071 \, u +$
$(3 \times 1.00867 \, u)$
Therefore: $235.995061 \, u \rightarrow 235.816292$
$\Delta m = 233.019559 \, u - 232.842946 \, u = 0.178769 \, u$ [1]

2. $\Delta E = 0.178769 \, u \times 931.5 \, MeV \, u^{-1} = 167 \, MeV = 170 \, MeV$ [1]
(*Ecf from **Answer 6–2–1***)

6–3 1. Mass of U-235 = $0.050 \times 13 \, kg = 0.65 \, kg$
Number of moles = $0.65 \, kg \div 0.235 \, kg \, mol^{-1}$
$= 2.77 \, mol$
Number of particles = $2.77 \, mol \times 6.0 \times 10^{23} \, mol^{-1}$
$= 1.66 \times 10^{24}$ [1]
Energy released = $1.66 \times 10^{24} \times 167 \times 10^6 \, eV \times 1.6 \times 10^{-19} \, J \, eV^{-1} = 4.44 \times 10^{13} \, J$
$= 4.4 \times 10^{13} \, J$ [1] (2 s.f.).

2. Mass of diesel fuel = $4.44 \times 10^{13} \, J \div 43 \times 10^6 \, J \, kg^{-1}$
$= 1.03 \times 10^6 \, kg$ (= 1030 tonnes) [1]

6–4 1. As the power plant is 40% efficient,
$E = 1500 \times 10^6 \, W \div 0.40 = 3.75 \times 10^9 \, J \, s^{-1}$ [1]
Mass used per second = $3.75 \times 10^9 \, J \div 6.8 \times 10^{13} \, J \, kg^{-1}$
$= 5.5 \times 10^{-5} \, kg \, s^{-1}$
Mass used per day = $5.5 \times 10^{-5} \, kg \, s^{-1} \times 86 \, 400 \, s$
$= 4.75 \, kg$ [1]

2. The total amount of fuel = $4.75 \, kg \div 0.05 = 95 \, kg$ [1]

7–1 [1]

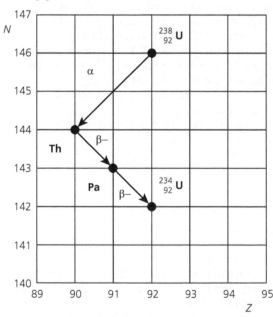

7–2 See grid in **Answer 7–1**. (*2 marks for all three correct lines, 1 mark for two correct lines*)

7–3 $^{238}_{92}U \rightarrow \, ^{234}_{90}Th + \, ^{4}_{2}He$

$^{234}_{90}Th \rightarrow \, ^{234}_{91}Pa + \, ^{0}_{-1}e^- + \bar{\nu}_e$

$^{234}_{91}Pa \rightarrow \, ^{234}_{92}U + \, ^{0}_{-1}e^- + \bar{\nu}_e$

(*2 marks for all three correct equations, 1 mark for two correct equations*)

7–4 1. $t_{\frac{1}{2}} = 4.468 \times 10^9 \, a \times 365 \, d \times 86\,400 \, s = 1.41 \times 10^{17} \, s$ [1]

$\lambda = \dfrac{\ln 2}{t_{\frac{1}{2}}} = 4.92 \times 10^{-18} \, s^{-1}$ [1]

2. Number of atoms in 1 kg = $(1.0 \, kg \div 0.238 \, kg \, mol^{-1}) \times 6.02 \times 10^{23} \, mol^{-1} = 2.52 \times 10^{24}$ [1]

$\dfrac{\Delta N}{\Delta t} = -\lambda N$

$\Delta N / \Delta t = -\lambda = 4.92 \times 10^{-18} \, s^{-1} \times 2.52 \times 10^{24} = 1.24 \times 10^6 \, Bq = 1.2 \times 10^6$ [1] Bq (2 s.f.) [1]

8–1 For gamma emission to occur, there has to be a radioactive decay event, e.g. alpha, beta-, beta+, or K-capture [1]. If the nucleus is left in an excited state, meaning it has excess energy, that energy is lost in the form of a gamma photon [1]. The photon is of a specific energy, as there is a discrete difference between the energy levels [1].

8–2 1. $^{60}_{27}Co \rightarrow \, ^{60}_{28}Ni + \, ^{0}_{-1}e + \bar{\nu}_e$ [1]

2. Gamma [1]

8–3 $X = 2.50 \, MeV - 1.33 \, MeV = 1.17 \, MeV$ [1]
$Y = (2.50 \, MeV + 0.31 \, MeV) - 1.33 \, MeV = 1.48 \, MeV$ [1]

8–4 1. $E = 1.33 \times 10^6 \, eV \times 1.60 \times 10^{-19} \, J \, eV^{-1} = 2.128 \times 10^{-13} \, J$ [1]

2. Equation: $\lambda = \dfrac{hc}{E}$ [1]

$\lambda = (6.63 \times 10^{-34} \, J \, s \times 3.00 \times 10^8 \, m \, s^{-1}) \div 2.128 \times 10^{-13} \, J = 9.35 \times 10^{-13} \, m$ [1]

8–5 1. $T_{\frac{1}{2}} = 5.272 \, a \times 365 \, d \, a^{-1} \times 86\,400 \, s \, d^{-1} = 1.66 \times 10^8 \, s$ [1]

$\lambda = \dfrac{\ln 2}{t_{\frac{1}{2}}}$

$\lambda = \ln 2 \div 1.66 \times 10^8 \, s = 4.17 \times 10^{-9} \, s^{-1}$ [1]

2. $T_{\frac{1}{2}} = 25 \, a \times 365 \, d \, a^{-1} \times 86\,400 \, s \, d^{-1} = 7.88 \times 10^8 \, s$ [1]

$A = A_0 e^{-\lambda t}$
$A = 1000 \, Bq \times e^{-(4.17 \times 10^{-9} \, s^{-1} \times 7.88 \times 10^8 \, s)}$
$A = 1000 \, Bq \times e^{-3.288} = 1000 \, Bq \times 0.0373 = 37.3 \, Bq$ [1]

9–1 1. $\Delta E = 142.6 \, keV - 2.1 \, keV = 140.5 \, keV$ [1]

2. $E = 2.1 \times 10^3 \, eV \times 1.6 \times 10^{-16} \, J \, eV^{-1} = 3.36 \times 10^{-16} \, J$
$\lambda = (6.63 \times 10^{-34} \, J \, s \times 3.0 \times 10^8 \, m \, s^{-1}) \div 3.36 \times 10^{-16} \, J = 5.92 \times 10^{-10} \, m$ [1]

3. It is in the X-ray region [1].

9–2 48 h = 8 half-lives
Fraction left = $2^{-8} = 3.9 \times 10^{-3}$ [1]

9–3 1. Proton number = 44; neutron number = $99 - 44 = 55$ [1]

2. $T_{\frac{1}{2}} = 211\,000 \, a \times 365 \, d \, a^{-1} \times 86\,400 \, s \, d^{-1} = 6.65 \times 10^{12} \, s$ [1]

$\lambda = \dfrac{\ln 2}{t_{\frac{1}{2}}}$

$\lambda = \ln (2) \div 6.65 \times 10^{12} \, s = 1.04 \times 10^{-13} \, s^{-1}$
The probability of any one nucleus decaying is $1.04 \times 10^{-13} \, s^{-1}$ [1].

3. The <u>extra dose</u> of radioactivity is very small [1]. If there are 10^{10} nuclei, the <u>probability of a decay event would be about 10^{-3}</u>, or once every 1000 s [1]. This is much lower than the background radiation. It is reduced further as Tc atoms are removed by <u>excretion</u> [1].

Telescopes (p.97)

Quick questions

1 Three parallel rays going to the lens; then going from the lens and meeting at the principal focus

2a) Image can be projected onto a screen.
b) Image cannot be projected onto a screen.
c) Image is bigger than the object.
d) Image is smaller than the object.
e) Right way up compared to the object.
f) Upside down compared to the object.

3a) Image is real, diminished, and inverted.

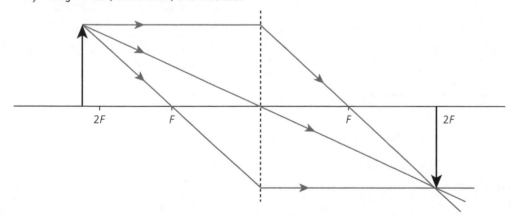

b) Image is real, the same size, and inverted.

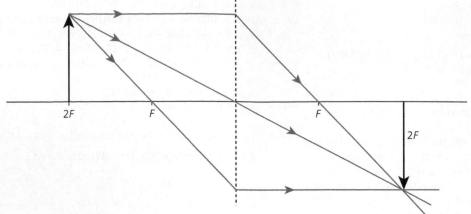

c) Image is magnified, inverted, and real.

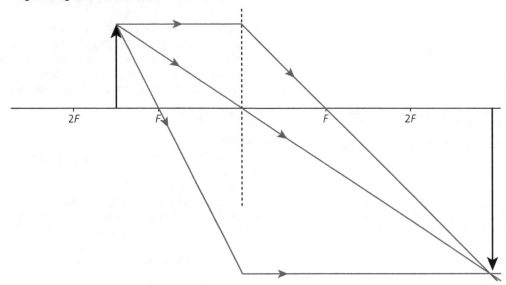

d) Image at infinity.

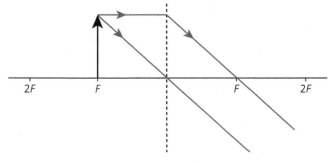

e) Image is virtual, upright, and magnified.

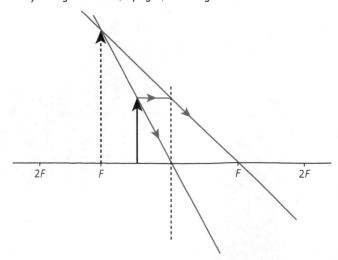

4a) The image from the objective lens is real, inverted, and diminished.

b) The image in the eyepiece lens is virtual, magnified, and upright.

5 The image of the bird would be upside down, since the image from the objective formed at the focal plane would be inverted.

6 Angular magnification =

$$\frac{\text{angle subtended by image at eye}}{\text{angle subtended by object to the unaided eye}}$$

7 Advantages:
- It is much easier to grind a parabolic mirror than a parabolic lens.
- There is no chromatic aberration on a mirror.
- Much larger diameter mirrors can be used than is possible for a lens.
- Large lenses can sag and go out of shape. Large mirrors can be supported to prevent them going out of shape.
- Wide apertures reduce diffraction effects.

Disadvantages:
- Mirrors are exposed to the atmosphere and collect dust.
- Mirrors may need to be recoated if they get scratched while cleaning.
- Mirrors can go out of alignment.
- Secondary mirror blocks some light.
- Secondary mirror causes diffraction which will reduce image quality.

8 $I \propto d^2$ OR $I = \pi d^2/4$

9 Ratio = $(35\,\text{cm})^2 \div (5\,\text{cm})^2 = 49$ (i.e. the intensity of the light collected by the large telescope is 49 times that of the smaller telescope)

The construction makes no difference.

Exam-style questions

10–1 $M = \dfrac{f_o}{f_e}$

$M = 45\,\text{cm} \div 2.5\,\text{cm} = 18$ [1]

10–2 $d = 45\,\text{cm} + 2.5\,\text{cm} = 47.5\,\text{cm}$ [1]

10–3 $M = \dfrac{f_o}{f_e} = \dfrac{\beta}{\alpha}$ [1]

$M = 85\,\text{cm} \div 2.5\,\text{cm} = 34$

$a = b \div M = 0.050\,\text{rad} \div 34 = 0.0015\,\text{rad}$ [1]

10–4 1. $\tan \alpha = \alpha = 3500\,\text{km} \div 410\,000\,\text{km} = 0.0085\,\text{rad}$ [1]

2. Focal length of the objective = $1.00\,\text{m} - 0.050\,\text{m} = 0.95\,\text{m}$
Magnification = $0.95\,\text{m} \div 0.050\,\text{m} = 19$
$\beta = 19 \times 0.0085 = 0.1615$ [1]

3. $d = 0.05\,\text{m} \times 0.1615 = 8.1 \times 10^{-3}\,\text{m}$ [1]

11–1 1. $M = 2.50\,\text{m} \div 0.050\,\text{m} = 50$ [1]

2. $\tan \alpha = 150\,\text{km} \div 20 \times 10^6\,\text{km} = 7.5 \times 10^{-6}$
$\tan \alpha = \alpha = 7.5 \times 10^{-6}\,\text{rad}$ [1]

3. $\beta = 50 \times 7.5 \times 10^{-6}\,\text{rad} = 3.75 \times 10^{-4}\,\text{rad}$ [1]

11–2 Single slit diffraction: $n\lambda = D \sin\theta$
In this case, $n = 1$, and θ is a <u>small</u> angle in radians, so
$\sin \theta = \theta$ [1].

Therefore: $\lambda = D\theta$

Rearrange to: $\theta = \dfrac{\lambda}{D}$ [1]

11–3 1. Cannot be resolved: $\theta < \dfrac{\lambda}{D}$ [1]

2. Can just be resolved: $\theta \approx \dfrac{\lambda}{D}$ [1]

3. Can easily be resolved: $\theta > \dfrac{\lambda}{D}$ [1]

11–4 1. Use: $\theta = \dfrac{\lambda}{D}$

$\theta = 550 \times 10^{-9}\,\text{m} \div 4.0 \times 10^{-3}\,\text{m} = 1.4 \times 10^{-4}\,\text{rad}$ [1]

2. $d = 2.0\,\text{m} \div 1.4 \times 10^{-4}\,\text{rad} = 14.5 \times 10^4\,\text{m} = 15\,\text{km}$
(2 s.f.) [1]

12–1 $\theta = 30\,000\,\text{km} \div 500 \times 10^6\,\text{km} = 6.0 \times 10^{-5}\,\text{rad}$ [1]

12–2 $\theta = 550 \times 10^{-9}\,\text{m} \div 0.20\,\text{m} = 2.75 \times 10^{-6}\,\text{rad}$ [1]

12–3 The telescope will resolve the two asteroids easily, as the angle subtended by the asteroids is $6.0 \times 10^{-5}\,\text{rad} \div 2.75 \times 10^{-6}\,\text{rad} = 22$ times greater than the angular resolution of the telescope [1].

13–1 $f = c/\lambda = 3.0 \times 10^8\,\text{m s}^{-1} \div 0.21\,\text{m} = 1.43 \times 10^9\,\text{Hz}$
$= 1.4 \times 10^9\,\text{Hz}$ (2 s.f.) [1]

13–2 $\theta = 0.21\,\text{m} \div 76\,\text{m} = 2.76 \times 10^{-3}\,\text{rad} = 2.8 \times 10^{-3}\,\text{rad}$
(2 s.f.) [1]

13–3 Separation = $1200\,\text{ly} \times 2.8 \times 10^{-3}\,\text{rad} = 3.4\,\text{ly}$ [1]

13–4 Collecting power = $(36\,000\,\text{m})^2 \div (76\,\text{m})^2 = 22\,400$
$= 22\,000$ times more powerful (2 s.f.) [1]

13–5 $\theta = 0.21\,\text{m} \div 36\,000\,\text{m} = 5.83 \times 10^{-6}\,\text{rad}$ [1]

13–6 Separation = $1200\,\text{ly} \times 5.83 \times 10^{-6}\,\text{rad} = 7.0 \times 10^{-3}\,\text{ly}$ [1]

14 Film (*any three from*):
Advantages:
- Extremely large photographic plates could be used to cover large areas of the night sky [1].
- Very high quality images possible [1].
- No noise (spurious signals) [1].

Disadvantages:
- Film has to be developed (which can go wrong) [1].
- Quantum efficiency is low (4–10%) [1].
- Astronomical film is monochrome (black and white) [1].

CCD (*any three from*):
Advantages:
- Has very high quantum efficiency (70–90%) [1].
- Can be very sensitive [1].
- Can connect directly to computers [1].
- Can detect radiations beyond the visible spectrum (e.g. IR, UV) [1].
- Can be made into arrays of millions of pixels [1].
- Can be made to record colours [1].
- Can be used many times [1].

Disadvantages:
- CCDs can suffer noise (spurious electrical signals) [1].
- Resolution of colour CCD is lower as one pixel is required for each of red, green and blue [1].
- CCDs are still expensive [1].
- CCDs need to be cooled to reduce noise [1].

15–1 *Diagram with:*

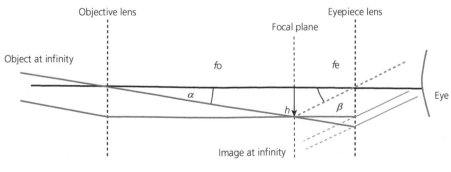

- Principal axis, focal plane, and lenses shown [1]
- Rays from object shown, and crossing at focal plane [1]
- Parallel rays from the eye [1]

15–2 1. $d = f_o + f_e = 65\,\text{cm} + 3.5\,\text{cm} = 68.5\,\text{cm}$ [1]

2. $M = \dfrac{f_o}{f_e} = \dfrac{65\,\text{cm}}{3.5\,\text{cm}} = 18.6$ [1]

15–3 1. $\theta = 3.0 \times 10^3\,\text{km} \div 1.5 \times 10^8\,\text{km} = 2.0 \times 10^{-5}\,\text{rad}$ [1]

2. $\theta = \lambda/D = 570 \times 10^{-9}\,\text{m} \div 0.10\,\text{m} = 5.7 \times 10^{-6}\,\text{rad}$.
($\theta = \lambda/D = 7.0 \times 10^6\,\text{rad}$) [1]
The telescope will be able to resolve the two objects, as the angle is greater than the angle of resolution possible with the telescope [1].

16–1 1. Spherical aberration is when parallel rays of light passing through a lens are refracted by different amounts so that they do not pass through the same point [1].

2. Chromatic aberration is the different wavelengths of light being refracted by different amounts when passing through a lens [1].

16–2 Spherical aberration arises because lenses are ground into a spherical shape, so the rays do not pass through the same focal point [1]. The problem can be reduced by using a lens ground into a parabolic shape [1].

16–3 The diameter of lenses is limited to less than 1 m [1]. Large lenses can go out of shape, since glass is a super-cooled fluid [1].

16–4 *Any two from:*
- Lenses are held in a metal tube, so the instrument is more robust [1].
- No mirrors to knock out of adjustment [1].
- Less expensive to produce [1].

16–5 The ratio of the diameters is $15:5 = 3:1$
Ratio of the collecting powers is $3^2:1^2 = 9:1$ [1]

17–1

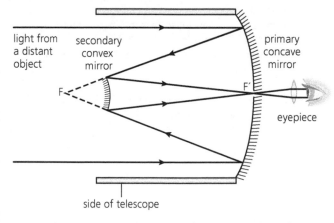

light from a distant object — secondary convex mirror — primary concave mirror — F — F′ — eyepiece — side of telescope

Primary concave mirror, secondary convex mirror, focal points, and eyepiece are all shown in the correct position [1].
At least two rays are shown reflected with angle of incidence = angle of reflection [1].

17–2 1. $M = \dfrac{f_o}{f_e} = \dfrac{225\,\text{cm}}{3.5\,\text{cm}} = 64.3$ [1]

2. $\tan \alpha = \alpha = 3500\,\text{km} \div 384\,000\,\text{km} = 9.11 \times 10^{-3}\,\text{rad}$ [1]

3. $\beta = 9.11 \times 10^{-3}\,\text{rad} \times 64 = 0.583\,\text{rad}$ [1]

4. $\beta = (0.583\,\text{rad} \times 360°) \div 2\pi = 33°$ [1]

17–3 1. $\theta = \dfrac{\lambda}{D}$
$\sin \theta = \theta = (570 \times 10^{-9}\,\text{m} \div 0.25\,\text{m}) = 2.28 \times 10^{-6}\,\text{rad}$ [1]

2. $d = 3.84 \times 10^8\,\text{m} \times 2.28 \times 10^{-6}\,\text{rad} = 875.52\,\text{m}$
$= 880\,\text{m}$ [1]

17–4 1. Use of a charge-coupled device (CCD) [1].

2. (*Any one advantage. Accept sensible alternatives.*)
Advantage:

- Clarity of images (no atmospheric distortion) [1].
- Images of IR or UV radiation can be observed [1].
(*Any one disadvantage. Accept sensible alternatives.*)
Disadvantage:
- Costly to launch [1].
- Hard to repair if there is a failure [1].

18–1 Collecting power ratio $= (76\,\text{m})^2 \div (38\,\text{m})^2 = 4:1$ [1]

18–2 1. $\lambda = c/f = 3.0 \times 10^8\,\text{Hz} \div 400 \times 10^6\,\text{Hz} = 0.75\,\text{m}$ [1]

2. A pulsar is a <u>spinning</u> [1] <u>neutron star</u> (or <u>white dwarf</u>) [1] that gives off pulses of electromagnetic radiation at regular intervals.
(*No marks for the last point as it's in the stem of the question.*)

3. $f = 1/T = (1.8 \times 10^{-3}\,\text{s})^{-1} = 556\,\text{Hz}$ [1]

18–3 1. $\theta = \lambda/D = 0.75\,\text{m} \div 76\,\text{m} = 9.9 \times 10^{-3}\,\text{rad}$ [1]

2. Distance $= 15\,\text{ly} \div \tan(9.9 \times 10^{-3}\,\text{rad}) = 15\,\text{ly} \div 9.9 \times 10^{-3}\,\text{rad} = 1515\,\text{ly} = 1500\,\text{ly}$
Minimum resolution angle, $\theta = \lambda/D = 0.75\,\text{m} \div 38\,\text{m} = 2.0 \times 10^{-2}\,\text{rad}$
Therefore the Mark 2 telescope would not be able to resolve the pulsars. [1]

18–4 Similarity: both have <u>parabolic</u> reflectors. The parabolic reflector ensures that the waves are <u>focused</u> perfectly onto the <u>focal point/principal focus</u> [1].
Difference: the radio telescope is much <u>larger</u>, as the radio wavelengths are much <u>longer than the wavelengths of light</u> [1].
OR The radio telescope does not have a secondary mirror as the antenna is at the principal focus [1].
(*For each mark, the point should be supported by an explanation.*)

18–5 Instead of having a larger telescope, we could use an array of many telescopes [1].

Classification of stars (p.100)

Quick questions

1. Difference in apparent magnitudes $= 2$
A is brighter by $2.51^2 = 6.30$ times than B.

2. It's not a fair test, because some stars are a lot further away than others. Therefore, a bright star at a long distance may appear dimmer than a comparatively dim star, which is closer to us.

3. $m - M = 5\log_{10}\left(\dfrac{d}{10}\right)$

Therefore: $M = m - 5\log_{10}\left(\dfrac{d}{10}\right)$

$M = +5.32 - 5 \times \log_{10}(8.7\,\text{pc} \div 10\,\text{pc})$
$= +5.32 - 5\log_{10}(0.87) = +5.32 - 5(-0.0605)$
$M = +5.32 + 0.3025 = +5.62$

4a) $1\,\text{AU} = 1.50 \times 10^{11}\,\text{m} \div 3.086 \times 10^{16}\,\text{m} = 4.86 \times 10^{-6}\,\text{pc}$

b) $M = m - 5\log_{10}\left(\dfrac{d}{10}\right)$
$M = -26.7 - (5 \times \log_{10}(4.86 \times 10^{-6}\,\text{pc} \div 10\,\text{pc}))$
$= -26.7 - (5 \times \log_{10}(4.86 \times 10^{-7}\,\text{pc}))$
$M = -26.7 - (5 \times -6.31) = -26.7 - -31.6 = +4.87$

5. $\lambda_{max} = 2.89 \times 10^{-3}\,\text{m K} \div 6500\,\text{K} = 4.45 \times 10^7\,\text{m} = 445\,\text{nm}$

6.

Term	Meaning	Unit
P	Power	Watt (W)
σ	**Stefan's constant**	$5.67 \times 10^{-8}\,\text{W m}^{-2}\,\text{K}^{-4}$
A	**Area**	m^2
T	Temperature	**K**

7. The Balmer series are hydrogen transitions that end at $n = 2$.

8 Both stars have the same luminosity. Therefore: $\dfrac{A_M}{A_R} = \dfrac{T_R^4}{T_M^4}$

$(A_M/A_R) = (4000\,\text{K})^4 \div (10\,000\,\text{K})^4 = 0.0256$
$(r_M/r_R) = (0.0256)^{0.5} = 0.16$
$(r_R/r_M) = 6.25$
The red giant is 6.25 times the diameter of the main sequence star.

9 *Relevant points to include:*
- Large stars have masses up to 100 times the solar mass.
- Large stars have short lives, less than 10^7 years.
- They burn fuel very quickly and at a high temperature.
- A Class G star like the Sun will last for about 10^{10} years.
- A star with about 0.5 solar mass will burn dimly, lasting 10^{12} years.
- Stars with masses less than 0.1 solar masses will not sustain nuclear fusion. They become brown dwarfs.

10a) A white dwarf.

b) Mass $= 2.0 \times 10^{30}\,\text{kg} \times 0.5 = 1.0 \times 10^{30}\,\text{kg}$
Volume $= 4/3 \times \pi \times (6.4 \times 10^6\,\text{m})^3 = 1.1 \times 10^{21}\,\text{m}^3$
Density $= 1.0 \times 10^{30}\,\text{kg} \div 1.1 \times 10^{21}\,\text{m}^3 = 9.1 \times 10^8\,\text{kg m}^{-3}$

11 Argon, $^{36}_{18}\text{Ar} + ^4_2\text{He} \rightarrow ^{40}_{20}\text{Ca}$

12 Neutrinos are the product of the interaction when electrons are forced into protons to make neutrons in the extreme conditions as a star collapses. The neutrinos can pass through the outer layers of the collapsing star as they only rarely interact with matter.
Gamma rays are the result of intensely energetic collisions between nuclei as the star collapses. The gamma rays are produced as the star collapses, which tells astronomers how long the collapse lasts. They travel along the axis of rotation of the star.

Exam-style questions

13-1 Difference is dimmest − brightest $= 1.25 - -0.04 = 1.29$
Ratio $= $ Arcturus/Deneb $= 2.51^{1.29} = 3.28$
Arcturus is 3.3 times brighter than Deneb. [1]

13-2 $1.29 = 2.5\,(\log\,(1.0 \times 10^{-15}\,\text{W m}^{-2}) - \log\,(I_B))$
$2.5\log I_B = (2.5 \times - 15) - 1.29 = -38.79$ [1]
$\log I_B = -38.79 \div 2.5 = -15.52$
$I_B = 10^{-15.52} = 3.048 \times 10^{-16}\,\text{W m}^{-2} = 3.0 \times 10^{-16}\,\text{W m}^{-2}$
(2 s.f.) [1]

14-1 1 degree $= 2\pi \div 360° = 0.0175\,\text{rad}$
1 arc second $= 0.0175\,\text{rad} \div 3600 = 4.85 \times 10^{-6}\,\text{rad}$ [1]

14-2 $\tan\theta = \dfrac{R}{d}$

Rearrange: $d = \dfrac{R}{\tan\theta}$ [1]

Since the angle in radians is so small, we can say that

$\tan\theta \approx \theta$. Therefore: $d = \dfrac{R}{\theta}$ [1]

14-3 $1\,\text{pc} = 1.50 \times 10^{11}\,\text{m} \div 4.85 \times 10^{-6}\,\text{rad} = 3.09 \times 10^{16}\,\text{m}$
$1\,\text{pc} = 3.09 \times 10^{16}\,\text{m} \div 9.46 \times 10^{15}\,\text{m ly}^{-1} = 3.27\,\text{ly}$ [1]
(Allow 3.26 ly if unrounded figures have been brought in from earlier.)

15-1 Equation: $P = \sigma A T^4$
$P = 5.67 \times 10^{-8}\,\text{W m}^{-2}\,\text{K}^{-4} \times (4 \times \pi \times (6.96 \times 10^8\,\text{m})^2) \times (5800\,\text{K})^4$
$P = 3.91 \times 10^{26}\,\text{W}$ [1]

15-2 1. $\lambda = 2.90 \times 10^{-3}\,\text{m K} \div 4660\,\text{K} = 6.22 \times 10^{-7}\,\text{m} = 622\,\text{nm}$.
It's red light. [1]

2. Rearrange: $A = \dfrac{P}{\sigma T^4}$

$A = 1.23 \times 10^{29}\,\text{W} \div (5.67 \times 10^{-8}\,\text{W m}^{-2}\,\text{K}^{-4} \times (4660\,\text{K})^4)$
$= 4.60 \times 10^{21}\,\text{m}^2$ [1]
$r^2 = 4.60 \times 10^{21}\,\text{m}^2 \div (4 \times \pi) = 3.66 \times 10^{20}\,\text{m}^2$
$r = (3.66 \times 10^{20}\,\text{m}^2)^{0.5} = 1.91 \times 10^{10}\,\text{m}$ [1]

3. Ratio $= $ radius of Dubhe $\div$ radius of the Sun
$= 1.91 \times 10^{10}\,\text{m} \div 6.96 \times 10^8\,\text{m} = 27.5$ times [1]

15-3 1. $\Delta E = -3.4\,\text{eV} - -13.6\,\text{eV} = 10.2\,\text{eV}$
$\Delta E = 10.2\,\text{eV} \times 1.60 \times 10^{-19}\,\text{J eV}^{-1} = 1.63 \times 10^{-18}\,\text{J}$ [1]

2. Equation: $\lambda = \dfrac{hc}{E}$

$\lambda = (6.63 \times 10^{-34}\,\text{J s} \times 3.00 \times 10^8\,\text{m s}^{-1}) \div 1.63 \times 10^{-18}\,\text{J} = 1.22 \times 10^{-7}\,\text{m} = 122\,\text{nm}$ [1]

3. The photons of this wavelength are in the UV region [1].

16-1 It has to be an electron neutrino to ensure that the lepton number is conserved at +1 [1]. *(If it were an electron antineutrino, the lepton number would be −1, so L would not be conserved.)*

16-2 Elements that have proton numbers greater than 26, and nucleon numbers greater than 56 [1].

16-3 $d = 11 \times 10^6\,\text{ly} \div 3.26\,\text{ly pc}^{-1} = 3.37 \times 10^6\,\text{pc}$ [1]

16-4 Equation: $M = m - 5\log_{10}\left(\dfrac{d}{10}\right)$

$M = +10.8 - 5\log\,(3.37 \times 10^6\,\text{pc} \div 10\,\text{pc}) = +10.8 - (5 \times 5.53) = +10.8 - 27.6$
$M = -16.8$ [1]

16-5 Time $= $ energy $\div$ power $= 10^{44}\,\text{J} \div 4.0 \times 10^{26}\,\text{J s}^{-1} = 2.5 \times 10^{17}\,\text{s}$
Time in years $= 2.5 \times 10^{17}\,\text{s} \div (365\,\text{d y}^{-1} \times 86\,400\,\text{s d}^{-1}) = 8 \times 10^9\,\text{y}$ [1]

17-1 Radius $= 12\,500\,\text{m}$
Volume $= (4 \div 3) \times \pi \times (12\,500\,\text{m})^3 = 8.2 \times 10^{12}\,\text{m}^3$
Density $= $ mass $\div$ volume $= (1.5 \times 2.0 \times 10^{30}\,\text{kg}) \div 8.2 \times 10^{12}\,\text{m}^3$
Density $= 3.7 \times 10^{17}\,\text{kg m}^{-3}$ [1]

17-2 A black hole is a region in space in which material is squashed so tightly that a singularity results, where the normal laws of physics do not apply [1]. The gravity is so intense that not even photons of light can escape. *(Or the escape velocity is greater than that of light.)* [1]

17-3 *Any one from:*
- Two white dwarfs can collide, and the combined mass results in a gravity field sufficiently powerful to squeeze the material together to form a singularity [1].
- A neutron star can attract sufficient material from a companion star for the combined mass to be sufficient to form a singularity [1].

17-4 Equation: $R_S = \dfrac{2GM}{c^2}$

$M = 1.99 \times 10^{30}\,\text{kg} \times 3.5 = 6.97 \times 10^{30}\,\text{kg}$
$R_S = (2 \times 6.67 \times 10^{-11}\,\text{N m}^2\,\text{kg}^{-2} \times 6.97 \times 10^{30}\,\text{kg}) \div (3.0 \times 10^8\,\text{m s}^{-1})^2$
$R_S = 1.03 \times 10^4\,\text{m} = 10\,\text{km}$ [1]

17-5 1. Radius of the photon sphere $= 30\,\text{km}$, therefore the radius of the event horizon $= 30 \div 1.5 = 20\,\text{km} = 2.0 \times 10^4\,\text{m}$ [1]

2. Rearrange: $M = \dfrac{R_S c^2}{2G}$

$M = (20\,000\,\text{m} \times (3.0 \times 10^8\,\text{m s}^{-1})^2) \div (2 \times 6.67 \times 10^{-11}\,\text{N m}^2\,\text{kg}^{-2})$
$M = 1.35 \times 10^{31}\,\text{kg}$. [1]
In solar masses, this is $1.35 \times 10^{31}\,\text{kg} \div 1.99 \times 10^{30}\,\text{kg} = 6.8$ solar masses [1].

18-1 Its luminosity/power and its distance [1].
(Both factors needed for the mark)

18-2 1. He catalogued all the stars on a scale of 1 (brightest) to 6 (dimmest) [1].
The brightest star was 100 times brighter than the dimmest [1].

2. From **Answer 18-2-1**, ratio, r, is given by [1]:

$\dfrac{I_1}{I_6} = r = 100$

Therefore: $r = 100^{\frac{1}{5}} = 2.512$

For each change in magnitude, the brightness increases by 2.51 times [1]

18–3 1. Betelgeuse, as its magnitude is less positive [1].

2. $\Delta m = 1.1 - 0.4 = 0.7$
Ratio of brightness $= 2.51^{\Delta m} = 2.51^{0.7} = 1.90$
Betelgeuse is 1.9 times as bright as Antares [1]

19–1 Astronomical unit: distance from the Sun to the Earth. [1]
Light year: the distance travelled in one year by light travelling at $3.0 \times 10^8\,\mathrm{m\,s^{-1}}$ [1] (*Both points needed for this mark.*)
Parsec: the distance to a star that subtends an angle of one arc second to the line from the Earth to the Sun [1].

19–2 Since the angle is very small and in rad, $\tan \theta = \theta$.
From the definition of the parsec:
$\theta = (1/3600)° = (2\pi \div 360) \div 3600 = 4.85 \times 10^{-6}\,\mathrm{rad}$ [1]
$1\,\mathrm{pc} = 1\,\mathrm{AU} \div 4.85 \times 10^{-6}\,\mathrm{rad} = 1.50 \times 10^{11}\,\mathrm{m} \div 4.85 \times 10^{-6}\,\mathrm{rad} = 3.09 \times 10^{16}\,\mathrm{m}$
$1\,\mathrm{pc} = 3.09 \times 10^{16}\,\mathrm{m} \div 9.46 \times 10^{15}\,\mathrm{m\,ly^{-1}} = 3.26\,\mathrm{ly}$ [1]

19–3 Difference in apparent magnitudes: $\Delta m = m_A - m_B$

Ratio of the intensities [1]: $\dfrac{I_A}{I_B} = 100^{\frac{\Delta m}{5}} = 100^{0.2\Delta m}$

Take logs [1]: $\log I_A - \log I_B = 2 \times 0.2\,\Delta m = 0.4\,\Delta m$

Rearrange: $\Delta m = 2.5\log\left(\dfrac{I_A}{I_B}\right)$

Using inverse square law [1]: $\left(\dfrac{I_A}{I_B}\right) = \left(\dfrac{d_B}{d_A}\right)^2$

Therefore [1]: $\Delta m = 2.5\log\left(\dfrac{d_B}{d_A}\right)^2 = 5\log\left(\dfrac{d_B}{d_A}\right)$

20–1 The absolute magnitude is the <u>brightness</u> of any star if it were placed at a <u>distance of 10 parsecs</u> from the Earth [1]. (*Mention of light years is a physics error.*)

20–2 1. Equation: $M = m - 5\log\left(\dfrac{d}{10}\right)$

$d = 10\,\mathrm{ly} \div 3.26\,\mathrm{ly\,pc^{-1}} = 3.07\,\mathrm{pc}$ [1]
$M = +5.6 - 5 \times (\log(3.07\,\mathrm{pc} \div 10\,\mathrm{pc}))$
$= +5.6 - (5 \times -0.513) = +5.6 - -2.57 = +8.2$ [1]

2. <u>Dimmer</u>, as the star has been <u>moved further out</u> [1].

20–3 1. Distance $= 300\,\mathrm{ly} \div 3.26\,\mathrm{ly\,pc^{-1}} = 92\,\mathrm{pc}$ [1]

2. Rearrange the equation: $m - M = 5\log\left(\dfrac{d}{10}\right)$

$m - -4.0 = 5 \times \log(92\,\mathrm{pc} \div 10\,\mathrm{pc}) = 4.82$ [1]
$m = 4.82 - 4.0 = +0.82$ [1]

3. The apparent magnitude is more negative, therefore brighter, because Star B is closer to the Earth than Star A [1].

21–1 1. Wien's law states: The <u>product</u> of the <u>wavelength</u> at which maximum energy is radiated and the <u>Kelvin temperature</u> is a constant. (*Or, the <u>wavelength</u> at which maximum energy is radiated is <u>inversely proportional</u> to the <u>Kelvin temperature</u>.*) [1]
(*Quote of the formula is not sufficient.*)

2. *Any two from:*
• A black body is a perfect absorber and a perfect emitter [1].
• A black body emits radiation across a wide range of wavelength [1].
• A black body emits a peak in intensity at a given wavelength [1].

• The hotter the object, the higher the peak [1].
• The hotter the object, the shorter the peak wavelength [1].

21–2 $\lambda_{max}T = 2.9 \times 10^{-3}\,\mathrm{m\,K}$
$\lambda_{max} = 2.9 \times 10^{-3}\,\mathrm{m\,K} \div 5600\,\mathrm{K} = 5.2 \times 10^{-7}\,\mathrm{m} = 520\,\mathrm{nm}$ [1]
Answer to 2 s.f. [1]

21–3 1. 100 nm is UV light; 1500 nm is infrared [1]. (*Both needed for the mark*)

2. Grey line (marked 4000 K) on the graph [1].

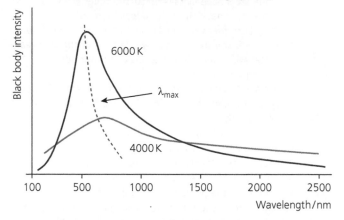

3. Dashed line on the graph [1].

22–1 The total energy per unit time (or power) radiated by a black body is proportional to the fourth power of its absolute temperature [1].

22–2 Luminosity is the total rate of production of energy (power) from the star.
Brightness is how much power is produced as visible light [1].
They are different because some of the energy is given off as IR/UV/X-rays [1].

22–3 1. $\lambda_{max} = 2.9 \times 10^{-3}\,\mathrm{m\,K} \div 5780\,\mathrm{K} = 502\,\mathrm{nm}$ [1]

2. Cyan [1]

3. $A = 4\pi r^2 = 4 \times \pi \times (6.96 \times 10^8\,\mathrm{m})^2 = 6.09 \times 10^{18}\,\mathrm{m}^2$ [1]
$P = 5.67 \times 10^{-8}\,\mathrm{W\,m^{-2}\,K^{-4}} \times 6.09 \times 10^{18}\,\mathrm{m}^2 \times (5780\,\mathrm{K})^4$
$= 3.85 \times 10^{26}\,\mathrm{W}$ [1]

22–4 1. Rearrange: $A = \dfrac{P}{\sigma T^4}$

$A = (3.85 \times 10^{26}\,\mathrm{W} \times 1.52) \div (5.67 \times 10^{-8}\,\mathrm{W\,m^{-2}\,K^{-4}} \times (5790\,\mathrm{K})^4) = 9.18 \times 10^{18}\,\mathrm{m}^2$ [1]
$r^2 = 9.18 \times 10^{18}\,\mathrm{m}^2 \div 4\pi = 7.31 \times 10^{17}\,\mathrm{m}^2$
$r = 8.55 \times 10^8\,\mathrm{m}$ [1]

2. Ratio of the radii $= 8.55 \times 10^8\,\mathrm{m} \div 6.96 \times 10^8\,\mathrm{m} = 1.23$ (or 1.23 : 1) [1]

23–1 $A = 4\pi r^2$
And: $P = \sigma A T^4$
Combine these to give: $P = 4\sigma\pi r^2 T^4$ [1]

23–2 Absolute magnitude considers that all stars are the same distance (10 pc) from the Earth [1]. The measured intensity (power per unit area) of the stars will be the same if the absolute magnitude is the same [1].

23–3 Use Stefan's law for star P: $P_P = 4\sigma\pi r_P^2 T_P^4$
For star Q: $P_Q = 4\sigma\pi r_Q^2 T_Q^4$

Since $P_P = P_Q$ [1]: $4\sigma\pi r_P^2 T_P^4 = 4\sigma\pi r_Q^2 T_Q^4$

Cancelling gives us: $r_P^2 T_P^4 = r_Q^2 T_Q^4$

By inspection, if $T_P = T_Q$ [1]: $r_P^2 = r_Q^2$
Therefore the radii of both stars are the same.

Answers

Spectral class	Surface temp/K	Colour	H Balmer series	Other elements
O	40 000	Blue	Weak	Ionised He
B	20 000	**Blue**	Medium	He atoms
A	10 000	Blue–white	Strong	Ionised metals
F	7 500	White	Medium	Ionised metals
G	**5 500**	**Yellow–white**	Weak	Medium ionised and neutral metals
K	4 500	**Orange**	Weaker	**Neutral metals**
M	3 000	Red	Very Weak	Neutral atoms, strong TiO

Table with:
- Spectral class column filled in [1].
- Class B and G rows complete [1].
- Class K and M rows complete [1].

24–2 1. The electrons in most of the hydrogen atoms do not have the energy to leave the ground state [1].
  2. The hydrogen has been ionised (*or has its electrons in very high energy levels*) [1], so few electrons will transition to $n = 2$ [1].

24–3 1. Molecules can only form by exchange of electrons between atoms [1]. In cooler stars, electrons can combine with nuclei to make atoms [1].
  2. In stars, only the strongest bonds can hold atoms together [1], only the most stable are likely to be found [1], e.g. titanium oxide (TiO).

24–4 1. $\lambda_{max}T = 2.9 \times 10^{-3}\,\text{m K}$
    $T = 2.9 \times 10^{-3}\,\text{m K} \div 530 \times 10^{-9}\,\text{m} = 5472\,\text{K} = 5500\,\text{K}$ (2 s.f.) [1]
  2. Class G [1]
  3. There is red light and blue light in the spectrum of a green star [1]. These will mix together to form white light [1].

25–1

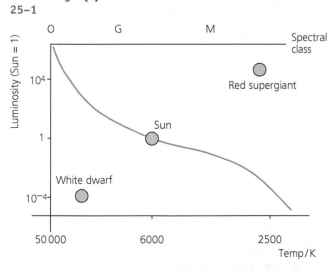

  1. The temperatures are decreasing. The scale is uneven [1].
  2. Spectral classes O, G and M as shown [1].
  3. Position of the Sun at $T = 6000\,\text{K}$, $L = 1$ [1].
  4. White dwarf and red supergiant in the correct place [1].
25–2 1. A main sequence star is one that fuelled by hydrogen [1].
  2. The <u>radiation pressure</u>/outward force of the explosion is <u>balanced</u> by the <u>force of gravity</u> pulling the star together [1].
25–3 1. Stefan's law: $P = \sigma A T^4$

    Rearrange: $\sigma = \dfrac{P}{AT^4}$

    Intensity = power per unit area, therefore: $\sigma = \dfrac{I}{T^4}$
    Rearrange to [1]: $I = \sigma T^4$
  2. Intensity of the Sun = $5.67 \times 10^{-8}\,\text{W m}^{-2}\,\text{K}^{-4} \times (5780\,\text{K})^4$
    = 6.33×10^7 [1] W m^{-2} [1]

25–4 1. Applying the inverse square law: $\Delta m = 2.5\log\left(\dfrac{I_{Star}}{I_{Sun}}\right)$
    $\Delta m = +4.8 - -5.1 = 9.9$
    $\log(I_{Star} / I_{Sun}) = 9.9 \div 2.5 = 3.96$ [1]
    $I_{Star} = 10^{3.96} \times I_{Sun} = 9120 \times I_{Sun}$ [1]
  2. $I = 9120 \times 6.33 \times 10^7\,\text{W m}^{-2} = 5.77 \times 10^{11}\,\text{W m}^{-2}$ [1]
  3. $\lambda_{max} = 2.9 \times 10^{-3}\,\text{m K} \div 25\,000\,\text{K} = 7.96 \times 10^{-9}\,\text{m}$ [1]
26–1 1. A nebula is a cloud of hydrogen gas [1].
  2. Processes of star formation:
    - Hydrogen molecules are very slowly attracted by gravity and come together [1].
    - As they accrete, they are warmed up by collisions [1].
    - In the centre of the cloud, the temperature becomes high enough to initiate nuclear fusion [1].
26–2 The amount of gas that has accreted is less than 0.25 solar masses [1]. Therefore, the temperature was not hot enough to initiate fusion [1].
26–3

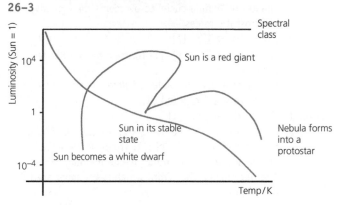

Graph with:
- labels correct [1] …
- … and in right positions [1]
- shape of path [1].

27–1 (*Any two from Part 1, any two from Part 2 and any two from Part 3.*)
Part 1 (End of hydrogen burning)
- When the hydrogen runs out, fusion will stop [1].
- The gravity will cause the Sun to collapse [1].
- The Sun will heat up enough to enable helium fusion [1] …
- … since the energy is sufficient to overcome electrostatic repulsion [1].
Part 2 (Formation of red giant)
- The Sun will re-light [1].
- The radiation pressure will make the star expand into a red giant [1].
- This has a cool outer envelope [1] …
- … but is much more luminous than the star was before [1].
Part 3 (Formation of white dwarf)
- When the helium fuel runs out, the star will collapse into a white dwarf [1].
- The surface temperature will be much hotter than the Sun [1].

- The white dwarf will not have a high luminosity [1] ...
- ... since it is so small [1].

27–2 1. Volume of Earth = $4/3 \times \pi \times (6.37 \times 10^6 \, \text{m})^3$
= $1.08 \times 10^{21} \, \text{m}^3$
Density = $1.99 \times 10^{30} \, \text{kg} \div 1.08 \times 10^{21} \, \text{m}^3$
= $1.84 \times 10^9 \, \text{kg m}^{-3}$ [1]

2. Equation: $g = -\dfrac{GM}{r^2}$
$g = (-6.67 \times 10^{-11} \, \text{N m}^2 \, \text{kg}^{-2} \times 1.99 \times 10^{30} \, \text{kg}) \div (6.37 \times 10^6 \, \text{m})^2 = (-)3.27 \times 10^6 \, \text{m s}^{-2}$ [1]

3. $v^2 = \dfrac{2GM}{r}$
$v^2 = 2 \times (6.67 \times 10^{-11} \, \text{N m}^2 \, \text{kg}^{-2} \times 1.99 \times 10^{30} \, \text{kg}) \div 6.37 \times 10^6 \, \text{m} = 4.17 \times 10^{13} \, \text{m}^2 \, \text{s}^{-2}$ [1]
$v = 6.46 \times 10^6 \, \text{m s}^{-1}$ [1]
$v = 2.2 \times 10^{-2} \, c$ [1]

27–3 Such stars may have 100 times the mass of the Sun and be 10^6 times as luminous [1]; Therefore they use up a finite amount of fuel at a much greater rate [1].

28–1 *Table completed:*
Helium fusion shell: C; low density envelope: A; heavier elements: D; hydrogen fusion shell: B. [1]

28–2 1. $r = 5.2 \times 1.50 \times 10^{11} \, \text{m} = 7.8 \times 10^{11} \, \text{m}$ [1]

2. Equation: $P = 4\sigma\pi r^2 T^4$
$P = 4 \times 5.67 \times 10^{-8} \, \text{W m}^{-2} \, \text{K}^{-4} \times \pi \times (7.8 \times 10^{11} \, \text{m})^2 \times (3500 \, \text{K})^4 = 6.51 \times 10^{31} \, \text{W}$ [1]
$P = 6.5 \times 10^{31} \, \text{W}$ (2 s.f.) [1]

3. Ratio = $6.51 \times 10^{31} \, \text{W} \div 3.83 \times 10^{26} \, \text{W} = 1.70 \times 10^5$ times the luminosity of the Sun [1].

28–3 1. The temperature needs to be about 10^8 K [1], to enable the helium nuclei to overcome the repulsive electrostatic forces [1].

2. $^4_2\text{He} + ^4_2\text{He} \rightarrow ^8_4\text{Be}$ [1]

3. $^8_4\text{Be} + ^4_2\text{He} \rightarrow ^{12}_6\text{C} \, (+\gamma)$ [1] (*Gamma ray not needed for the mark.*)

4. Beryllium is very unstable (*or almost immediately decays to <u>smaller nuclei</u>*) [1].

29–1 1. $^{36}_{18}\text{Ar} + ^4_2\text{He} \rightarrow ^{40}_{20}\text{Ca}$ [1]

2. The conditions in the dying star are not hot enough to fuse two sulfur nuclei [1]. The result of a fusion of two sulfur nuclei would give germanium which is a larger nuclide than iron, which is the most stable nuclide [1].

29–2
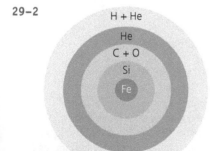

Diagram with:
- 5 layers [1]
- each layer labelled correctly [1].

29–3 1. Iron, it has the highest binding energy per nucleon [1].
2. *Any five from:*
- As the fusion stops, gravity is no longer balanced by radiation pressure [1].
- The outer layers collapse into the core [1].
- The temperature rises to 10^{11} K [1].
- The iron nuclei dissociate into helium nuclei, protons and neutrons [1].
- The pressure is so intense that electrons are forced into protons to make neutrons [1].

- The outer layers collide with the neutron core and rebound [1].
- The shockwave tears the star apart [1].
- The high temperatures start fusion reactions in the outer layers to produce a huge amount of energy [1].
- The energy produces elements larger than iron [1].

29–4 1. The core of neutrons [1].
2. $^1_1\text{p}^+ + ^0_{-1}\text{e}^- \rightarrow ^1_0\text{n} + v_e$ [1] (*Any reference to electron antineutrino is a physics error.*)

Cosmology (p.108)

Quick questions

1 Equation: $\dfrac{\Delta f}{f} = \dfrac{v}{c}$
$(\Delta f \div 512 \, \text{Hz}) = 40 \, \text{m s}^{-1} \div 340 \, \text{m s}^{-1}$
$\Delta f = 0.118 \times 512 \, \text{Hz} = 60 \, \text{Hz}$

a) Frequency approaching = $512 \, \text{Hz} + 60 \, \text{Hz} = 572 \, \text{Hz}$
b) Frequency going away = $512 \, \text{Hz} - 60 \, \text{Hz} = 452 \, \text{Hz}$

2 Equation: $\dfrac{\Delta\lambda}{\lambda} = -\dfrac{v}{c}$
$\Delta\lambda = --5000 \times 10^3 \, \text{m s}^{-1} \times 350 \times 10^{-9} \, \text{m} \div 3 \times 10^8 \, \text{m s}^{-1}$
$\Delta\lambda = 5.83 \times 10^{-9} \, \text{m}$
As the star is receding, we add the difference to the wavelength.
New wavelength = $355.83 \, \text{nm}$

3a) The negative difference indicates that the spectral line is blue-shifted, so the star is approaching the Earth.

b) Equation: $\dfrac{\Delta\lambda}{\lambda} = -\dfrac{v}{c}$
$(-0.264 \, \text{nm} \div 547 \, \text{nm}) = -(v \div 3.0 \times 10^8 \, \text{m s}^{-1})$
$-v = -4.83 \times 10^{-4} \times 3.0 \times 10^8 \, \text{m s}^{-1}$
$v = 1.45 \times 10^5 \, \text{m s}^{-1}$

4a) $\Delta\lambda = 428.6 \, \text{nm} - 422.3 \, \text{nm} = 6.3 \, \text{nm}$
$v = (6.3 \, \text{nm} \div 422.3 \, \text{nm}) \times 3.0 \times 10^8 \, \text{m s}^{-1}$
$= 4.48 \times 10^6 \, \text{m s}^{-1} \, (= 4480 \, \text{km s}^{-1})$

b) Hubble's law: $d = \dfrac{v}{H}$
$d = 4480 \, \text{km s}^{-1} \div 65 \, \text{km s}^{-1} \, \text{Mpc}^{-1} = 69 \, \text{Mpc}$

5a) Units for H in SI units are $\text{m s}^{-1} \, \text{m}^{-1} = \text{s}^{-1}$

b) Assume a constant rate of expansion. We know that:
$d = vt$
We know from Hubble's law that: $d = \dfrac{v}{H}$
Therefore: $vt = \dfrac{v}{H}$
The v terms cancel to give: $t = \dfrac{1}{H}$

6a) $H = 65 \, \text{km s}^{-1} \, \text{MPc}^{-1}$
$H = 6.5 \times 10^4 \, \text{m s}^{-1} \div 3.08 \times 10^{22} \, \text{m} = 2.1 \times 10^{-18} \, \text{s}^{-1}$

b) $t = \dfrac{1}{H}$
$t = (2.1 \times 10^{-18} \, \text{s}^{-1})^{-1} = 4.7 \times 10^{17} \, \text{s}$
In years: $t = 4.7 \times 10^{17} \, \text{s} \div (365 \, \text{d y}^{-1} \times 86\,400 \, \text{s d}^{-1})$
$= 15 \times 10^9 \, \text{y}$

7 *Relevant points to include:*
- The universe is expanding.
- The further the galaxies are away from the Earth, the faster they are moving away.
- As the universe cooled, the temperature fell to 2.7 K, which corresponds to a microwave wavelength of 1.8 mm. The cosmic background radiation is fairly uniform.
- Theoretical calculations suggest that the earliest universe consisted of 75% H, 25% He.
- Observations of very distant (hence very old) objects show that ratio of hydrogen and helium.

8 $E = \dfrac{hc}{\lambda}$
$E = (6.63 \times 10^{-34} \, \text{J s} \times 3.0 \times 10^8 \, \text{m s}^{-1}) \div 1.8 \times 10^{-3} \, \text{m}$
$= 1.11 \times 10^{-22} \, \text{J}$

9 Quasars:
- are very distant objects
- have a very high absolute magnitude
- emit jets of matter
- are smaller than a galaxy
- are thought to be super-massive black holes that are gobbling up stars.

10a) Exoplanets are planets that are outside the solar system, and orbit a star or a binary system.

b) The only exoplanets that have been directly observed are gas giants at least the size of Jupiter or even larger.

11 *Relevant points to include:*
- Such planets are too small for direct observation, even by the best telescopes.
- They reflect light from stars, so are much dimmer.
- The light given out is swamped by the star(s) around which they orbit.

12 The planet exerts a gravitational pull on its star, which means that the system centre of mass is not at the centre of the star. The star moves around the system centre of mass.

As the star comes towards us, it is blue-shifted. As it moves away, it is red-shifted. The planet has to be large to have a noticeable effect. This method does not allow the mass of the planet to be worked out, as we do not know the distance from the star.

Exam-style questions

13–1 1. Red shift [1]. (*'Doppler effect' on its own is not enough for the mark.*)
   2. The star is moving away, so the waves are of a longer wavelength [1].

13–2 Equation: $\dfrac{\Delta\lambda}{\lambda} = \dfrac{v}{c}$

$\Delta\lambda = 434.2\,\text{nm} - 439.7\,\text{nm} = -5.5\,\text{nm} = -5.5 \times 10^{-9}\,\text{m}$

$v = (-5.5 \times 10^{-9}\,\text{m} \div 4.342 \times 10^{-7}\,\text{m}) \times 3.0 \times 10^{8}\,\text{m s}^{-1}$ [1]

$v = -3.8 \times 10^{6}\,\text{m s}^{-1}\ (= 3800\,\text{km s}^{-1})$

Minus sign means the star is moving away [1]. (*This must be included for mark.*)

13–3 $v = H_0 d$

Rearrange: $d = \dfrac{v}{H_0}$

$d = 3800\,\text{km s}^{-1} \div 65\,\text{km s}^{-1}\,\text{Mpc}^{-1} = 58\,\text{Mpc}$ [1]

13–4 The red shift would be less [1].
The recession speed would be less because $v \propto d$ (Hubble's law) [1].
(*Hubble's law should be quoted for the mark.*)

14–1 The Steady State Theory stated that the Universe had always been the way it was and would remain so for ever [1]. The Expansion Theory suggested that the Universe had started in a single point and was expanding all the time [1].

14–2 Proportion of hydrogen to helium. The universe was so hot that fusion of hydrogen nuclei formed helium [1] in a proportion of 75% H and 25% He. Distant objects in the Universe have shown this proportion [1].
Cosmic Background Radiation. As the universe expanded, it cooled and is currently at a temperature of 2.7 K [1]. This is shown by microwave radiation of about 1.8 mm wavelength [1].
(*Each piece of evidence must be given.*)

14–3 1. The SI units for km s^{-1} and Mpc^{-1} are m s^{-1} and m^{-1}, respectively.
   Therefore m s^{-1} combined with m^{-1} results in s^{-1}. (Or use of $H = v/d$.) [1]
   2. $1\,\text{Mpc} = 1 \times 10^{6} \times 3.08 \times 10^{16}\,\text{m} = 3.08 \times 10^{22}\,\text{m}$
   $65\,\text{km s}^{-1} = 6.5 \times 10^{4}\,\text{m s}^{-1}$
   $H = 6.5 \times 10^{4}\,\text{m s}^{-1} \div 3.08 \times 10^{22}\,\text{m} = 2.1 \times 10^{-18}\,\text{s}^{-1}$ [1]
   3. The age of the universe $= 1/H = (2.1 \times 10^{-18}\,\text{s}^{-1})^{-1} = 4.7 \times 10^{17}\,\text{s}$ [1]

In years: $t = 4.7 \times 10^{17}\,\text{s} \div (365\,\text{dy y}^{-1} \times 86\,400\,\text{s dy}^{-1})$
$= 1.5 \times 10^{10}\,\text{y}$ [1]
   4. The expansion is constant [1].

15–1 The spectra were red-shifted considerably, suggesting a high recession speed [1].

15–2 $d = \dfrac{v}{H}$

$d = (0.15 \times 3.0 \times 10^{8}\,\text{m s}^{-1}) \div 2.2 \times 10^{-18}\,\text{s}^{-1} = 2.05 \times 10^{25}\,\text{m}$ [1]

$d = 2.05 \times 10^{25}\,\text{m} \div 3.08 \times 10^{22}\,\text{m Mpc}^{-1} = 660\,\text{Mpc}$ [1]

15–3 1. Equation: $\dfrac{\Delta\lambda}{\lambda} = \dfrac{v}{c}$

Rearranging: $\Delta\lambda = \dfrac{v\lambda}{c}$

$\Delta\lambda = 0.15 \times 425 \times 10^{-9}\,\text{m} = 63.8 \times 10^{-9}\,\text{m}$

Wavelength $= 425 \times 10^{-9}\,\text{m} + 63.8 \times 10^{-9}\,\text{m} = 489 \times 10^{-9}\,\text{m} = 489\,\text{nm}$ [1]

   2. $f = \dfrac{c}{\lambda}$

$f = 3.0 \times 10^{8}\,\text{m s}^{-1} \div 0.21\,\text{m} = 1.43 \times 10^{9}\,\text{Hz}$ [1]

$-\dfrac{\Delta f}{f} = \dfrac{v}{c}$

Rearranging: $-\Delta f = \dfrac{vf}{c}$

$\Delta f = -0.15 \times 1.43 \times 10^{9}\,\text{Hz} = -2.14 \times 10^{8}\,\text{Hz}$

Frequency $= 1.43 \times 10^{9}\,\text{Hz} - 2.14 \times 10^{8}\,\text{Hz} = 1.22 \times 10^{9}\,\text{Hz}$ [1]

16–1 1. A supermassive black hole in the centre of a galaxy [1].
   2. Quasars are very distant objects. It is thought that they represent what the universe was like shortly after the Big Bang [1].

16–2 *Any three from:*
- The black holes are consuming material at a great rate [1].
- Energy is released as electromagnetic waves [1].
- The EM waves have a range of wavelengths from radio waves to gamma rays [1].
- This occurs at a faster rate than fusion [1].

16–3 1. Equation: $M = m - 5\log\left(\dfrac{d}{10}\right)$

$M = +14.6 - 5\log(7.8 \times 10^{8}\,\text{pc} \div 10\,\text{pc}) = +14.6 - 5 \times \log(7.8 \times 10^{7})$ [1]

$M = +14.6 - 5 \times 7.89 = +14.6 - 39.5 = -24.9$ [1]

   2. $\Delta M = 2.5\log\left(\dfrac{I_Q}{I_S}\right)$

$\Delta M = 4.8 - -24.9 = 29.7$

$\log(I_Q/I_S) = 29.7 \div 2.5 = 11.88$

$I_Q = 7.59 \times 10^{12}$ times that of the Sun [1]

$I_Q = 7.59 \times 10^{11} \times 7.3 \times 10^{7}\,\text{W m}^{-2} = 5.5 \times 10^{19}\,\text{W m}^{-2}$ [1]

   3. $R_s = \dfrac{2GM}{c^2}$

Mass of quasar $= 1.99 \times 10^{30}\,\text{kg} \times 5.0 \times 10^{8}$
$= 9.95 \times 10^{38}\,\text{kg}$ [1]

$R_S = (2 \times 6.67 \times 10^{-11}\,\text{N m}^2\,\text{kg}^{-2} \times 9.95 \times 10^{38}\,\text{kg}) \div (3.0 \times 10^{8}\,\text{m s}^{-1})^2 = 1.50 \times 10^{12}\,\text{m}$ [1]

   4. Area of the event horizon $= 4 \times \pi \times (1.50 \times 10^{12}\,\text{m})^2$
$= 2.83 \times 10^{25}\,\text{m}^2$
Luminosity $= 5.5 \times 10^{19}\,\text{W m}^{-2} \times 2.83 \times 10^{25}\,\text{m}^2$
$= 1.6 \times 10^{45}\,\text{W}$ [1]
Ratio of quasar to the Sun $= 1.6 \times 10^{45}\,\text{W} \div 4.47 \times 10^{26}\,\text{W} = 3.6 \times 10^{18}$ times [1]

17–1 *Any two from:*
- infinitely dense [1]
- laws of physics do not apply [1]
- contained all the mass (and space–time) of the universe [1].

(*No reference to space–time needed*)

17–2 1. The Big Bang. [1]
2. *Any three from:*
- There was an enormous explosion and the universe was extremely hot [1].
- There was a very rapid expansion in the first second [1].
- In the first second, exchange particles, quarks and leptons were formed [1].
- Within minutes, the temperature had fallen to allow fusion, and the first simple nuclei were formed [1].

17–3 [1]

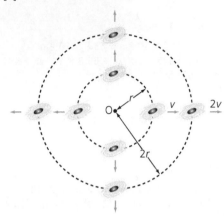

The drawing should show that at r, the recession speed should be v, and at $2r$, the recession speed is $2v$ [1].

17–4 A – the Universe is above the critical density. It will collapse back to a single point [1].

B – the Universe is at the critical density. Its rate of expansion will slow and it will end up at a constant size [1].

C – the Universe is slightly below the critical density. It will continue to expand slowly [1].

D – the Universe is well below the critical density. It will expand at an increasing rate [1].

Topic review: astrophysics (p.111)

1–1 1 ly is the <u>distance</u> travelled by light (at a speed of $3.0 \times 10^8\,\text{m s}^{-1}$) in a <u>period of 1 year</u> [1].

1–2 1. $A = 4 \times \pi \times (6.96 \times 10^8\,\text{m})^2 = 6.09 \times 10^{18}\,\text{m}^2$ [1]
$I = P/A = 3.83 \times 10^{26}\,\text{W} \div 6.09 \times 10^{18}\,\text{m}^2$
$= 6.29 \times 10^7\,\text{W m}^{-2}$ [1]

2. Equation: $\dfrac{I_B}{I_A} = \left(\dfrac{d_A}{d_B}\right)^2$

Rearrange: $I_B = \left(\dfrac{d_A}{d_B}\right)^2 \times I_A$

$d_B = 50\,\text{ly} \times 9.46 \times 10^{15}\,\text{m} = 4.73 \times 10^{17}\,\text{m}$
$I_B = (6.96 \times 10^8\,\text{m} \div 4.73 \times 10^{17}\,\text{m})^2 \times 6.29 \times 10^7\,\text{W m}^{-2}$
$= 1.36 \times 10^{-10}\,\text{W m}^{-2}$ [1]

1–3 1. Equation (*from photon energy*): $E = \dfrac{hc}{\lambda}$

$E = (6.63 \times 10^{-34}\,\text{J s} \times 3.0 \times 10^8\,\text{m s}^{-1}) \div 483 \times 10^{-9}\,\text{m}$
$= 4.12 \times 10^{-19}\,\text{J}$ [1]

2. $A = \pi d^2/4 = (\pi \times (1.50\,\text{m})^2) \div 4 = 1.77\,\text{m}^2$
Number of photons per second = (intensity ÷ photon energy) × area
$N = (1.36 \times 10^{-10}\,\text{W m}^{-2} \div 4.12 \times 10^{-19}\,\text{J}) \times 1.77\,\text{m}^2 = 5.84 \times 10^8\,\text{s}^{-1}$ [1]

1–4 1. $QE = \dfrac{\text{number of electrons produced every second}}{\text{number of photons absorbed every second}}$ [1]

2. Number of electrons per second = $0.87 \times 5.84 \times 10^8\,\text{s}^{-1}$
$= 5.08 \times 10^8\,\text{s}^{-1}$ [1]

1–5 1. They do not reflect except at shallow angles [1].

2. They can be reflected at shallow angles from highly reflective surfaces [1].

1–6 1. Use: $\dfrac{\Delta\lambda}{\lambda} = \dfrac{v}{c}$

$\Delta\lambda = 488.0265\,\text{nm} - 488.0265\,\text{nm} = 2.8 \times 10^{-3}\,\text{nm}$ [1]
$v/c = 2.8 \times 10^{-3}\,\text{nm} \div 488.0265\,\text{nm} = 5.74 \times 10^{-6}$
$v = 5.74 \times 10^{-6} \times 3.0 \times 10^8\,\text{m s}^{-1} = 1721\,\text{m s}^{-1}$ [1]
$= 1.7\,\text{km s}^{-1}$

2. Use Hubble's law: $d = \dfrac{v}{H}$

$d = 1.7\,\text{km s}^{-1} \div 65\,\text{km s}^{-1}\,\text{Mpc}^{-1} = 0.262\,\text{Mpc} = 26\,000\,\text{pc}$
[1] (2 s.f.) [1]

1–7 Use: $m - M = 5\log\left(\dfrac{d}{10}\right)$

$m - M = 5\log(26\,000\,\text{pc} \div 10\,\text{pc}) = 5\log(2640)$ [1]
$+10.3 - M = 5 \times 3.415 = 17.1$
$-M = 6.8$
$M = -6.8$ [1]

1–8 The answer has a <u>wide amount of uncertainty</u>, as the figure given for Hubble's constant has a <u>wide range of values</u> [1].

2–1

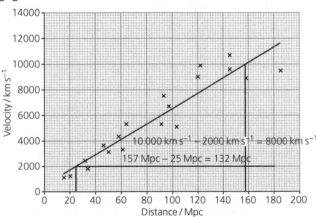

Graph with:
- axes labelled: velocity on y-axis, distance on x-axis [1]
- sensible scales [1]
- points plotted accurately [1]
- line of best fit [1].

2–2

E.g. (depending on the line of best fit) rise = $10\,000\,\text{km s}^{-1} - 2000\,\text{km s}^{-1} = 8000\,\text{km s}^{-1}$
Run = $157\,\text{Mpc} - 25\,\text{Mpc} = 132\,\text{Mpc}$ [1].
Gradient = $8000\,\text{km s}^{-1} \div 132\,\text{Mpc} = 61\,\text{km s}^{-1}\,\text{Mpc}^{-1}$ [1]
(*Correct unit needed for the second mark.*)
Answer will depend on the line of best fit. Credit for
Answers 2–2 and 2–3 *are awarded for answers that are consistent with the graphs from* ***Answer 2–1***.

2–3 1.

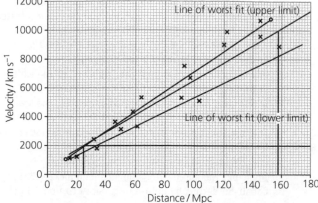

Lines of worst fit shown for upper and lower limits [1].

2. Upper limit gradient, e.g. Rise = 1000 km s⁻¹ – 2000 km s⁻¹ = 8000 km s⁻¹

   Run = 141 Mpc – 25 Mpc = 116 Mpc

   Gradient = 8000 km s⁻¹ ÷ 116 Mpc = 69 km s⁻¹ Mpc⁻¹ [1]

   Lower limit, e.g. Gradient = 6000 km s⁻¹ ÷ 120 Mpc = 50 km s⁻¹ Mpc⁻¹ [1]

3. $\Delta G = \dfrac{G_{max} - G_{min}}{2}$

   $\Delta G = (69\,\text{km s}^{-1}\,\text{Mpc}^{-1} \div 50\,\text{km s}^{-1}\,\text{Mpc}^{-1}) \div 2$

   $= 9.5\,\text{km s}^{-1}\,\text{Mpc}^{-1}$ [1]

2–4 Hubble initially looked at stars that were close by (*or some of the objects he called stars were more distant star clusters*) [1].

3–1 Direct observation is difficult because: such planets have to be very large to be seen from such distances [1]; The planets do not produce their own light, but reflect light from their star [1].

3–2 1. • The planet has to be large enough for the joint centre of mass of the star and the planet to be outside the star [1].
   • Blue shift indicates that the star is moving towards us; red shift that the star is moving away [1].
   (*Both blue and red shift need to be mentioned.*)

2. Equation: $\Delta\lambda = \dfrac{v\lambda}{c}$

   $\Delta\lambda = (140\,\text{m s}^{-1} \times 436.3 \times 10^{-9}\,\text{m}) \div 3.0 \times 10^{8}\,\text{m s}^{-1}$

   $= 2.0 \times 10^{-13}\,\text{m}$ [1]

3. Use $v = Hd$

   $d = 20\,\text{m s}^{-1} \div 2.2 \times 10^{-18}\,\text{s}^{-1} = 9.1 \times 10^{18}\,\text{m}$ [1]

   $d = 9.1 \times 10^{18}\,\text{m} \div 3.08 \times 10^{16}\,\text{m pc}^{-1} = 300\,\text{pc}$ [1]

3–3 The intensity of light coming from a star is reduced [1] by the same ratio as the planet's area is to the star's area [1].
Diagram [1]

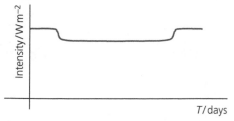

3–4 Goldilocks zone means that the stars have to be in a specific orbital zone to ensure:
   • presence of liquid water
   • temperature about 300 K.

   [2] (1 mark for the mention of the *orbital zone*, 1 mark for the *two* bulleted points)

4–1 1. An absorption spectrum shows the light from stars as seen through a diffraction grating [1].

2. Dark lines are seen across a continuous visible spectrum [1]. These are characteristic of the elements that are in the observed object [1].

3. The patterns of lines show what elements are present in a star [1].

4–2 1. Transitions that absorb/emit photons starting at energy level $n = 2$ [1].

2. A photon of exactly the right energy ($E = hf$) raises an electron to a higher energy level [1]. The electron falls from the higher level to the lower level emitting a photon of energy $E = hf$ as it falls back to $n = 2$ [1]. The photons are radiated in random directions, so the intensity is much lower (hence leading to dark lines) [1]. (*For full marks, mention must be made of E = hf, photons and electrons.*)

3. The transitions would be formed in the atmosphere/ photosphere/outer envelope [1]. This region is energetic enough to excite the atoms, but not sufficiently energetic to ionise the atoms [1].

4–3 1. Energy change = –1.51 eV – –3.41 eV = 1.90 eV

   Energy change (J) = 1.90 eV × 1.60 × 10⁻¹⁹ J eV⁻¹ = 3.04 × 10⁻¹⁹ J [1]

   $\lambda = hc/E = 6.63 \times 10^{-34}\,\text{J s} \times 3.0 \times 10^{8}\,\text{m s}^{-1} \div 3.04 \times 10^{-19}\,\text{J} = 6.54 \times 10^{-7}\,\text{m} = 654\,\text{nm}$ [1]

2. This is in the red region [1].

5–1 Planetary nebula ticked [1].

5–2 They have a higher temperature than the Sun (*or T = 10 000 K*) [1]. They are the same size as a planet [1].

5–3 $P = 4\sigma\pi r^2 T^4$

   Assume $r = 6.4 \times 10^6\,\text{m}$; $T = 10\,000\,\text{K}$;

   $P = 4 \times 5.7 \times 10^{-8}\,\text{W m}^{-2}\,\text{K}^{-4} \times \pi \times (6.4 \times 10^6\,\text{m})^2 \times (10\,000\,\text{K})^4 = 2.9 \times 10^{23}\,\text{W}$ [1]

   Ratio = 2.9 × 10²³ W ÷ 4 × 10²⁶ W = 7.3 × 10⁻⁴ [1]

   This is about 1400 times less luminous than the Sun [1]. (*Answer should be to no more than 2 s.f.*)

5–4 The white dwarf remains hot as gravity pulls it together and heat is given off as it squashes together [1]. Eventually it will cool to become a brown dwarf [1]. Finally, it will end up as an inert dark mass [1].

6–1 1. d = 250 000 ly ÷ 3.26 ly pc⁻¹ = 76 700 pc [1]

2. $M = m - 5\log\left(\dfrac{d}{10}\right)$

   $M = +3.25 - 5\log(7670) = +3.25 - 19.42$ [1]

   $M = -16.2$ [1]

6–2 *Any two from:*
   • The gravity is very intense [1].
   • The electrons are forced into protons to make neutrons [1].
   • This makes dense neutron material [1].

6–3 Mass = 1.5 × 2.0 × 10³⁰ kg = 3.0 × 10³⁰ kg [1]

   Equation: $g = -\dfrac{GM}{r^2}$

   $g = -(6.67 \times 10^{-11}\,\text{N m}^2\,\text{kg}^{-2} \times 3.0 \times 10^{30}\,\text{kg}) \div (12\,000\,\text{m})^2 = -1.4 \times 10^{12}\,\text{N kg}^{-1}$ [1]

6–4 1. A is a beam of radio waves; B is a magnetic field [1].

2. A pulsar [1].

7–1 A black hole is so dense that even light cannot escape [1]. (*'Heavy' or 'massive' does not get the mark.*)

7–2 1. The event horizon is the boundary below which light cannot escape.
   (*Alternative correct answer is 'The boundary at which the escape speed is the speed of light.'*) [1]

2. Consider a spacecraft, mass m, orbiting at the event horizon. To get away from the black hole, it has to have potential energy: $E_p = -\dfrac{GMm}{r}$

   To do the work required to move to infinity, it needs to have kinetic energy: $E_k = \dfrac{1}{2}mv^2$

   By definition, $v = c = 3 \times 10^8\,\text{m s}^{-1}$. Therefore, at the Schwarzschild radius [1]: $\dfrac{1}{2}mc^2 = \dfrac{GMm}{R_S}$

The mass cancels out and the equation rearranges [1]:

$$R_s = \frac{2GM}{c^2}$$

(*For this last mark, you need to demonstrate evidence that the mass terms are cancelled and the equation is rearranged.*)

3. Rearrange: $M = \frac{R_s c^2}{2G}$

   $M = (30 \times 10^3 \, \text{m} \times (3.0 \times 10^8 \, \text{m s}^{-1})^2) \div 2 \times 6.67 \times 10^{-11} \, \text{N m}^2 \, \text{kg}^{-2}$ [1]

   $M = 2.0 \times 10^{31} \, \text{kg}$ [1]

4. Ratio = $2.0 \times 10^{31} \, \text{kg} \div 2.0 \times 10^{30} \, \text{kg} = 10$ times bigger than the Sun [1]

7–3 Gravitational lensing *OR evidence of stellar material being attracted into a black hole* [1].

7–4 Gamma ray bursts last for a few seconds (or minutes), indicating that the supernova collapse takes place in the same period of time [1]. They are formed by the intense energy of the interactions taking place at the core of the star [1].

8–1 1. Absolute magnitude is the <u>apparent magnitude</u> if the star was <u>10 pc from the Earth</u> [1].

2. Standard candle is an <u>absolute apparent magnitude</u> of a <u>known value</u> [1].

8–2 *Any five from:*

- These supernovae form from binary star systems [1].
- One star is bigger than the other, so evolves at a faster rate [1].
- The big star has become a white dwarf, while the smaller star has become a red giant [1].
- Gravity pulls material from the red giant to the white dwarf [1].
- If the mass becomes over 1.4 solar masses, the star collapses [1].
- The resulting fusion results in a supernova explosion [1].

8–3 1. Nickel decay at the peak; cobalt decay between 100 and 200 days [1]

2. <u>Iron</u> (-52) was <u>fused with helium</u> to form nickel (-56) [1].

3. Cobalt-56 decays by beta-minus decay to iron-56 [1].

8–4 Equation: $M = m - 5\log\left(\dfrac{d}{10}\right)$

$-19.5 = +2.1 - 5 \log (d/10)$

$5 \times (\log d - 1) = 21.6$

$5 \log d = 26.6$ [1]

$\log d = 5.32$

$d = 2.1 \times 10^5 \, \text{pc}$ [1]

Practice exam papers

Practice paper 1 (p.116)

Question 1

Question	Answer	Guidance	Marks	ID
1–1	An isotope is an element with the same number of protons, but a different number of neutrons in its nucleus.	Allow analogous definition.	1	AO1
1–2	$^{10}_{6}C \rightarrow\ ^{10}_{5}B +\ ^{0}_{1}e^+ + v_e$	1 mark for correct boron and 1 mark for e^+ and neutrino.	2	AO2
1–3		All particles from the diagram (p, n and W^+) should be shown for full marks.	3	AO2
1–4	Weak interaction	Accept weak force	1	AO1

Question 2

Question	Answer	Guidance	Marks	ID
2–1	Individual photons are interacting with individual electrons. [1] Photon energy is absorbed by an individual electron. [1]	Using intense light of lower frequency not inducing photoelectricity is also a good argument.	2	AO3
2–2	Convert 520 nm to 5.2×10^{-7} m and $0.078\,eV = 0.078 \times 1.6 \times 10^{-19}\,J$ $= 1.25 \times 10^{-20}\,J$ [1] $E_{k(max)} = 1.25 \times 10^{-20}\,J$ [1] Rearrange $hf = E_{k(max)} + \Phi\ \rightarrow$ $\Phi = hc/\lambda - E_{k(max)}$ [1] $= 6.63 \times 10^{-34}\,Js \times 3.0 \times 10^8\,ms^{-1} / 5.20 \times 10^{-7}\,m - 1.25 \times 10^{-20}\,J$ $= 3.7 \times 10^{-19}\,J$ [1]		4	AO2 MS 2.3
2–3	Electrons from the metal surface are emitted with max E_k, because no energy is used to move them to the surface. [1] Electrons deeper in the potassium block have less kinetic energy, because some energy was used to move the electrons to the surface. [1]		2	AO3 MS 2.2 MS 2.3
2–4	Increase the intensity of the incident light.		1	AO1
2–5	The new metal has a higher work function [1], which would require photons of shorter wavelength to carry enough energy to win the work function [1].		2	AO1

Question 3

Question	Answer	Guidance	Marks	ID
3–1	$\sin\upsilon_{2r} = \dfrac{\sin55°}{n_{2r}} = \dfrac{\sin55°}{1.50917} = 0.543$ [1] $\sin\upsilon_{2v} = \dfrac{\sin55°}{n_{2v}} = \dfrac{\sin55°}{1.52136} = 0.538$ [1] $\upsilon_{2r} = a\sin\upsilon_{2r} = 32.87°$ and $\upsilon_{2v} = a\sin\upsilon_{2v} = 32.58°$ [1]	Accept answers to 4 s.f. only	3	AO2 MS 0.6 MS 4.1
3–2	$0.140\,\text{m} \times \tan\upsilon_{2r} - 0.140\,\text{m} \times \tan\upsilon_{2v}$ [1] $= 1 \times 10^{-3}\,\text{m}$ [1]	ECF from previous answer	2	AO1 MS 0.6 MS 4.1
3–3			1	AO1
3–4	Because of the different angles of refraction for different wavelengths, due to their different speeds in the material of the core. [1] This could lead to pulse broadening, which limits the maximum frequency of pulses, i.e. the bandwidth. [1]		2	AO3
3–5	Use a semicircle of borosilicate glass attached to a glass of lower refractive index. [1] Vary the angle of incidence between the borosilicate and the test glass. [1] Measure and record the critical angle and change the test glass (using different refractive indexes). [1] A large critical angle is preferable in optical fibres to minimise modal dispersion. [1]	Alternative methods that achieve similar outcomes are acceptable.	4	AO3

Question 4

Question	Answer	Guidance	Marks	ID
4–1	$W_p = m\,g\,\sin12° = 233\,\text{kg} \times 9.81\,\text{m s}^{-2} \times \sin12°$ $= 475\,\text{N}$ [1] $R = W_p + F = 475\,\text{N} + 140\,\text{N} = 615\,\text{N}$ [1]		2	AO2 MS 0.6, 4.2, 4.4, 4.5
4–2	$v^2 = u^2 + 2as$ $a = \dfrac{R}{m} = \dfrac{615\,\text{kg m s}^{-2}}{233\,\text{kg}} = 2.64\,\text{m s}^{-2}$ [1] $v = \sqrt{2as} = \sqrt{2 \times 2.64\,\text{m s}^{-2} \times 5\,\text{m}} = 5.14\,\text{m s}^{-1}$ [1]		2	AO1 AO2 MS 0.5, 2.2, 2.3, 2.4
4–3	Acceleration after A: $a = \dfrac{W_p}{m} = \dfrac{476\,\text{N}}{233\,\text{kg}} = 2.04\,\text{m s}^{-2}$ [1] Velocity at B: $v = \sqrt{u^2 + 2as} = \sqrt{5.14^2\,\text{m}^2\,\text{s}^{-2} + 2 \times 2.04\,\text{m s}^{-2} \times 23\,\text{m}}$ $= 10.97\,\text{m s}^{-1}$ [1] From B, v becomes initial velocity u: $s = \dfrac{v+u}{2}t = \dfrac{0+u}{2}t = \dfrac{10.97\,\text{m s}^{-1}}{2} \times 3.2\,\text{s} = 17.55\,\text{m}$	Allow ECF from **Question 4–2**. Allow alternative method with energy analysis.	3	AO2 AO3 MS 0.5, 2.2, 2.3, 2.4
4–4	Frictionless track [1] Constant braking force [1]		2	AO1
4–5	The cart needs to be lifted so its centre of mass is beyond the front wheel (contact point). [1] On the slope, the cart is already partially tilted. [1]		2	AO2

Question 5

Question	Answer	Guidance	Marks	ID
5–1	$V = \dfrac{m}{\rho} = \dfrac{8.15 \times 10^7\,\text{kg}}{7.7 \times 10^3\,\text{kg m}^{-3}} = 10584\,\text{m}^3$ [1] $A_{\text{cable}} = \dfrac{V}{l} = \dfrac{10584\,\text{m}^3}{2332\,\text{m}} = 4.54\,\text{m}^2$ [1] $A_{\text{wire}} = \dfrac{A_{\text{cable}}}{27\,572} = 1.65 \times 10^{-4}\,\text{m}^2$ [1]		3	AO1 AO2 MS 0.2, 4.3
5–2	Young modulus $(M) = \dfrac{FL}{A_{\text{cable}}\Delta L} \rightarrow \Delta L = \dfrac{FL}{A_{\text{cable}}M}$ [1] $= \dfrac{1.39 \times 10^6\,\text{N} \times 2332\,\text{m}}{4.5\,\text{m}^2 \times 195 \times 10^9\,\text{N m}^2}$ [1 mark for substitution and conversion] $= 3.7 \times 10^{-3}\,\text{m}$ [1]	ECF from A of cable; do not grant mark if A of wire is used.	3	AO2 MS 3.1
5–3	So that the integrity of the bridge is not compromised if individual wires fail.		1	AO1

Question 6

Question	Answer	Guidance	Marks	ID
6 (circuit diagram)		Award a maximum of 2 marks for the circuit diagram.	6	AO1 AO3 MS 3.2, 4.3 PS 1.2
6 (method)	Answers could include the following points: • Add wires in parallel to circuit. • Measure V across and I through the wires. • Calculate R of combined wires in parallel. • Use a micrometer to measure the diameter of individual wires. • Measure the diameter across 3 or 4 sections of each wire and calculate the mean diameter. • Calculate the cross-sectional area A of the individual wires. • Add A_1, A_2 … for wires in parallel. • L = constant • Plot R against 1/A. • ρ = gradient/L			

Question 7

Question	Answer	Guidance	Marks	ID
7–1	Resistance of $T_1 = 8000\,\Omega$ [1] $I_{\text{tot}} = \dfrac{V_{T_1}}{R_{30°}} = \dfrac{7.5\,\text{V}}{8000\,\Omega} = 9.4 \times 10^{-4}\,\text{A}$ [1] $V_{R_2} = V_{\text{Load}} = 12\,\text{V} - 7.5\,\text{V} = 4.5\,\text{V}$ [1] $I_{R_2} = \dfrac{V_{R_2}}{R_2} = \dfrac{4.5\,\text{V}}{10 \times 10^3\,\Omega} = 4.5 \times 10^{-4}\,\text{A}$ [1] $R_{\text{Load}} = \dfrac{V_{\text{Load}}}{I_{\text{tot}} - I_{R_2}} = \dfrac{4.5\,\text{V}}{4.9 \times 10^{-4}\,\text{A}} = 9184\,\Omega$ [1]		5	AO1 AO3 MS 3.2 PS 4.1
7–2	Drop 2.5 kΩ from 8 kΩ to 5.5 kΩ. Corresponding temperature is 42°.	Accept 40°.	1	AO1

Practice paper 2 (p.127)

Question 1

Question	Answer	Guidance	Marks	ID
1–1	Specific heat capacity is the <u>quantity of heat</u> required to raise the <u>temperature</u> of a <u>unit mass</u> through a <u>unit temperature rise</u>.	The underlined points must be included. Allow quotes from units, e.g. 1 kg, 1 K or °C (but **not** °K).	1	AO1
1–2	Temperature rise per second: $$\Delta\theta = \frac{\Delta Q}{mc}$$ $\Delta\theta = 35\,000\,W \div (5.0\,kg\,s^{-1} \times 4200\,J\,kg^{-1}K^{-1})$ $\Delta\theta = 1.67\,K$	Allow °C.	1	AO2 MS 0.3, 2.2
1–3	1 day = 86400 s Energy supplied by the boiler = $35\,000\,W \times 86\,400\,s$ $\Delta Q = 3.024 \times 10^9\,J$ [1] $\Delta\theta = 3.024 \times 10^9\,J \div (9.0 \times 10^4\,kg \times 4200\,J\,kg^{-1}K^{-1})$ $\Delta\theta = 8.0\,K$ (°C) Temperature after 1 day = 10°C + 8.0°C = 18°C [1]	Answer must be in °C. ECF from ΔQ.	2	AO2 MS 0.1, 2.3
1–4	Rate of evaporation = $9.0\,kg \div 3600\,s$ $= 2.5 \times 10^{-2}\,kg\,s^{-1}$. Using $\Delta Q = mL$ Power = $2.5 \times 10^{-3}\,kg\,s^{-1} \times 2.26 \times 106\,J\,kg^{-1}$ $= 5700\,W$	Accept 5650 W, 5.7 kW or 5.65 kW.	1	AO2 MS 0.3, 2.3
1–5	Any one of: <ul><li>Humidity of the air</li><li>Wind speed</li><li>Temperature difference between the water and the air.</li></ul>		1	AO1

Question 2

Question	Answer	Guidance	Marks	ID
2–1	Accept either graph:	Graphs must have axes correctly labelled with units. The line must be straight. If Celsius is used, absolute zero must be shown as −273°C. No mark if temperature units are mixed up (e.g. K used on a Celsius graph) or for °K.	1	AO1 MS 3.1
2–2	$$n = \frac{pV}{RT}$$ n = $(1.01 \times 10^5\,Pa \times 1.5 \times 10^{-4}\,m^3) \div$ $(8.31\,J\,K^{-1}mol^{-1} \times 318\,K)$ n = $5.73 \times 10^{-3}\,mol$	T in Celsius is a physics error, no mark.	1	AO1 MS 2.2, 2.3
2–3	N = $5.73 \times 10^{-3}\,mol \times 6.0 \times 10^{23}\,mol^{-1} = 3.4 \times 10^{21}$	Accept ECF from **2–2**, unless T is in Celsius.	1	AO1 MS 2.2, 2.3

Question	Answer	Guidance	Marks	ID
2–4	$$\frac{V_1}{T_1} = \frac{V_2}{T_2}$$ $V_2 = (1.5 \times 10^{-4}\,m^3 \times 318K) \div 273K = 1.75 \times 10^{-4}\,m^3$ $\Delta V = 2.5 \times 10^{-5}\,m^3$ [1] $$W = p\Delta V$$ $W = 1.01 \times 10^5\,Pa \times 2.5 \times 10^{-5}\,m^3 = 2.5\,J$ [1]	Accept ECF from first answer.	2	AO2 MS 2.2, 2.3
2–5	Any one of: • All collisions between molecules and the walls of the container are perfectly elastic. • Intermolecular forces are negligible, as is gravity. • Molecules move in straight lines and at constant speed between collisions. • Collision times are negligible compared with the time between collisions. • The volume of the gas molecules is negligible compared to that of the gas. • Newton's laws of motion are applicable.		1	AO1
2–6	$$(c_{rms})^2 = \frac{3pV}{Nm}$$ $Nm = 5.73 \times 10^{-3}\,mol \times 4.00 \times 10^{-2}\,kg\,mol^{-1}$ $= 2.3 \times 10^{-4}\,kg$ [1] $(c_{rms})^2 = (3 \times 1.01 \times 10^5\,Pa \times 1.75 \times 10^{-4}\,m^3)$ $\div 2.3 \times 10^{-4}\,kg$ $= 1.54 \times 10^3\,m^2\,s^{-2}$ $c_{rms} = (1.54 \times 10^3\,m^2\,s^{-2})^{0.5} = 392\,m\,s^{-1} = 390\,m\,s^{-1}$ [1] Answer to 2 s.f. as data is to 2 s.f. [1]	Alternatively: $Nm = 40u \times 1.661 \times 10^{-27}\,kg\,u^{-1} \times 3.4 \times 10^{21} = 2.3 \times 10^{-4}\,kg$ [1] Accept any correct method of calculating c_{rms} for full credit. ECF from **Questions 2–2** and **2–3**. The significant figures mark is independent of the final answer.	3	AO2 MS 1.1, 2.1
2–7	Temperature		1	AO1

Question 3

Question	Answer	Guidance	Marks	ID
3–1	Gravitational field strength at a point is the <u>gravitational force</u> per <u>unit mass</u> at that point.	Both underlined points are needed. Do not accept $N\,kg^{-1}$ or ms^{-2} as the whole answer.	1	AO1
3–2	Radius of orbit $= 6.5 \times 10^6\,m + 1.5 \times 10^6\,m = 8.0 \times 10^6\,m$ $g = (-) \dfrac{(6.67 \times 10^{-11}\,N\,m^2\,kg^{-1} \times 5.51 \times 10^{24}\,kg)}{(8.0 \times 10^6\,m)^2}$ $g = 5.7\,N\,kg^{-1}$	Use of $r = 6.5 \times 10^6\,m$ is a physics error. Minus sign is not needed. Allow ms^{-2} for the unit.	1	AO2 MS 0.1, 2.2
3–3	Gravitational field strength is acceleration (Newton II), therefore: $$a = \frac{GM}{r^2}$$ Centripetal acceleration: $$a = \omega^2 r = \frac{4\pi^2}{T^2} r$$ Combine these two [1]: $$\frac{4\pi^2}{T^2} = \frac{GM}{r^3}$$ Invert the equation: $$\frac{T^2}{4\pi^2} = \frac{r^3}{GM}$$ Rearrange [1]: $$T^2 = \frac{4\pi^2 r^3}{GM}$$	Minus signs left out of the argument: they cancel out when the equations are combined. Second mark is dependent on the student showing the steps of inverting the equation and rearranging. A simple statement: '...therefore: $T^2 = \dfrac{4\pi^2 r^3}{GM}$...' is not sufficient.	2	AO3 MS 2.2
3–4	$$T^2 = \frac{4\pi^2 \times (8.0 \times 10^6\,m)^3}{6.67 \times 10^{-11}\,N\,m^2\,kg^{-2} \times 5.51 \times 10^{24}\,kg}$$ [1] $T = (5.45 \times 10^6\,s^2)^{0.5} = 7416\,s = 7400\,s$ (2 s.f.) [1]	Allow ECF for radius from **Question 3–2**.	2	AO2 MS 2.3, 2.4
3–5	$v^2 = 2 \times 5.7\,N\,kg^{-1} \times 8.0 \times 10^6\,m$ $v = (9.12 \times 10^7\,m^2\,s^{-2})^{0.5} = 9549\,m\,s^{-1} = 9500\,m\,s^{-1}$	ECF from **Question 3–2**.	1	AO2 MS 2.4

Question	Answer	Guidance	Marks	ID
4–1	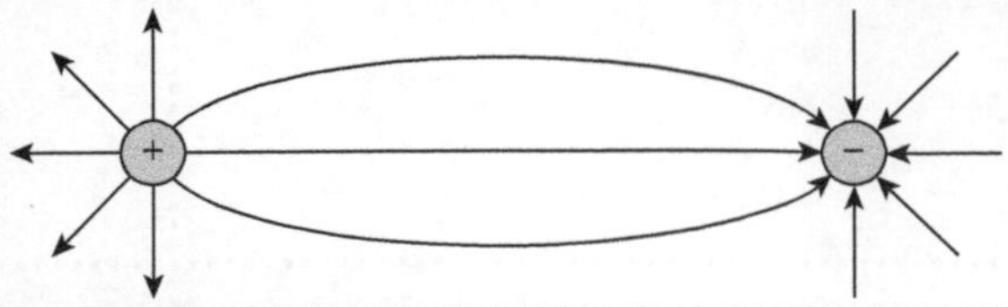	Field lines should be shown between the point sources and around them.	1	AO1
4–2	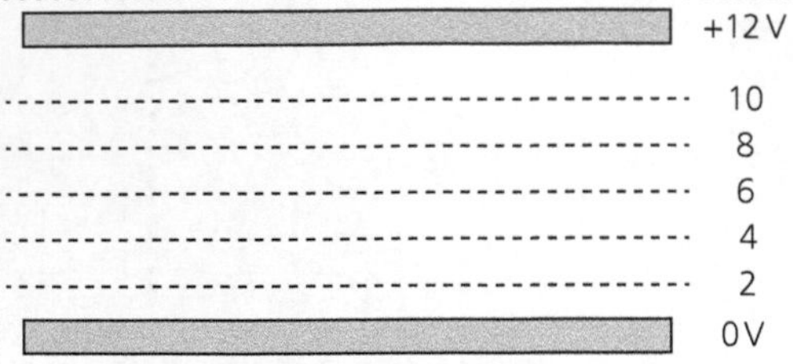	Potentials must be shown for the mark.	1	AO1
4–3	Area under the graph shown between 0.4 m and 1.0 m: 132 ± 5 squares counted [1] Each square = $8.0\,J\,C^{-1}$ $V = 1056\,J\,C^{-1} = 1100\,J\,C^{-1}$ [1]	Evidence that squares were counted for the first mark. The second mark is independent of the first.	2	AO2 MS 3.9
4–4	Equation: $$E = \frac{1}{4\pi\varepsilon_0}\frac{Q_1 Q_2}{r_1} - \frac{1}{4\pi\varepsilon_0}\frac{Q_1 Q_2}{r_2}$$ $E = (8.99 \times 10^9\,F\,m^{-1} \times 80 \times 10^{-9}\,C \times 12 \times 10^{-9}\,C)\ldots$ $\ldots(1/0.4\,m - 1/1.4\,m)$ [1] $E = 1.54 \times 10^{-6}\,J = 1.5 \times 10^{-6}\,J$ [1]		2	AO2 MS 2.2, 2.3
4–5	Combine uniform electric field equation with definition of electric field as force per unit charge: $$E = \frac{V}{d} = \frac{F}{Q}$$ Therefore $F = \dfrac{QV}{d}$ Using Newton's second law [1]: $$a = \frac{F}{m} = \frac{QV}{dm}$$ Horizontal velocity remains the same, i.e. v. Vertical velocity v_V obeys the equation of motion: $v^2 = u^2 + 2as$ In this derivation, $s = \dfrac{d}{2}$ Therefore: $$v_V^2 = ad = \frac{QVd}{dm}$$ The d terms cancel out to give [1]: $$v_V^2 = \frac{QV}{m}$$ By Pythagoras [1]: $$v_R^2 = v^2 + \frac{QV}{m}$$ Therefore: $v_R = \left(v^2 + \dfrac{QV}{m}\right)^{0.5}$	The only equation that will work is: $v^2 = u^2 + 2as$ as the length of the plates is not specified.	3	AO3 MS 0.6, 2.5, 4.4

Question 5

Question	Answer	Guidance	Marks	ID
5–1	Magnetic flux density is the <u>number of lines of magnetic flux</u> passing <u>through a unit area</u>.	Both the underlined points are needed.	1	AO1
5–2	They move downwards in a circular path.		1	AO1
5–3	Force on a charged particle in a magnetic field: $F = Bqv$ From circular motion: $$F = \frac{mv^2}{r}$$ Therefore [1]: $$Bqv = \frac{mv^2}{r}$$ Rearranging [1]: $$r = \frac{mv^2}{Bqv} = \frac{mv}{Bq}$$		2	AO3 MS 2.2, 2.4
5–4	$$r = \frac{9.11 \times 10^{-31}\,\text{kg} \times 3.5 \times 10^6\,\text{m s}^{-1}}{1.5 \times 10^{-4}\,\text{T} \times 1.6 \times 10^{-19}\,\text{C}}\ [1]$$ $r = 0.13\,\text{m}$ to 2 s.f. [1]	Answer must be to 2 s.f. for the second mark.	2	AO2 MS 2.3
5–5	First case: $$r = \frac{mv}{Bq}$$ Second case: $$r' = \frac{2m \times 2v}{\frac{B}{4} \times 2q}$$ $r' = 8\,r$		1	AO2
5–6	A spiral path starting from the entry point and going to the exit point.	The spiral does not have to be perfect.	1	AO2
5–7	Equation rearranged for m: $$m = \frac{Bq}{2\pi f}$$ $m = (2.4\,\text{T} \times 3.2 \times 10^{-19}\,\text{C}) \div (2 \times \pi \times 3.6 \times 10^6\,\text{Hz})$ $m = 3.4 \times 10^{-26}\,\text{kg}$ [1] Mass compared with proton: Ratio $= 3.4 \times 10^{-26}\,\text{kg} \div 1.67 \times 10^{-27}\,\text{kg}$ $= 20.3$ Ratio: 20.3 : 1 or 20 : 1 [1]	If 36 is inserted, the first mark is lost. 20 : 2 can be accepted for the second mark.	2	AO2 MS 1.1

Question 6

Question	Answer	Guidance	Marks	ID
6–1	$^{60}_{27}\text{Co} \rightarrow\ ^{60}_{28}\text{Ni} +\ ^{0}_{-1}e +\ ^{0}_{0}\bar{\nu}_e$	Both parts are needed for the mark. Equation must be in nuclide notation, except for the antineutrino. If a neutrino is written, this is a physics error.	1	AO2
6–2	y-axis: Distance^2 / m^2 [1] x-axis: Count rate^{-1} / s	Allow quantities to be written in appropriate symbol form, e.g. x for distance. Units for activity are s^{-1} or Bq. Therefore, activity^{-1} would have units of s. Allow Bq^{-1}.	2	AO3
6–3	Sketch graph with time on x-axis and activity on y-axis. Curve should show exponential decay. Half-life shown at about 5.25 years.	For the mark, the following must be shown: • Half-life is about 5¼ y (5 y is not sufficiently precise) • The line must be exponential.	1	AO1 MS 3.1
6–4	Equation: $$\lambda = \frac{\ln 2}{t_{1/2}}$$ $\lambda = 0.693 \div (5.27\,\text{y} \times 365\,\text{d y}^{-1} \times 86400\,\text{s d}^{-1})$ $\lambda = 4.17 \times 10^{-9}$ [1] s^{-1} [1]	1 mark for using the correct unit, independent of the answer.	2	AO2 MS 1.3

Question	Answer	Guidance	Marks	ID
6–5	Equation: $$A = A_0 e^{-\lambda t}$$ Therefore: $$\ln A - \ln A_0 = -\lambda t$$ $t = (\ln (150\,\text{Bq}) - \ln (2500\,\text{Bq})) \div -4.17 \times 10^{-9}\,\text{s}^{-1}$ $t = 5.011 - 7.824 \div -4.17 \times 10^{-9}\,\text{s}^{-1}$ $t = 6.745 \times 10^8\,\text{s}$ [1] $t\,(\text{y}) = 6.745 \times 10^8\,\text{s} \div (365\,\text{d}\,\text{y}^{-1} \times 86400\,\text{s}\,\text{d}^{-1})$ $t\,(\text{y}) = 21.4\,\text{y} = 21\,\text{y}$ [1]	Alternatively, use: $\lambda = 0.132\,\text{y}^{-1}$ [1] This gives $21.3\text{y} = 21\,\text{y}$ [1]	2	AO2 MS 0.5, 2.2, 2.3
6–6	$\Delta E = 0.33\,\text{MeV} + 1.173\,\text{MeV} + 1.333\,\text{MeV}$ $\Delta E = 2.836\,\text{MeV}$ [1] $\Delta m = 2.836\,\text{MeV} \div 931.5\,\text{MeV}\,\text{u}^{-1}$ $\Delta m = 3.04 \times 10^{-3}\,\text{u}$ [1]		2	AO3 MS 0.1

Question 7

Question	Answer	Guidance	Marks	ID
7–1	A: coolant B: moderator C: pressure vessel D: fuel rods E: control rods	All five correct to get the mark.	1	AO2
7–2	High level: • Contents of spent fuel rods are stored in deep ponds. • Material is added to sand and the sand is melted to vitrify it. Medium level: • Spent fuel rod cladding is set in concrete and stored in deep repositories. Low level: • Protective clothing is buried in landfill.	Student should choose one only. If more than one, mark the first and ignore the others. For the mark, the type of material should be stated, and one method of disposal given. The list here is not exhaustive. The mark should be awarded for any reasonable answer.	1	AO2
7–3	Advantages: • Much reduced CO_2 emissions; no fossil fuels are burned. • Constant energy available; power stations are running all the time (OR renewables may not work all the time/wind farms are off when it's calm). • High energy density; only a small mass of fuel is needed to give the same energy as a large amount of coal. Disadvantages: • Waste from nuclear power stations is very dangerous and has to be handled very carefully. • Start-up costs; power plants are very expensive and take many years to build. • Decommissioning is very expensive and potentially hazardous due to the amount of radioactive materials. • Accidents; although rare, accidents have caused catastrophic damage.	3 advantages and 3 disadvantages should be listed (not, for example, 2 advantages and 4 disadvantages). Each advantage and disadvantage must have an explanation for the mark. Give marks for valid points that are not included here.	6	AO3

Section B

8	B	13	A	18	D	23	B	28	C
9	C	14	D	19	C	24	D	29	D
10	D	15	B	20	B	25	B	30	D
11	C	16	C	21	A	26	B	31	C
12	A	17	B	22	D	27	A	32	A

Paper 3 Part A (p.141)

Question 1

Question	Answer	Guidance	Marks	ID
1–1	Time to jump to highest point: $\dfrac{1s}{1200} = \dfrac{t}{946} \rightarrow t = \dfrac{946}{1200}s = 0.79\,s$ [1] Height of jump: $s = 55\,cm - (3.0 + 2.5)\,cm = 49.5\,cm = 0.495\,m$ [1] $s = \dfrac{u+v}{2}t = \dfrac{u+0}{2}t \rightarrow u = \dfrac{2s}{t} = 1.25\,m\,s^{-1}$ [1]		3	AO2 MS 0.5, 2.2, 2.3, 2.4
1–2	Identify the time, t, from the frame count of the camera. [1] Identify the height of base for each value of t. [1] Subtract $s = ut + \dfrac{gt^2}{2}$ from the value of height to obtain a displacement from the centre of oscillation for each value of t. [1]		3	AO2 MS 0.5, 2.2, 2.3, 2.4
1–3	Derive $T = 0.05\,s$ from graph. [1] Convert 23 g to 0.023 kg. Use $T = 2\pi\sqrt{\dfrac{m}{2k}}$ Rearrange $k = \dfrac{2\pi^2 m}{T^2} = \dfrac{2\pi^2 \times 0.023\,kg}{2.5 \times 10^{-3}\,s^2} = 182\,kg\,s^{-2}$ [1]		2	AO1 AO2 MS 4.6 AT b, c
1–4	The toy is in free fall as soon as it loses contact with the desk [1], so the mass of the whole system is considered [1] because both ends will oscillate [1].	Accept a similar correct argument.	3	AO3

Question 2

Question	Answer	Guidance	Marks	ID
2–1	Diagram showing: • Any two socket pairs connected in series with a power supply, an ammeter [1] and a variable resistor [1] • A voltmeter connected in parallel with the selected socket pair [1].		3	AO1
2–2	All points correctly plotted. [2] Curve of best fit correctly drawn. [1]		3	AO3
2–3	The current cannot be changed directly and it varies as a result of changes we make to the p.d. [1] OR The p.d. is the independent variable and the current the dependent variable. [1]		1	AO2
2–4	The resistance of component 1 is constant across all values of V. [2] The resistance of component 2 and components 1 and 2 together increases as the p.d. increases. [1]		3	AO3
2–5	Component 1 is a fixed resistor. [1] Component 2 could be a filament lamp. [1]	Accept just 'resistor' for component 1.	2	AO1
2–6	Diagram showing: • Components 1 and 2 connected in parallel with each other between the bottom two sockets [1] • Correct symbols for resistor and filament lamp [1].		2	AO2 AO3

Question 3

Question	Answer	Guidance	Marks	ID
3–1	Take three (or more) repeat readings for each height. [1] Start the car from the same height for each repeated set of readings. [1] Calculate a mean stopping distance. [1] Use light gates to take repeat readings of velocity and calculate the mean velocity. [1]		4	AO2
3–2	This method is not correct [1] because the kinetic energy is dependent on v^2 [1] (so the curve is not a straight line).		2	AO1 AO3
3–3	The centre of mass of the car is actually higher than the height mark, so the GPE should be greater than that they calculated.		1	AO3
3–4	Start the car by positioning its centre of mass in line with the height mark for each drop.		1	AO3
3–5	Values for the GPE (J) column: 0.035 0.047 0.059 0.071 0.082	Correct values are calculated using 0.024 kg for mass and 9.81 N kg^{-1} for g. All values must be correct for 2 marks. 1 mark if 1 or 2 values are incorrect. 1 mark if values have been calculated using 24 g for the mass.	2	AO2
3–6	Graph with: • Correct scales and labels [1] • Correct plots drawn [1] • Line of best fit extended to origin of graph [1].		3	AO1
3–7	Use a large triangle to calculate the gradient of the line of best fit. [1] Gradient $= \dfrac{\Delta E_k}{\Delta s_{mean}}$ [1] $= \dfrac{0.100\,\text{J}}{0.810\,\text{m}} = 0.123\,\text{N}$ [1]	Allow other values of $\dfrac{\Delta E_k}{\Delta s_{mean}}$	3	AO2
3–8	The rough surface might cause the toy car's wheels to have uneven contact with the surface [1] and prevent it from following a straight path [1].	Accept equally reasonable suggestions.	2	AO3
3–9	Record a video of the toy car travelling along the carpet. [1] Use the video to plot a distance–time graph. [1] Use maths software to derive a velocity–time graph from the gradient of the d–t graph, and an acceleration–time graph from the gradient of the v–t graph. [1]		3	AO3

Paper 3 Part B (p.145)

Question 1

Question	Answer	Guidance	Marks	ID
1–1		For the mark: • Lines to be drawn with a ruler • Ray diagram as shown, extending beyond the eyepiece • Angles of incidence and reflection should be equal (use your judgement) • Directional arrows • Principal focus marked.	1	AO1
1–2	$M = \dfrac{f_o}{f_e}$ $M = 1.25\,\text{m} \div 0.05\,\text{m} = 25$		1	AO2 MS 2.3
1–3	$\theta \approx \dfrac{\lambda}{D}$ $\theta = 550 \times 10^{-9}\,\text{m} \div 0.20\,\text{m} = 2.8 \times 10^{-6}\,\text{rad}$		1	AO2 MS 2.1

Question	Answer	Guidance	Marks	ID
1–4	Advantages and reasons: • No chromatic aberration; which means light is refracted by different amounts in lenses • Larger apertures can be used in reflectors, whereas maximum lens size is about 1 m • Easier to make magnification greater, whereas maximum focal length for a lens is about 20 m • More light can be transmitted; some light is lost in a lens.	Any two of these for one mark each. Each point needs to be supported with a reason.	2	AO1
1–5	Photon energy: $$E = \frac{hc}{\lambda}$$ $E = \dfrac{6.63 \times 10^{-34}\,\text{J s} \times 3.00 \times 10^{8}\,\text{m s}^{-1}}{550 \times 10^{-9}\,\text{m}}$ $E = 3.62 \times 10^{-19}\,\text{J}$ Power $= 2.2 \times 10^{-10}\,\text{W m}^{-2} \times 3.5 \times 10^{-2}\,\text{m}^2$ $= 7.7 \times 10^{-12}\,\text{W}$ [1] Number of photons per second $=$ power $\div$ photon energy $N = 7.7 \times 10^{-12}\,\text{W} \div 3.62 \times 10^{-19}\,\text{J} = 2.1 \times 10^{7}\,\text{s}^{-1}$ [1]	Error can be carried forward from the first calculation to the second.	2	AO3 MS 2.2, 2.3
1–6	No. of electrons $= 2.13 \times 10^{7}\,\text{s}^{-1} \times 0.7 = 1.5 \times 10^{7}\,\text{s}^{-1}$	Error in **Question 1–5** can be carried forward.	1	
1–7	These radiations <u>are blocked by the atmosphere</u>, therefore <u>space telescopes</u> are needed to observe them.	Both underlined points are needed for the mark.	1	AO1

Question 2

Question	Answer	Guidance	Marks	ID
2–1	Equation: $$m - M = 5\log\left(\frac{d}{10}\right)$$ $(-0.01 - +4.4) \div 5 = \log(d/10)$ $\log d - \log 10 = -0.878$ $\log d = +0.122$ $d = 1.3(2)$ [1] pc [1]	Unit must be parsec (pc). Do not accept light years, even if quoted correctly (4.3 ly).	2	AO2 MS 0.3, 0.5
2–2	$$\lambda_{\text{max}} = \frac{2.9 \times 10^{-3}}{T}$$ $\lambda_{\text{max}} = (2.9 \times 10^{-3}\,\text{m K}) \div 5790\,\text{K}$ $\lambda_{\text{max}} = 5.0 \times 10^{-7}\,\text{m}\ (= 500\,\text{nm})$	Accept 500 nm, as long as the unit is consistent.	1	AO2 MS 2.2, 2.3
2–3	$$\frac{A_S}{A_A} = \left(\frac{T_A}{T_S}\right)^4 \times \frac{P_S}{P_A}$$ Substituting: $\dfrac{A_S}{A_A} = \left(\dfrac{5790\,\text{K}}{5778\,\text{K}}\right)^4 \times \dfrac{1}{1.519} = 0.664$ [1] Therefore: $\dfrac{d_A}{d_S} = 1.51^{0.5} = 1.23$ Diameter of Alpha Centauri $= 1.2$ times that of the Sun [1]	ECF from first part to second.	2	AO2 MS 2.2, 2.3
2–4	Spectral class G		1	AO3 MS 3.1
2–5	Alpha Centauri shown at 6000 K, with a luminosity of 1.		1	AO3 MS 3.1
2–6	A Type 1a supernova: 1 Is formed when a <u>white dwarf</u> in a <u>binary star system</u> attracts material from its <u>companion star which is a red giant</u>. [1] 2 Mass exceeds 1.4 solar masses… [1] 3 … leading to the collapse of the white dwarf into a <u>neutron star</u>. [1]	Statement 1 must be given, followed by either 2 or 3 for the second mark.	2	AO1
2–7	It is used as a standard candle because it has absolute magnitude.		1	AO1

Question 3

Question	Answer	Guidance	Marks	ID
3–1	$\Delta\lambda = 553.13\,nm - 553.00\,nm = 0.13\,nm$ $z = 0.13\,nm \div 553.00\,nm = 2.350 \times 10^{-5}$ [1] $v = zc = 2.350 \times 10^{-5} \times 3.00 \times 10^{8}\,m\,s^{-1}$ $v = 7052\,m\,s^{-1}$ [1]	Negative sign is not needed.	2	AO2 MS 2.2, 2.3
3–2	A <u>large exoplanet</u> [1] is <u>attracting</u> Star B by <u>gravity</u> [1].	Accept 'large planet'. The underlined parts are needed for each mark.	2	AO2
3–3	An exoplanet is passing across the face of the star [1], shading a small fraction of the disc [1].		2	AO2
3–4	Life is most likely to be found where: • The temperature is between 260 and 310 K [1] • There is water [1] • There is a rocky surface [1].	Allow credit for any other valid suggestion.	3	AO3

Question 4

Question	Answer	Guidance	Marks	ID
4–1	Any two of: • Intense radio sources • Very bright • Supermassive black holes • Move away at very high speeds • Emit powerful jets of matter • Consume about 10 solar masses of matter each year • Have strange spectral lines.	Any two of these for 1 mark each. Allow credit for any other valid suggestion. No credit for 'smaller than galaxies' or 'very distant', as these are in the stem of the question.	2	AO1
4–2	$\Delta\lambda = 550\,nm - 660\,nm = (-)\,110\,nm$ $z = 110\,nm \div 550\,nm = 0.20$ $v = zc = 0.20 \times 3.00 \times 10^{8}\,m\,s^{-1}$ $v = 6.0 \times 10^{7}\,m\,s^{-1} = 60\,000\,km\,s^{-1}$	Signs are not important. No marks for: $110 \div 660 = 0.167$ but allow ECF. Watch out for the conversion to $km\,s^{-1}$. Failure to do so is a physics error.	1	AO3 MS 1.1, 2.1
4–3	Hubble's law: $$d = \frac{v}{H}$$ $d = 60\,000\,km\,s^{-1} \div 65\,km\,s^{-1}\,Mpc^{-1}$ $d = 923\,Mpc$ [1] $= 920\,Mpc$ (to 2 s.f.) [1]	1 mark for using the correct unit, independent of the answer.	2	AO3 MS 1.1, 2.1
4–4	Cosmic background radiation: the temperature (2.7 K) of the universe suggests that it expanded from something much hotter. [1] Hydrogen and helium ratio: it is 3:1 by mass, indicating cooling to allow quarks to form baryons. [1] OR Proton to neutron ration: it is 7:1, indicating cooling to allow quarks to form baryons. [1]	For the mark, the point needs to be made with an explanation. A hydrogen to helium ratio of 7:1 is incorrect, as is a proton to neutron ratio of 3:1. Alternative mark for dark energy: distant objects are still accelerating when the expectation is that gravitational attraction should slow them down.	2	AO1
4–5	$1\,Mpc = 3.08 \times 10^{16}\,m \times 1 \times 10^{6} = 3.08 \times 10^{22}\,m$ $1\,Mpc = 3.08 \times 10^{19}\,km$ $H = 65\,km\,s^{-1} \div 1\,Mpc$ $H = 65\,km\,s^{-1} \div 3.08 \times 10^{19}\,km = 2.11 \times 10^{-18}\,s^{-1}$ $Age = H^{-1} = (2.11 \times 10^{-18}\,s^{-1})^{-1} = 4.74 \times 10^{17}\,s$ $Age = 1.4 \times 10^{10}\,y\ (= 14\,000\ million\,y)$		1	AO3 MS 0.1, 2.4